Experimentation in Mathematics

Experimentation in Mathematics
Computational Paths to Discovery

Jonathan Borwein
David Bailey
Roland Girgensohn

A K Peters
Natick, Massachusetts

Editorial, Sales, and Customer Service Office

A K Peters, Ltd.
63 South Avenue
Natick, MA 01760
www.akpeters.com

Library of Congress Cataloging-in-Publication Data

Borwein, Jonathan M.
  *Experimentation in Mathematics :computational paths to discovery /
Jonathan M.*
 Borwein, David H. Bailey, Roland Girgensohn.
   p. cm.
  Includes bibliographical references and index.
  ISBN 1-56881-136-5
  1. Mathematics--Research. I. Bailey, David H. II. Girgensohn, Roland, 1964- III. Title.

QA12.B67 2003
510'.72--dc22

2003062324

Printed in Canada
08 07 06 05 04   10 9 8 7 6 5 4 3 2 1

Contents

Preface

Moreover a mathematical problem should be difficult in order to entice us, yet not completely inaccessible, lest it mock our efforts. It should be to us a guidepost on the mazy path to hidden truths, and ultimately a reminder of our pleasure in the successful solution. ...

Besides it is an error to believe that rigor in the proof is the enemy of simplicity.
David Hilbert, Paris International Congress, 1900

As we recounted in the first volume of this work, *Mathematics by Experiment: Plausible Reasoning in the 21st Century* [43], when we started our collaboration in 1985, relatively few mathematicians employed computations in serious research work. In fact, there appeared to be a widespread view in the field that "real mathematicians don't compute." In the ensuing years, computer hardware has skyrocketed in power and plummeted in cost, thanks to the remarkable phenomenon of Moore's law. In addition, numerous powerful mathematical software products, both commercial and noncommercial, have become available. But just as importantly, a new generation of mathematicians is eager to use these tools, and consequently numerous new results are being discovered.

The experimental methodology described in these books provides a compelling way to generate understanding and insight; to generate and confirm or confront conjectures; and generally to make mathematics more tangible, lively, and fun for both the professional researcher and the novice. Furthermore, the experimental approach helps broaden the interdisciplinary nature of mathematical research: a chemist, physicist, engineer, and a mathematician may not understand each others' motivation or technical language, but they often share an underlying computational approach, usually to the benefit of all parties involved.

A typical scenario of using this experimental methodology is the following. Note the "dialogue" between human and computer, which is very typical of this approach to mathematical research:

1. Studying a mathematical problem to identify aspects that need to be better understood.

2. Using a computer to explore these aspects, by working out specific examples, generating plots, etc.

3. Noting patterns or other phenomena evident in the computer-based results that relate to the problem under study.

4. Using computer-based tools to identify or "explain" these patterns.

5. Formulating a chain of credible conjectures that, if true, would resolve the question under study.

6. Deciding if the potential result points in the desired direction and is worth a full-fledged attempt at formal proof.

7. Performing additional computer-based experiments to gain greater confidence in the key conjectures.

8. Confirming these conjectures by rigorous proof.

9. Using symbolic computing software to double-check analytical derivations.

Our goal in these books is to present a variety of *accessible* examples of modern mathematics where intelligent computing plays a significant role (along with a few examples showing the limitations of computing). We have concentrated primarily on examples from analysis and number theory, as this is where we have the most experience, but there are numerous excursions into other areas of mathematics as well (see the Table of Contents). For the most part, we have contented ourselves with outlining reasons and exploring phenomena, leaving a more detailed investigation to the reader. There is, however, a substantial amount of new material, including numerous specific results that have not yet appeared in the mathematical literature, as far as we are aware.

This work is divided into two volumes, each of which can stand by itself. The first volume, *Mathematics by Experiment: Plausible Reasoning in the 21st Century* [43], presents the rationale and historical context of experimental mathematics, and then presents a series of examples that exemplify the experimental methodology. We include in the first volume a reprint of an article co-authored by two of us that complements this material. This second volume, *Experimentation in Mathematics: Computational Paths to Discovery*, continues with several chapters of additional examples. Both volumes include a chapter on numerical techniques relevant to experimental mathematics.

Each volume is targeted to a fairly broad cross-section of mathematically trained readers. Most of the first volume should be readable by anyone with solid undergraduate coursework in mathematics. Most of this volume should be readable by persons with upper-division undergraduate or graduate-level coursework. None of this material involves highly abstract or esoteric mathematics.

Some programming experience is valuable to address the material in this book. Readers with no computer programming experience are invited to try a few of our examples using commercial software such as *Mathematica* and *Maple*. Happily, much of the benefit of computational-experimental mathematics can be obtained on any modern laptop or desktop computer— a major investment in computing equipment and software is not required.

Each chapter concludes with a section of commentary and exercises. This permits us to include material that relates to the general topic of the chapter, but which does not fit nicely within the chapter exposition. This material is not necessarily sorted by topic nor graded by difficulty, although some hints, discussion and answers are given. This is because mathematics in the raw does not announce, "I am solved using such and such a technique." In most cases, half the battle is to determine how to start and which tools to apply.

We are grateful to our colleagues Victor Adamchik, Heinz Bauschke, Peter Borwein, David Bradley, Gregory Chaitin, David and Gregory Chudnovsky, Robert Corless, Richard Crandall, Richard Fateman, Greg Fee, Helaman Ferguson, Steven Finch, Ronald Graham, Andrew Granville, Christoph Haenel, David Jeffrey, Jeff Joyce, Adrian Lewis, Petr Lisonek, Russell Luke, Mathew Morin, David Mumford, Andrew Odlyzko, Hristo Sendov, Luis Serrano, Neil Sloane, Daniel Rudolph, Asia Weiss, and John Zucker who were kind enough to help us prepare and review material for this book; to Mason Macklem, who helped with material, indexing (note that in the index figures are marked by "‡," and quotes with "†"), and more; to Jen Chang and Rob Scharein, who helped with graphics; to Janet Vertesi who helped with bibliographic research; to Will Galway, Xiaoye Li, and Yozo Hida, who helped with computer programming; and to numerous others who have assisted in one way or another in this work. We owe a special debt of gratitude to Klaus Peters for urging us to write this book and for helping us nurse it into existence. Finally, we wish to acknowledge the assistance and the patience exhibited by our spouses and family members during the course of this work.

Borwein's work is supported by the Canada Research Chair Program and the Natural Sciences and Engineering Council of Canada. Bailey's work is supported by the Director, Office of Computational and Technology Research, Division of Mathematical, Information, and Computational Sciences of the U.S. Department of Energy, under contract number DE-AC03-76SF00098; also by the National Science Foundation, award number DMS-0342255.

Experimental Mathematics Web Site

The authors have established a web site containing an updated collection of links to many of the URLs mentioned in the two volumes, plus errata, software, tools, and other web useful information on experimental mathematics. This can be found at the following URL: http://www.expmath.info.

Jonathan M. Borwein   August 2003
David H. Bailey
Roland Girgensohn

1 | Sequences, Series, Products and Integrals

Several years ago I was invited to contemplate being marooned on the proverbial desert island. What book would I most wish to have there, in addition to the Bible and the complete works of Shakespeare? My immediate answer was: Abramowitz and Stegun's *Handbook of Mathematical Functions*. If I could substitute for the Bible, I would choose Gradsteyn and Ryzhik's *Table of Integrals, Series and Products*. Compounding the impiety, I would give up Shakespeare in favor of Prudnikov, Brychkov and Marichev's *Tables of Integrals and Series.*∴. On the island, there would be much time to think about waves on the water that carve ridges on the sand beneath and focus sunlight there; shapes of clouds; subtle tints in the sky... With the arrogance that keeps us theorists going, I harbor the delusion that it would be not too difficult to guess the underlying physics and formulate the governing equations. It is when contemplating how to solve these equations—to convert formulations into explanations—that humility sets in. Then, compendia of formulas become indispensable.

Michael Berry, "Why Are Special Functions Special?", 2001

In *Mathematics by Experiment* [43], we presented numerous examples of experimental mathematics in action. In particular, we examined how a computational-experimental approach could be used to identify constants and sequences, evaluate definite integrals and infinite series, discover new identities involving fundamental constants and functions of mathematics, provide a more intuitive approach to mathematical proofs, and formulate conjectures that can lead to important advances in the field. In this chapter, we introduce our discussion with a number of additional intriguing examples in the realm of sequences, series, products and integrals.

1.1 Pi Is Not 22/7

We first consider an example from the early history of π, as described in Chapter 3 of [43].

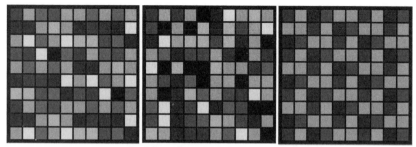

Archimedes: 223/71 < π < 22/7

Figure 1.1. A pictorial proof of Archimedes' inequality.

Even *Maple* or *Mathematica* "knows" $\pi \neq 22/7$, since

$$0 \ < \ \int_0^1 \frac{(1-x)^4 x^4}{1+x^2}\, dx \ = \ \frac{22}{7} - \pi, \tag{1.1}$$

though it would be prudent to ask "why" it can perform the evaluation and "whether" we should trust it?

Assume we trust it. Then the integrand is strictly positive on the interior of the interval of integration, and the answer in (1.1) is necessarily an area and thus strictly positive, despite millennia of claims that π is 22/7. Of course, 22/7 is one of the early continued fraction approximations to π. The first four are $3, 22/7, 333/106, 355/113$.

In this case, computing the indefinite integral provides immediate reassurance. We obtain

$$\int_0^t \frac{x^4 (1-x)^4}{1+x^2}\, dx \ = \ \frac{1}{7} t^7 - \frac{2}{3} t^6 + t^5 - \frac{4}{3} t^3 + 4t - 4 \arctan(t) \tag{1.2}$$

This is easily confirmed by differentiation, and the Fundamental Theorem of Calculus substantiates (1.1).

In fact, one can take this idea a bit further. We note that

$$\int_0^1 x^4 (1-x)^4\, dx \ = \ \frac{1}{630}, \tag{1.3}$$

and we observe that

$$\frac{1}{2} \int_0^1 x^4 (1-x)^4\, dx \ < \ \int_0^1 \frac{(1-x)^4 x^4}{1+x^2}\, dx < \int_0^1 x^4 (1-x)^4\, dx. \tag{1.4}$$

On combining this with (1.1) and (1.3), we straightforwardly derive $223/71 < 22/7 - 1/630 < \pi < 22/7 - 1/1260 < 22/7$, and so re-obtain Archimedes' famous computation

$$3\frac{10}{71} < \pi < 3\frac{10}{70} \tag{1.5}$$

(illustrating that it is sometimes better not to fully reduce a fraction to lowest terms).

This derivation of the estimate above seems first to have been written down in *Eureka*, the Cambridge student journal in 1971 [103]. The integral in (1.1) was apparently shown by Kurt Mahler to his students in the mid-1960s, and it had appeared in a mathematical examination at the University of Sydney in November, 1960. Figure 1.1 (see Color Plate I) shows the estimate graphically illustrated. The three 10×10 arrays color the digits of the first hundred digits of $223/71$, π, and $22/7$. One sees a clear pattern on the right $(22/7)$, a more subtle structure on the left $(223/71)$, and a "random" coloring in the middle (π).

It is tempting to ask if there is a clean general way to mimic (1.1) for more general rational approximations, or even continued fraction convergents. This is indeed possible to some degree, as discussed by Beukers in [24]. The most satisfactory result is

$$a_n\pi - \frac{b_n}{c_n} = \int_0^1 \frac{t^{2n}\left(1-t^2\right)^{2n}\left((1+it)^{3n+1} + (1-it)^{3n+1}\right)}{\left(1+t^2\right)^{3n+1}}dt, \tag{1.6}$$

for $n \geq 1$, where the integers a_n, b_n and c_n are implicitly defined by the integral in (1.6). The first three integrals evaluate to $14\pi - 44$, $968\pi - 45616/15$, and $75920\pi - 1669568/7$, so again we start with $\pi - 22/7$.

Unlike Beukers' preliminary attempts in [24], such as the seemingly promising

$$\int_0^1 \frac{t^n\left(1-t\right)^n}{\left(t^2+1\right)^{n+1}}dt,$$

this set of approximates p_n/q_n actually produces an explicit if weak *irrationality estimate* [44, 24]: for large n,

$$\left|\pi - \frac{p_n}{q_n}\right| \geq \frac{1}{q_n^{1.0499}}.$$

As Beukers sketches, one consequence of this explicit sequence is

$$\left|\pi - \frac{p}{q}\right| \geq \frac{1}{q^{21.04\ldots}}.$$

for all integers p, q with sufficiently large q. (Here $21.04\ldots = 1 + 1/0.0499$. In fact, in 1993 Hata by different methods had improved the number 21.04 to 8.02.)

While it is easy to discover "natural" results like

$$\frac{1}{5} \int_0^1 \frac{x\,(1-x)^2}{(1+x)^3}\,dx = \frac{7}{10} - \log(2), \tag{1.7}$$

the fact that $7/10$ is again a convergent to $\log 2$ seems to be largely a happenstance. For example,

$$\int_0^1 \frac{x^{12}\,(1-x)^{12}}{16\,(1+x^2)}\,dx = \frac{431302721}{137287920} - \pi$$

$$\int_0^1 \frac{x^{12}\,(1-x)^{12}}{16}\,dx = \frac{1}{1081662400}$$

leads to the true, if inelegant, estimate that $5902037233/1878676800 < \pi < 224277414953/71389718400$, where the interval is of size $1.39 \cdot 10^{-9}$.

In contrast to this easy symbolic success, *Maple* and *Mathematica* struggle with the following version of the *sophomore's dream*:

$$\int_0^1 \frac{1}{x^x}\,dx = \sum_{n=1}^{\infty} \frac{1}{n^n}. \tag{1.8}$$

When students are asked to confirm this, they most typically mistake numerical validation for symbolic proof: $1.291285997 = 1.291285997$. One seems to need to nurse a computer system, starting with integrating

$$x^{-x} = \exp\left(-x \log x\right) = \sum_{n=0}^{\infty} \frac{(-x \log x)^n}{n!},$$

term by term. See Exercise 3 at the end of this chapter.

1.2   Two Products

Consider the product

$$\prod_{n=2}^{\infty} \frac{n^3 - 1}{n^3 + 1} = \frac{2}{3}, \tag{1.9}$$

which has a rational value, and the seemingly simpler one (squares instead of cubes)

$$\prod_{n=2}^{\infty} \frac{n^2 - 1}{n^2 + 1} = \frac{\pi}{\sinh(\pi)}, \tag{1.10}$$

which evaluates to a transcendental number. *Mathematica* and *Maple* successfully evaluate such products, although not always in the same form. In this case, *Mathematica* produces expressions involving the Gamma function, while *Maple* returns the values shown above. In either case, we learn little or nothing from the results, since the software typically cannot recreate the steps of validation. In such a situation, it often pays to ask our software to evaluate the finite products and then take limits. Note that in earlier versions of *Maple* or *Mathematica*, the infinite products would have been returned unevaluated, so that we may have been led directly to the finite products. Nowadays the system knows more, but we often learn less! To use a modern educational term, we are not led to "unpack" the concepts.

When asked to evaluate the finite products, *Maple* returns expressions involving Gamma function values. For the first product (1.9), this expression can be simplified to

$$\prod_{n=2}^{N} \frac{n^3 - 1}{n^3 + 1} = \frac{2}{3} \frac{N^2 + N + 1}{N(N+1)},$$

and from this, one may get the idea that the evaluation can be done by telescoping. This directly leads to the following proof, which just consists of filling in the intermediate steps *Maple* still does not care to tell us:

$$\prod_{n=2}^{N} \frac{n^3 - 1}{n^3 + 1} = \prod_{n=2}^{N} \frac{(n-1)(n^2 + n + 1)}{(n+1)(n^2 - n + 1)} = \frac{\displaystyle\prod_{n=0}^{N-2}(n+1) \ \prod_{n=2}^{N}(n^2 + n + 1)}{\displaystyle\prod_{n=2}^{N}(n+1) \ \prod_{n=1}^{N-1}(n^2 + n + 1)}$$

$$= \frac{2}{N(N+1)} \cdot \frac{N^2 + N + 1}{3} \to \frac{2}{3}.$$

The second finite product does not simplify in any helpful way; however, the Gamma function expression, together with the *Maple* evaluation of the infinite product, gives us the hint that the sin-product formula

$$\sin(\pi x) = \pi x \prod_{n=1}^{\infty} \left(1 - \frac{x^2}{n^2}\right), \tag{1.11}$$

which is introduced in Chapter 5 of [43], plays a role here. With this idea, the proof of the evaluation is simple: By complexification (and holomorphy), it follows from (1.11) that

$$\frac{\sinh(\pi)}{\pi} = 2 \prod_{n=2}^{\infty} \frac{n^2 + 1}{n^2},$$

and we get

$$\frac{\sinh(\pi)}{\pi} \cdot \prod_{n=2}^{\infty} \frac{n^2 - 1}{n^2 + 1} = 2 \prod_{n=2}^{\infty} \frac{n^2 - 1}{n^2} = 1,$$

since the final product is again telescoping.

Do these evaluations generalize in a useful manner? For example, does the product $\prod_{n=2}^{\infty}(n^4 - 1)/(n^4 + 1)$ have an evaluation in terms of basic constants? *Maple* tells us that indeed

$$\prod_{n=2}^{\infty} \frac{n^4 - 1}{n^4 + 1} = \frac{\pi \sinh(\pi)}{\cosh(\sqrt{2}\,\pi) - \cos(\sqrt{2}\,\pi)}, \tag{1.12}$$

and it again produces a Gamma function expression for the finite product. For analogous products with fifth powers, *Maple* fails to return an evaluation. However, we now have enough hints to try our own hands at these products: Apparently we have to use properties of the Gamma function. In fact, setting $\omega = \exp(\pi i/r)$ and using the relations $\prod_{j=1}^{2r}(n - z\omega^j)^{(-1)^j} = (n^r - z^r)/(n^r + z^r)$, as well as $\sum_{j=1}^{2r} \omega^j(-1)^j = 0$ and $\prod_{j=1}^{2r}(\omega^j)^{(-1)^j} = -1$, it follows from the product representation of the Gamma function

$$\Gamma(x) = \lim_{n \to \infty} \frac{n!\, n^x}{x(x+1)\cdots(x+n)} \tag{1.13}$$

that, for $r \in \mathbb{N}$, $r > 1$, and $z \in \mathbb{C} \setminus \mathbb{N}$,

$$\prod_{n=1}^{\infty} \frac{n^r - z^r}{n^r + z^r} = \prod_{n=1}^{\infty} \prod_{j=1}^{2r} \left(1 - \frac{z\omega^j}{n}\right)^{(-1)^j} = -\prod_{j=1}^{2r} \Gamma(-z\omega^j)^{-(-1)^j}.$$

Hence, for $m \in \mathbb{N}$,

$$\prod_{n=1,\ n \neq m}^{\infty} \frac{n^r - z^r}{n^r + z^r} = -\frac{m^r + z^r}{(m^r - z^r)\Gamma(-z)} \prod_{j=1}^{2r-1} \Gamma(-zw^j)^{-(-1)^j},$$

where as $z \to m$,

$$(m^r - z^r)\Gamma(-z) = \frac{m^r - z^r}{(m-z)} \frac{\Gamma(m+1-z)}{(m-1-z)\cdots(1-z)(-z)} \to rm^{r-1}\frac{1}{m!(-1)^m}.$$

This gives the finite evaluation

$$P_r(m) \quad = \quad \prod_{n=1,\ n\neq m}^{\infty} \frac{n^r - m^r}{n^r + m^r}$$

$$= \quad (-1)^{m+1} \frac{2m(m!)}{r} \prod_{j=1}^{2r-1} \Gamma(-m\omega^j)^{-(-1)^j}. \qquad (1.14)$$

When $r = 2s$ is even, this can in a few steps be further reduced to

$$-(-1)^m \frac{2^\epsilon \pi m}{s} (\sinh \pi m)^{(-1)^s}$$

$$\times \prod_{j=1}^{s-1} \left(\cosh \left(2\pi m \sin \left(\tfrac{j\pi}{2s} \right) \right) - \cos \left(2\pi m \cos \left(\tfrac{j\pi}{2s} \right) \right) \right)^{(-1)^j},$$

where ϵ is 0 or 1 as s is respectively odd or even. From this, evaluations (1.10) and (1.12) immediately follow as special cases.

Interestingly, for odd $r \geq 5$, these products do not seem to have a closed form "nicer" than (1.14). In particular, they do not seem to be rational numbers like $P_3(1)$ (and in fact $P_3(m)$). The use of integer relation algorithms—to 400 digits—shows that $P_5(1)$ satisfies no integer polynomial with degree less than 21 and Euclidean norm less than $5 \cdot 10^{18}$.

1.3 A Recursive Sequence Problem

The following problem on a recursively defined sequence appeared in *American Mathematical Monthly* (Problem 10901, [58]). We will describe here how the problem, which really is a problem about functional equations, can be solved via the experimental approach.

Problem: Let $a_1 = 1$,

$$a_2 = \frac{1}{2} + \frac{1}{3}, \quad a_3 = \frac{1}{3} + \frac{1}{7} + \frac{1}{4} + \frac{1}{13},$$

$$a_4 = \frac{1}{4} + \frac{1}{13} + \frac{1}{8} + \frac{1}{57} + \frac{1}{5} + \frac{1}{21} + \frac{1}{14} + \frac{1}{183},$$

and continue the sequence, constructing a_{n+1} by replacing each fraction $1/d$ in the expression for a_n with $1/(d+1) + 1/(d^2 + d + 1)$. Compute $\lim_{n\to\infty} a_n$.

Solution: We first observe that if $s_0(x) = 1/x$ and $s_{n+1}(x) = s_n(x + 1) + s_n(x^2 + x + 1)$ for $n \geq 0$, $x > 0$, then $a_n = s_{n+1}(1)$. What do

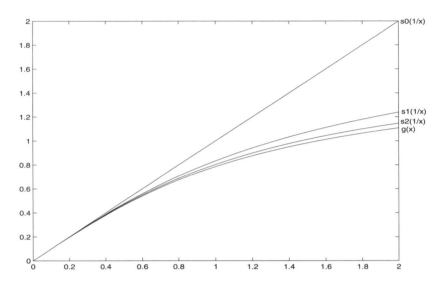

Figure 1.2. Convergence of $s_n(1/x)$ to $g(x)$.

these functions $s_n(x)$ look like? Like $s_0(x)$, the plots of successive $s_n(x)$ resemble reciprocal functions. If we instead examine the functions $s_n(1/x)$, we find that these are fairly well behaved, appearing to converge quickly to a smooth, monotone increasing function $g(x)$ (see Figure 1.2). Indeed, we find fairly good convergence (to roughly four decimal places) for $n = 25$, by comparing $s_{24}(1/x)$ with $s_{25}(1/x)$. What is this function $g(x)$?

Examining the sequence of calculated numerical values used for plotting, we find that while $g(x) = \lim_n s_n(1/x)$ is not defined at zero, it appears that $\lim_{x \to 0} g(x) = 0$. Further, it appears that $g'(0) = 1$, $g(1) \approx 0.7854$ and $g'(1) = 1/2$. Needless to say, the value 0.7854 is an approximation to $\pi/4$. These observations suggest that perhaps $g(x) = \arctan x$.

Let $f(x) = \arctan(1/x)$ for $x > 0$. By applying the addition formula for the tangent, we note that

$$\tan[f(x+1) + f(x^2 + x + 1)] = \cfrac{\cfrac{1}{x+1} + \cfrac{1}{x^2 + x + 1}}{1 - \cfrac{1}{x+1} \cdot \cfrac{1}{x^2 + x + 1}}$$

$$= \frac{1}{x} = \tan[f(x)].$$

This means that $f(x)$ satisfies $f(x) = f(x+1) + f(x^2 + x + 1)$, confirming that we are on the right track. In fact, we are finished if we can show that $s_n(x)$ converges pointwise to $f(x)$.

To demonstrate this, we first verify that the function $E(x) = 1/(x\,f(x))$ decreases strictly to 1 as $x \to \infty$. By differentiation, it suffices to show that $-\arctan(x) + x/(x^2+1) < 0$. But this follows since $-\arctan(x) + x/(x^2+1)$ is strictly decreasing (its derivative is $-2x^2/(x^2+1)^2$), and it starts at 0 for $x = 0$.

The second step is to show that for all $x > 0$, we have

$$f(x) \leq s_n(x) \leq f(x) \cdot E(x+n). \tag{1.15}$$

For $n = 0$ this is merely the condition $xf(x) \leq 1$. Now if (1.15) holds for some $n > 0$, then we infer

$$f(x+1) \leq s_n(x+1) \leq f(x+1) \cdot E(x+n+1),$$

and (using the monotonicity of E),

$$f(x^2+x+1) \leq s_n(x^2+x+1) \leq f(x^2+x+1) \cdot E(x^2+x+1+n)$$
$$\leq f(x^2+x+1) \cdot E(x+n+1).$$

Adding (and using the functional equation for f), we obtain (1.15) for $n+1$. These facts together imply $\lim_{n \to \infty} s_n(x) = f(x)$ for each $x > 0$.

Thus we have demonstrated here that $a_n \to \pi/4$ [58]. $\qquad\square$

An extension of these methods leads to the following theorem:

Theorem 1.1. *Let $\mathcal{A} = \{s : \mathrm{R}_+ \to \mathrm{R} : \lim_{x \to \infty} x\,s(x) = 1\}$ and define a mapping $T : \mathcal{A} \to \mathcal{A}$ by $(Ts)(x) = s(x+1) + s(x^2+x+1)$ for $s \in \mathcal{A}$. Then the sequence (s_n) defined by the iteration $s_{n+1} = Ts_n$ converges pointwise to $f(x) = \arctan(1/x)$, for every $s_0 \in \mathcal{A}$. Equivalently, every orbit of T converges pointwise to f, which is the unique fixed point of T in $\mathcal{A}$.*

Proof. Define $e(x) = \inf_{y \geq x} s_0(y)/f(y)$ and $E(x) = \sup_{y \geq x} s_0(y)/f(y)$. Then $f(x)\,e(x) \leq s_0(x) \leq f(x)\,E(x)$ for all $x > 0$, while $e(x)$ increases to 1 and $E(x)$ decreases to 1 as $x \to \infty$. Now the same induction as in Step 2 above gives us

$$f(x)\,e(x+n) \leq s_n(x) \leq f(x)\,E(x+n) \quad \text{for all } n \geq 0,\ x > 0.$$

This implies $s_n(x) \to f(x)$ for $x > 0$. This argument, slightly modified, also shows that $f(x) = \arctan(1/x)$ is the unique fixed point of T in $\mathcal{A}$. $\qquad\square$

The same procedure works for $1/x \to 1/(x+y) + y/(x^2+xy+1)$, for $0 < y \leq 1$. More generally, there are similar functional equations for other inverse functions. Thus, for $l(x) = \log(1 + 1/x)$ the equation is

$$l(x) = l(2x+1) + l(2x).$$

The corresponding iteration, for $x = 1$, starting with $s_0(x) = 1/x$, produces the classical result

$$\sum_{k=2^n}^{2^{n+1}-1} 1/k \to \log(2) \quad \text{as } n \to \infty.$$

Similarly, for $\tau(x) = \text{arctanh}(1/x) = \frac{1}{2}\log((x-1)/(x+1))$, the functional equation is

$$\tau(x) \quad = \quad \tau(x+1) + \tau(x^2 + x - 1),$$

for $x > 1$. Likewise, for $\sigma = x \mapsto \text{arcsinh}(1/x)$, we have

$$\sigma(x) \quad = \quad \sigma\left(\sqrt{x^2+1}\right) + \sigma\left(x\sqrt{x^2+1}\left(\sqrt{x^2+1} + \sqrt{x^2+2}\right)\right).$$

And, for $\rho(x) = \arcsin(1/x)$, we have

$$\rho(x) \quad = \quad \rho\left(\sqrt{x^2+1}\right) + \rho\left(x\sqrt{x^2+1}\left(x + \sqrt{x^2-1}\right)\right),$$

for all $x \geq 1$.

For these four functional equations, the result corresponding to Theorem 1.1 can be established. In fact, the basic inequality corresponding to Step 2 above for the last two functions would read

$$f(x) \cdot e(\sqrt{x^2+n}) \quad \leq \quad s_n(x) \quad \leq \quad f(x) \cdot E(\sqrt{x^2+n}),$$

for $x \geq 1$, where s_n is defined, as before, by $s_0 \in \mathcal{A}$ and

$$s_{n+1}(x) \quad = \quad s_n(\sqrt{x^2+1}) + s_n\left(x\sqrt{x^2+1}\left(\sqrt{x^2+1} + \sqrt{x^2+2}\right)\right)$$

in the case of σ, and

$$s_{n+1}(x) \quad = \quad s_n(\sqrt{x^2+1}) + s_n\left(x\sqrt{x^2+1}\left(x + \sqrt{x^2-1}\right)\right)$$

in the case of ρ. By contrast,

$$\rho(x) \quad = \quad \rho\left(\frac{x^2}{(x-1)\sqrt{x^2-1} + \sqrt{2x-1}}\right) - \rho\left(\frac{x}{x-1}\right)$$

is another functional equation for $\arcsin(1/x)$ which does not have convergent orbits.

1.4   High Precision Fraud

Consider the sums

$$\sum_{n=1}^{\infty} \frac{\lfloor n \tanh(\pi) \rfloor}{10^n} \stackrel{?}{=} \frac{1}{81},$$

an evaluation that is wrong, but valid to 268 decimal places, and

$$\sum_{n=1}^{\infty} \frac{\lfloor n \tanh(\pi/2) \rfloor}{10^n} \stackrel{?}{=} \frac{1}{81},$$

which is valid to "only" 12 places. Both series actually evaluate to transcendental numbers.

What underlies these "fraudulent" evaluations? The "quick" reason is that $\tanh(\pi)$ and $\tanh(\pi/2)$ are almost integers, with, e.g., $0.99 < \tanh(\pi) < 1$. Therefore, $\lfloor n \tanh(\pi) \rfloor$ will be equal to $n - 1$ for many n; precisely for $n = 1, \cdots, 268$. Since

$$\sum_{n=1}^{\infty} \frac{n - 1}{10^n} = \frac{1}{81},$$

this explains the evaluations. Looking more closely at this argument, one is directly led to *continued fractions* as the deeper reason behind the frauds. For any irrational positive α, we can write

$$\begin{aligned}
\alpha &= [a_0, a_1, \cdots, a_n, a_{n+1}, \cdots] \\
&= a_0 + \cfrac{1}{a_1 + \cfrac{1}{a_2 + \cfrac{1}{a_3 + \cdots}}},
\end{aligned}$$

with integral a_n and $a_0 \geq 0$, $a_n \geq 1$ for $n \geq 1$. This is hard to compute by hand, but easy even on a small computer or calculator. For the parameters in our series, we get

$$\tanh(\pi) = [0, 1, \mathbf{267}, 4, 14, 1, 2, 1, 2, 2, 1, 2, 3, 8, 3, 1, \cdots] \tag{1.16}$$

and

$$\tanh\left(\frac{\pi}{2}\right) = [0, 1, \mathbf{11}, 14, 4, 1, 1, 1, 3, 1, 295, 4, 4, 1, 5, 17, 7, \cdots]. \tag{1.17}$$

It cannot be a coincidence that the integers 267 and 11 (each equal to the number of places of agreement with $1/81$ in the respective formula) appear in these expansions! There must be a connection between series of the type $\sum \lfloor n\alpha \rfloor z^n$ and the continued fraction expansion of an irrational α. In fact, consider the infinite continued fraction approximations for α generated by

$$
\begin{aligned}
p_{n+1} &= p_n a_{n+1} + p_{n-1}, & p_0 = a_0 = \lfloor \alpha \rfloor, & \quad p_{-1} = 1, \\
q_{n+1} &= q_n a_{n+1} + q_{n-1}, & q_0 = 1, & \quad q_{-1} = 0.
\end{aligned}
$$

Then for $n \geq 0$, p_{2n}/q_{2n} increases to α, while p_{2n+1}/q_{2n+1} decreases to α and

$$
\frac{1}{q_n (q_n + q_{n+1})} < \left| \alpha - \frac{p_n}{q_n} \right| < \frac{1}{q_n q_{n+1}}.
$$

Let further $\epsilon_n = q_n \alpha - p_n$. Then from the above, it follows that

$$
|\epsilon_{n+1}| < \frac{1}{q_n + q_{n+1}} < |\epsilon_n| < \frac{1}{q_{n+1}} \leq 1.
$$

All of this is standard and may be found in [137], [212], or [179]. Our aim now is to show a relationship between the above series and the continued fraction expansion of α. A first key is the following lemma, which we will not prove here since it requires some knowledge about linear Diophantine equations (see [47], from which this material is taken).

Lemma 1.2. *For any irrational $\alpha > 0$ and $n, N \in \mathbb{N}$, we have*

$$
\begin{aligned}
\lfloor n\alpha + \epsilon_N \rfloor &= \lfloor n\alpha \rfloor & \textit{for } n < q_{N+1}, \\
\lfloor n\alpha + \epsilon_N \rfloor &= \lfloor n\alpha \rfloor + (-1)^N & \textit{for } n = q_{N+1}.
\end{aligned}
$$

Theorem 1.3. *For irrational $\alpha > 0$,*

$$
\sum_{n=1}^{\infty} \lfloor n\alpha \rfloor z^n = \frac{p_0 \, z}{(1-z)^2} + \sum_{n=0}^{\infty} (-1)^n \frac{z^{q_n} z^{q_{n+1}}}{(1 - z^{q_n})(1 - z^{q_{n+1}})}.
$$

Proof. Let

$$
G_\alpha(z, w) = \sum_{n=1}^{\infty} z^n w^{\lfloor n\alpha \rfloor}, \tag{1.18}
$$

for $|z|, |w| < 1$. Then for $N > 0$,

$$(1 - z^{q_N} w^{p_N}) G_\alpha(z, w) - \sum_{n=1}^{q_N} z^n w^{\lfloor n\alpha \rfloor}$$

$$= \sum_{n=1}^{\infty} z^{n+q_N} \left(w^{\lfloor (n+q_N)\alpha \rfloor} - w^{\lfloor n\alpha \rfloor + p_N} \right)$$

$$= \sum_{n=1}^{\infty} z^{n+q_N} w^{\lfloor n\alpha \rfloor + p_N} \left(w^{\lfloor n\alpha + \epsilon_N \rfloor - \lfloor n\alpha \rfloor} - 1 \right)$$

$$= z^{q_{N+1}+q_N} w^{\lfloor q_{N+1}\alpha \rfloor + p_N} \left(w^{(-1)^N} - 1 \right) + O(z^{q_{N+1}+q_N+1})$$

$$= z^{q_{N+1}+q_N} w^{p_{N+1}+p_N} (-1)^N \frac{w-1}{w} + O(z^{q_{N+1}+q_N+1}), \qquad (1.19)$$

since $\lfloor q_{N+1}\alpha \rfloor = \lfloor \epsilon_{N+1} \rfloor + p_{N+1} = p_{N+1}$ if N is odd, and $= p_{N+1} - 1$ if N is even.

Now write $P_N = \sum_{n=1}^{q_N} z^n w^{\lfloor n\alpha \rfloor}$ and $Q_N = 1 - z^{q_N} w^{p_N}$. Then $A_N = Q_N P_{N+1} - Q_{N+1} P_N$ is a polynomial of degree at most $q_N + q_{N+1}$ in z, and therefore it follows from (1.19) that

$$A_N = Q_{N+1}(Q_N G_\alpha - P_N) - Q_N(Q_{N+1} G_\alpha - P_{N+1})$$

$$= (-1)^N \frac{w-1}{w} z^{q_N} w^{p_N} z^{q_{N+1}} w^{p_{N+1}}.$$

This in turn implies

$$\frac{P_{N+1}}{Q_{N+1}} - \frac{P_N}{Q_N} = \frac{A_N}{Q_N Q_{N+1}} = (-1)^N \frac{w-1}{w} \frac{z^{q_N} w^{p_N} z^{q_{N+1}} w^{p_{N+1}}}{Q_N Q_{N+1}}.$$

Next summing from zero to infinity, and noting that (1.19) implies that $G_\alpha - P_N/Q_N$ tends to 0 as N tends to infinity, shows that

$$G_\alpha(z, w) = \frac{zw^{p_0}}{1 - zw^{p_0}} - \frac{1-w}{w} \sum_{n=0}^{\infty} (-1)^n \frac{z^{q_n} w^{p_n} z^{q_{n+1}} w^{p_{n+1}}}{(1 - z^{q_n} w^{p_n})(1 - z^{q_{n+1}} w^{p_{n+1}})}.$$

Now differentiating with respect to w and then letting w tend to 1 proves the assertion. □

This theorem was first proved (for $\alpha \in (0, 1)$) by Mahler in [162].

Example 1.4. $\alpha = \tanh(\pi)$.

In this case, $q_n = 1, 1, 268, 1073, \cdots$ for $n = 0, 1, 2, 3, \cdots$, and thus

$$\sum_{n=1}^{\infty} \lfloor n \tanh(\pi) \rfloor z^n = \frac{z^2}{(1-z)^2} - \frac{z^{269}}{(1-z)(1-z^{268})} + \cdots.$$

Therefore,

$$\frac{1}{81} - 2 \cdot 10^{-269} \leq \sum_{n=1}^{\infty} \frac{\lfloor n \tanh(\pi) \rfloor}{10^n} \leq \frac{1}{81} + 2 \cdot 10^{-269},$$

and similarly for $\alpha = \tanh(\frac{\pi}{2})$.   □

Example 1.5. $\alpha = e^{\pi \sqrt{163/9}}$.

With one of our favorite transcendental numbers, $\alpha = e^{\pi \sqrt{163/9}} = [640320,$
$1653264929, \cdots]$, we get the incorrect evaluation

$$\sum_{n=1}^{\infty} \frac{\lfloor n e^{\pi \sqrt{163/9}} \rfloor}{2^n} \stackrel{?}{=} 1280640,$$

which is, however, correct to at least half a billion digits.   □

Example 1.6. $\alpha = \log_{10}(2)$.

In this case, $\lfloor n\alpha \rfloor + 1$ is the number of decimal digits of 2^n. Then $q_n = 1, 3, 10, 93, \cdots$ for $n = 0, 1, 2, 3, \cdots$, and as a result the transcendental number $\sum \lfloor n \log_{10}(2) \rfloor / 2^n$ is equal to $146/1023$ to 30 decimal digits. Interestingly, if $e(n)$, respectively $o(n)$, count the number of even, respectively odd, decimal digits of n, then

$$\sum_{n=1}^{\infty} \frac{o(2^n)}{2^n} = \frac{1}{9}$$

is rational, while

$$\sum_{n=1}^{\infty} \frac{e(2^n)}{2^n} = \sum_{n=1}^{\infty} \frac{\lfloor n \log_{10}(2) \rfloor + 1}{2^n} - \sum_{n=1}^{\infty} \frac{o(2^n)}{2^n}$$

is transcendental. We will not prove the transcendency result here, but the evaluation for the sum with $o(2^n)$ follows in the next theorem.   □

Theorem 1.7. *If $o(n)$ counts the odd decimal digits of n, then*

$$\sum_{n=1}^{\infty} \frac{o(2^n)}{2^n} = \frac{1}{9}.$$

Proof. Let $0 < q < 1$ and $m \in \mathbb{N}$, $m > 1$, and consider the base-m expansion of q,

$$q = \sum_{n=1}^{\infty} \frac{a_n}{m^n} \quad \text{with } 0 \le a_n < m,$$

where when ambiguous we take the terminating expansion. Then a_n is the remainder of $\lfloor m^n q \rfloor$ modulo m, and therefore we can just as well write

$$q = \sum_{n=1}^{\infty} \frac{\lfloor m^n q \rfloor \ (\text{mod } m)}{m^n}. \tag{1.20}$$

Now let $F(q) = \sum_{k=1}^{\infty} c_k q^k$ be a power series with radius of convergence 1. Then for $0 < q < 1$, from (1.20) we obtain, by exchanging the order of summation (as is valid within the radius of convergence),

$$F(q) = \sum_{k=1}^{\infty} c_k q^k = \sum_{k=1}^{\infty} c_k \sum_{n=1}^{\infty} \frac{\lfloor m^n q^k \rfloor \bmod m}{m^n} = \sum_{n=1}^{\infty} \frac{f(n)}{m^n},$$

with $f(n) = \sum_{k \ge 1} c_k \left(\lfloor m^n q^k \rfloor \bmod m \right)$. Now if $q = 1/b$, where b is an integer multiple of m, then $\lfloor m^n / b^k \rfloor \bmod m$ is the k-th digit (mod m) of the base-b expansion of the integer m^n. (Here we start the numbering of the digits with 0, e.g., the 0-th digit of 1205 is 5.) Thus for $F(q) = q/(1-q)$ and $m = 2$ (and b even), $f(n)$ counts the odd digits in the base-b expansion of 2^n. For $b = 10$, we have $f(n) = o(2^n)$, and we obtain

$$\frac{1}{9} = F\left(\frac{1}{10}\right) = \sum_{n=1}^{\infty} \frac{o(2^n)}{2^n}. \qquad \square$$

1.5   Knuth's Series Problem

We give an account here of the solution, by one of the present authors (Borwein), to a problem recently posed by Donald E. Knuth of Stanford University in the *American Mathematical Monthly* (Problem 10832, Nov. 2000):

Problem: Evaluate

$$S = \sum_{k=1}^{\infty} \left(\frac{k^k}{k! e^k} - \frac{1}{\sqrt{2\pi k}} \right).$$

Solution: We first attempted to obtain a numerical value for S. Using *Maple*, we produced the approximation

$$S \approx -0.08406950872765599646.$$

Based on this numerical value, the Inverse Symbolic Calculator, available at the URL http://www.cecm.sfu.ca/projects/ISC, with the "Smart Lookup" feature, yielded the result

$$S \approx -\frac{2}{3} - \frac{1}{\sqrt{2\pi}}\zeta\left(\frac{1}{2}\right). \tag{1.21}$$

Calculations to even higher precision (50 decimal digits) confirmed this approximation. Thus within a few minutes we "knew" the answer.

Why should such an identity hold? One clue was provided by the surprising speed with which *Maple* was able to calculate a high-precision value of this slowly convergent infinite sum. Evidently, the *Maple* software knew something that we did not. Peering under the covers, we found that *Maple* was using the Lambert W function, which is the functional inverse of $w(z) = ze^z$.

Another clue was the appearance of $\zeta(1/2)$ in the above experimental identity, together with an obvious allusion to Stirling's formula in the original problem. This led us to conjecture the identity

$$\sum_{k=1}^{\infty}\left(\frac{1}{\sqrt{2\pi k}} - \frac{P(1/2, k-1)}{(k-1)!\sqrt{2}}\right) = \frac{1}{\sqrt{2\pi}}\zeta\left(\frac{1}{2}\right), \tag{1.22}$$

where $P(x, n)$ denotes the Pochhammer function $x(x+1)\cdots(x+n-1)$, and where the binomial coefficients in the LHS of (1.22) are the same as those of the function $1/\sqrt{2-2x}$. *Maple* successfully evaluated this summation, as shown on the RHS. We now needed to establish that

$$\sum_{k=1}^{\infty}\left(\frac{k^k}{k!e^k} - \frac{P(1/2, k-1)}{(k-1)!\sqrt{2}}\right) = -\frac{2}{3}.$$

Guided by the presence of the Lambert W function

$$W(z) = \sum_{k=1}^{\infty}\frac{(-k)^{k-1}z^k}{k!},$$

an appeal to Abel's limit theorem suggested the conjectured identity

$$\lim_{z\to 1}\left(\frac{dW(-z/e)}{dz} + \frac{1}{2-2z}\right) = 2/3.$$

Here again, *Maple* was able to evaluate this summation and establish the identity.   □

As can be seen from this account, the above manipulation took considerable human ingenuity, in addition to computer-based symbolic manipulation. We include this example to highlight a challenge for the next generation of mathematical computing software—these tools need to more completely automate this class of operations, so that similar derivations can be accomplished by a significantly broader segment of the mathematical community.

1.6   Ahmed's Integral Problem

The same comments apply to an "experimental" solution to a problem posed by Zafar Ahmed in the *American Mathematical Monthly* [3]:

Problem: Evaluate

$$F \;=\; \int_0^1 \frac{\arctan\left(\sqrt{x^2+2}\right)}{\sqrt{x^2+2}\,(x^2+1)}\,dx.$$

Solution: Since presently available symbolic computing software is unable to produce a closed-form evaluation, we try to identify the integral via its numerical value,

$$F \;\approx\; 0.51404189589900707613976297395768828.$$

The Inverse Symbolic Calculator (with the integer relations algorithm clicked on) declares that this number matches $5\pi^2/96$. A test to higher precision confirms this evaluation.

It remains to prove the experimentally well-founded conjecture

$$F \;=\; \frac{5}{96}\,\pi^2.$$

An idea is needed, which, as always, takes more human insight than any computer algebra system (at present) has built in. However, such a system can help to quickly identify promising starting points—and it can then carry out the symbolic manipulations which, taken together, constitute a proof. One possible starting point for this problem is to generalize: Using *Maple*, set

```
>  assume(x>0,p>0); interface(showassumed=0);
>  g := arctan(p*sqrt(x^2+2))/sqrt(x^2+2)/(x^2+1);
```

$$g = \frac{\arctan(p\sqrt{x^2+2})}{\sqrt{x^2+2}\,(x^2+1)}$$

```
>  G := Int(g(x,p),x=0..1);
```

$$G = \int_0^1 \frac{\arctan(p\sqrt{x^2+2})}{\sqrt{x^2+2}\,(x^2+1)}\,dx,$$

so that $F = G(1)$. On the other hand,

```
>  diff(g,p);
```

$$\frac{1}{(1+p^2\,(x^2+2))\,(x^2+1)}$$

is a rational function in x and p, so that it may pay to write

$$F = G(1) = \int_0^1 \int_0^1 \partial g(x,p)/\partial p \, dp \, dx = \int_0^1 \int_0^1 \partial g(x,p)/\partial p \, dx \, dp,$$

where the exchange of order of integration is justified by Fubini's (Tonelli's) theorem. Thus we are led to

```
>  int(diff(g,p),x=0..1);
```

$$\frac{1}{4} \frac{-4p\arctan\left(\dfrac{p}{\sqrt{1+2p^2}}\right) + \pi\sqrt{1+2p^2}}{(p^2+1)\sqrt{1+2p^2}}$$

```
>  map(int,expand(int(diff(g,p),x=0..1)),p=0..1);
```

$$\int_0^1 -\frac{p\arctan\left(\dfrac{p}{\sqrt{1+2p^2}}\right)}{(p^2+1)\sqrt{1+2p^2}}\,dp + \frac{1}{16}\pi^2$$

```
>  G1 := map(int,expand(int(diff(g,p),x=0..1)),p=0..1)-Pi^2/16;
```

$$G1 = \int_0^1 -\frac{p\arctan\left(\dfrac{p}{\sqrt{1+2p^2}}\right)}{(p^2+1)\sqrt{1+2p^2}}\,dp$$

and $F = G1 + \pi^2/16$. Now using $\arctan(y) + \arctan(1/y) = \pi/2$ for $y > 0$ and then doing a change of variables $x = 1/p$, we obtain

```
>   G2:=subs(arctan(p/sqrt(1+2*p^2))
        =Pi/2-arctan(sqrt(1+2*p^2)/p),G1);
```

$$G2 = \int_0^1 -\frac{p\left(\frac{1}{2}\pi - \arctan\left(\frac{\sqrt{1+2p^2}}{p}\right)\right)}{(p^2+1)\sqrt{1+2p^2}}\,dp$$

```
>   G3:=map(int,expand(op(1,G2)),p=0..1);
```

$$G3 = -\frac{1}{24}\pi^2 + \int_0^1 \frac{p\arctan\left(\frac{\sqrt{1+2p^2}}{p}\right)}{(p^2+1)\sqrt{1+2p^2}}\,dp$$

```
>   G4:=simplify(student[changevar](x=1/p,G3,x));
```

$$G4 = -\frac{1}{24}\pi^2 + \int_1^\infty \frac{\arctan\left(\sqrt{x^2+2}\right)}{(x^2+1)\sqrt{x^2+2}}\,dx$$

```
>   H:=G4+Pi^2/24;
```

$$H = \int_1^\infty \frac{\arctan\left(\sqrt{x^2+2}\right)}{(x^2+1)\sqrt{x^2+2}}\,dx$$

so that $F = G1 + \pi^2/16 = G4 + \pi^2/16 = H + \pi^2/48$. This evaluates $F - H = \pi^2/48$. This suggests we also evaluate

$$F + H = \int_0^\infty \int_0^1 \partial g(x,p)/\partial p\,dp\,dx = \int_0^1 \int_0^\infty \partial g(x,p)/\partial p\,dx\,dp.$$

Indeed,

```
>   int(diff(g,p),x=0..infinity);
```

$$-\frac{1}{2}\frac{\pi\left(p + \sqrt{1-2p^2}\right)}{(p^2+1)\sqrt{1+2p^2}}$$

```
>   FpH := int(int(diff(g,p),x=0..infinity),p=0..1);
```

$$FpH = \frac{1}{12}\pi^2.$$

Now the result is proved by

```
> solve({f-h=Pi^2/48,f+h=Pi^2/12},{f,h});
```

$$\{f = \frac{5}{96}\pi^2, h = \frac{1}{32}\pi^2\}.$$

□

Generalization: We note that in the same fashion we can evaluate

$$\int_0^1 \frac{\arctan(\sqrt{x^2+b^2})}{\sqrt{x^2+b^2}\,(x^2+1)}\,dx - \int_1^\infty \frac{\arctan(\sqrt{x^2+b^2})}{\sqrt{x^2+b^2}\,(-1+x^2+b^2)}\,dx$$

$$= \frac{3}{4}\frac{\pi\arctan(\sqrt{b^2-1})}{\sqrt{b^2-1}} - \frac{1}{2}\frac{\pi\arctan(\sqrt{b^4-1})}{\sqrt{b^2-1}},$$

$$\int_0^1 \frac{\arctan(\sqrt{x^2+b^2})}{\sqrt{x^2+b^2}\,(x^2+1)}\,dx + \int_1^\infty \frac{\arctan(\sqrt{x^2+b^2})}{\sqrt{x^2+b^2}\,(x^2+1)}\,dx$$

$$= \frac{\pi\arctan(\sqrt{b^2-1})}{\sqrt{b^2-1}} - \frac{1}{2}\frac{\pi\arctan(\sqrt{b^4-1})}{\sqrt{b^2-1}},$$

and for $b = \sqrt{2}$, this yields the previous closed form. Moreover, for $b = 1$,

$$\int_0^1 \frac{\arctan(\sqrt{x^2+1})}{(x^2+1)^{3/2}}\,dx = \left(\frac{1}{4} - \frac{\sqrt{2}}{2}\right)\pi + \frac{3}{2}\sqrt{2}\arctan(\sqrt{2}),$$

and for $b = 0$,

$$\int_0^1 \frac{\arctan(x)}{x\,(x^2+1)}\,dx = \frac{G}{2} + \frac{1}{8}\pi\log(2),$$

where G is Catalan's constant.

1.7 Evaluation of Binomial Series

A classical binomial series, derived from the arctan series, and given in
[44], is

$$\sum_{n\geq1} \frac{-9n+18}{\binom{2n}{n}} = 2\frac{\pi}{\sqrt{3}}. \tag{1.23}$$

A more modern sum, due to Bill Gosper [129], is

$$\sum_{n\geq0} \frac{50n-6}{\binom{3n}{n}2^n} = \pi. \tag{1.24}$$

In [7], whole classes of formulas for π of this type are proved, such as

$$\sum_{n \geq 0} \frac{S_k(n)}{\binom{8kn}{4kn}(-4)^{kn}} = \pi,$$

where $S_k(n)$ is a polynomial in n of degree $4k$ with rational coefficients, explicitly computable (for fixed k).

Motivated by such results, we shall consider here the following two families of sums:

$$b_2(k) = \sum_{n \geq 1} \frac{n^k}{\binom{2n}{n}},$$

$$b_3(k) = \sum_{n \geq 1} \frac{n^k}{\binom{3n}{n} 2^n},$$

for $k \in \mathbb{Z}$. We shall record closed forms for the sums $b_2(k)$ and recursion formulas for the sums $b_3(k)$, both in the case of positive k. These were discovered with the help of integer relation and similar methods, described below. The case of negative k is more complex; we shall finish with some primarily experimental results. This material is taken from [54] (for $b_2(k)$ with negative k) and from [59].

The key observation is that the sums have integral representations involving the *polylogarithms* $L_p(z) = \sum_{n>0} z^n/n^p$ (see [61]). Using the following properties of the β-function (see Section 5.4 of [43]):

$$\frac{1}{\binom{2n}{n}} = (2n+1)\,\beta(n+1, n+1) = n\,\beta(n, n+1) \quad \text{and}$$

$$\frac{1}{\binom{3n}{n}} = (3n+1)\,\beta(2n+1, n+1) = n\,\beta(2n+1, n) = 2n\,\beta(2n, n+1),$$

we find that

$$b_2(k) = \int_0^1 L_{-k}(x(1-x)) + 2\,L_{-k-1}(x(1-x))\,dx \qquad (1.25)$$

$$= \int_0^1 \frac{L_{-k-1}(x(1-x))}{x}\,dx,$$

and

$$b_3(k) = \int_0^1 L_{-k}\left(\frac{x^2(1-x)}{2}\right) + 3\,L_{-k-1}\left(\frac{x^2(1-x)}{2}\right)\,dx \qquad (1.26)$$

$$= \int_0^1 \frac{L_{-k-1}\left(\frac{x^2(1-x)}{2}\right)}{1-x}\,dx = 2\int_0^1 \frac{L_{-k-1}\left(\frac{x^2(1-x)}{2}\right)}{x}\,dx.$$

For fixed $k \geq 0$, these integrals are easy to compute symbolically in *Maple* and (with some additional effort) in *Mathematica*.

1.7.1   The Nonnegative Case

For integer $k \geq 0$, $L_{-k}(x)$ is clearly a rational function, and it is useful to write it as a partial fraction

$$L_{-k}(x) \;=\; \sum_{j=1}^{k+1} \frac{c_j^k}{(x-1)^j}.$$

Since $x \, dL_{-k}(x)/dx = L_{-k-1}(x)$, we may obtain the recursion

$$c_j^k \;=\; -\left(j \, c_j^{k-1} + (j-1) \, c_{j-1}^{k-1} \right). \tag{1.27}$$

Let

$$\begin{aligned}
M_2(k, x) &= L_{-k}(x) + 2\, L_{-k-1}(x) \\
M_3(k, x) &= L_{-k}(x) + 3\, L_{-k-1}(x).
\end{aligned}$$

We may then easily verify that the coefficients of the partial fraction of M_2 and M_3 are governed by recursion (1.27) with initial conditions given by $c_1^0(2) = 1, c_2^0(2) = 2$, otherwise $c_j^0(2) = 0$, and $c_1^0(3) = 2, c_2^0(3) = 3$, otherwise $c_j^0(3) = 0$, respectively.

This in turn is easily verified—by hand or in a computer algebra system—to yield

$$c_j^k(2) \;=\; \frac{(-1)^{k+j}}{j} \sum_{m=1}^{j} (-1)^m \, (2m-1) \, m^{k+1} \binom{j}{m}, \tag{1.28}$$

$$c_j^k(3) \;=\; \frac{(-1)^{k+j}}{j} \sum_{m=1}^{j} (-1)^m \, (3m-1) \, m^{k+1} \binom{j}{m}.$$

As is often the case, this is somewhat easier to verify than to find.

Next we observe that the values of the integrals in (1.25) and (1.26) are of the form

$$b_2(k) \;=\; \sum_{j} c_j^k(2) \int_0^1 (1 - x(1-x))^{-j} \, dx$$

and

$$b_3(k) \;=\; \sum_{j} c_j^k(3) \int_0^1 \left(1 - x^2(1-x)/2\right)^{-j} \, dx, \tag{1.29}$$

respectively. So we set ourselves to the task of hunting for a recursion for

$$d_2(j) = \int_0^1 (1 - x(1 - x))^{-j} \, dx$$

and for

$$d_3(j) = \int_0^1 \left(1 - x^2(1 - x)/2\right)^{-j} \, dx.$$

By computing the first few cases, we determine that $d_2(j)$ is a rational combination of 1 and $\pi/\sqrt{3}$, while $d_3(j)$ is a rational combination of $1, \log 2$ and π. Thus it is reasonable to hunt for two-term recursions for d_2 and three-term recursions for d_3.

Now integer relation computations come to the rescue. We look for relations between $d_2(p), d_2(p+1)$, and $d_2(p+2)$, say for $0 \le p \le 4$, and are rewarded by the relations $[2, 2, -3]$, $[-2, 9, -6]$, $[6, -16, 9]$, $[-10, 23, -12]$, $[14, -30, 15]$. By inspection, we have $d_2(0) = 1, d_2(1) = 2\pi/(3\sqrt{3})$ and

$$(4p - 10) \, d_2(p - 2) - (7p - 12) \, d_2(p - 1) + (3p - 3) \, d_2(p) \quad = \quad 0 \quad (1.30)$$

for $p \ge 2$.

For d_3, we look for four-term relations between $d_3(p), d_3(p+1), d_3(p+2)$ and $d_3(p+3)$, and we return $[-3, -24, 78, -50]$, $[-24, 183, -310, 150]$, $[-105, 500, -696, 300]$, $[240, -975, 1236, -500]$, $[-429, 1608, -1930, 750]$. A little more intense pattern matching leads to $d_3(0) = 1, d_3(1) = 3\log(2)/5 + \pi/5$, $d_3(2) = 9/25 + 48\log(2)/125 + 37\pi/250$, while $d_3(3) = 627/1250 + 972\log(2)/3125 + 843\pi/6250$, as is predicted by

$$3(3p - 10)(3p - 8) \, d_3(p - 3) \qquad\qquad\qquad\qquad (1.31)$$
$$- (79(p - 2)(p - 3) + 21 + p) \, d_3(p - 2)$$
$$+ (77p - 153)(p - 2) \, d_3(p - 1) - 25(p - 1)(p - 2) \, d_3(p) = 0$$

for $p \ge 3$.

In each case, once the recursion has been discovered, it can be proven by considering the indefinite integral from 0 to t, which *Mathematica* and *Maple* can perform. One may then verify that the integral has a zero at $t = 1$. We illustrate this for the case $N = 2$. We combine the integrals in (1.30) and consider

$$\int_0^t \left(\frac{4p - 10}{(1 - x(1 - x))^{p-2}} - \frac{7p - 12}{(1 - x(1 - x))^{p-1}} + \frac{3p - 3}{(1 - x(1 - x))^p}\right) dx =$$
$$= -\frac{(2t - 1)(t - 1)t}{(1 - t + t^2)^{p-1}}. \qquad\qquad (1.32)$$

If we differentiate this last expression back and simplify, we recover the integrand as required. Since the right-hand side of (1.32) has a zero at $t = 1$, we are done. Similarly for (1.31), and it is to assure the zero at $t = 1$ that the factor of $1/2^n$ is needed.

The quantities $d_2(j)$ (and therefore also the $b_2(k)$) can in fact be computed explicitly (but this realization for us came after having found the recursion). To find the explicit formula, substitute $y = 2x - 1$ in the integral $d_2(j)$ to get

$$d_2(j) \quad = \quad 4^j \int_0^1 \frac{1}{(3 + x^2)^j}\, dx.$$

This satisfies the recursion $d_2(1) = 2\pi/(3\sqrt{3})$ and

$$d_2(j + 1) \quad = \quad \frac{2}{3j}\left(1 + (2j - 1)\, d_2(j)\right).$$

This leads to the explicit representation

$$d_2(j) \quad = \quad \frac{1}{3^j}\binom{2j - 2}{j - 1} \cdot \left(\sum_{i=1}^{j-1} \frac{3^i}{(2i - 1)\binom{2i-2}{i-1}} + \frac{2}{\sqrt{3}}\,\pi\right).$$

Putting this together with formula (1.28) and simplifying as much as possible gives

$$b_2(k - 1) \quad = \quad \frac{(-1)^k}{2}\sum_{j=1}^{k}(-1)^j j!\, S_k^{(j)} \frac{\binom{2j}{j}}{3^j}\left(\sum_{i=0}^{j-1}\frac{3^i}{(2i + 1)\binom{2i}{i}} + \frac{2}{3\sqrt{3}}\,\pi\right),$$

for $k \geq 1$, where $S_k^{(j)}$ are the *Stirling numbers of the second kind*,

$$S_k^{(j)} \quad = \quad \frac{(-1)^j}{j!}\sum_{m=0}^{j}(-1)^m m^k \binom{j}{m}.$$

For a similar explicit formula for $b_3(k)$, we would need to evaluate the integrals $d_3(j)$ explicitly. This would involve doing the partial fraction decomposition of the integrand $1/[(1+x)^j(x^2 - 2x + 2)^j]$, and while possible in principle, it leads to such unwieldy recursions that we have refrained from doing so.

To recapitulate, we have now proven that

$$b_2(k) \quad = \quad p_k + q_k\frac{\pi}{\sqrt{3}},$$

with explicitly given rationals p_k, q_k, and

$$b_3(k) \quad = \quad r_k + s_k\,\pi + t_k\,\log 2,$$

with certain rationals r_k, s_k, t_k, for which we have very efficient iterations. Explicitly,

$$\sum_{n\geq 1} \frac{1}{\binom{2n}{n}} \quad = \quad \frac{1}{3} + \frac{2}{9}\frac{\pi}{\sqrt{3}},$$

$$\sum_{n\geq 1} \frac{n}{\binom{2n}{n}} \quad = \quad \frac{2}{3} + \frac{2}{9}\frac{\pi}{\sqrt{3}},$$

$$\sum_{n\geq 1} \frac{n^2}{\binom{2n}{n}} \quad = \quad \frac{4}{3} + \frac{10}{27}\frac{\pi}{\sqrt{3}},$$

$$\sum_{n\geq 1} \frac{1}{\binom{3n}{n} 2^n} \quad = \quad \frac{2}{25} - \frac{6}{125}\log(2) + \frac{11}{250}\pi,$$

$$\sum_{n\geq 1} \frac{n}{\binom{3n}{n} 2^n} \quad = \quad \frac{81}{625} - \frac{18}{3125}\log(2) + \frac{79}{3125}\pi,$$

$$\sum_{n\geq 1} \frac{n^2}{\binom{3n}{n} 2^n} \quad = \quad \frac{561}{3125} + \frac{42}{15625}\log(2) + \frac{673}{31250}\pi.$$

In particular, we can deduce, by elimination, that (1.23) and (1.24) hold, and we can deduce the corresponding formulas

$$\sum_{n\geq 1} \frac{-150n^2 + 230n - 36}{\binom{3n}{n} 2^n} \quad = \quad \pi,$$

$$\sum_{n\geq 1} \frac{575n^2 - 965n + 273}{\binom{3n}{n} 2^n} \quad = \quad 6\log 2.$$

Moreover, the recursions derived for $b_2(k)$ and $b_3(k)$, namely the expressions $b_2(k) = \sum_j c_j^k(2)d_2(j)$ and $b_3(k) = \sum_j c_j^k(3)d_3(j)$, are sufficiently concise that we can compute symbolic values such as those of $b_3(200)$ or of $b_2(300)$ in a few seconds.

Without the factor of $1/2^n$, the sum b_3 would not have such an evaluation in terms of simple constants. In general, the position of the poles of $L_{-k}(ax^2(1-x))$, i.e., of the zeroes of $f_a(x) = 1 - ax^2(1-x)$, determines the evaluation of $\sum a^n n^k / \binom{3n}{n}$. If f_a has a sufficiently simple factorization, then we can expect an evaluation of the sum in terms of more basic constants. For example, with $a = -1/4$, we get (using *Maple* to evaluate the integral)

$$\sum_{n\geq 1} \frac{(-1)^n}{\binom{3n}{n} 4^n} = -\frac{1}{28} - \frac{3}{32}\log(2) + \frac{13}{112}\frac{\arctan\left(\frac{\sqrt{7}}{5}\right)}{\sqrt{7}},$$

$$\sum_{n\geq 1} \frac{(-1)^n n}{\binom{3n}{n} 4^n} = -\frac{81}{1568} - \frac{9}{256}\log(2) + \frac{17}{6272}\frac{\arctan\left(\frac{\sqrt{7}}{5}\right)}{\sqrt{7}},$$

while for $a = 1$, *Maple* returns an expression that can only be simplified to

$$\sum_{n\geq 1} \frac{1}{\binom{3n}{n}} = \frac{4}{23} + \frac{2}{23}\sum_{23r^3 + 55r + 23 = 0} r\log(1987 - 598r + 621r^2).$$

Similarly, the sums $\sum a^n n^k / \binom{2n}{n}$ lead to similar recursions; for example, the classical

$$\sum_{n\geq 1}(-1)^n \frac{n^k}{\binom{2n}{n}} = r_k + s_k \frac{\operatorname{arctanh}(1/\sqrt{5})}{\sqrt{5}},$$

with appropriate rationals r_k, s_k. Another tractable example is the sum $\sum n^k / \binom{4n}{2n}$, which has

$$\sum_{n\geq 1} \frac{n^k}{\binom{4n}{2n}} = r_k + s_k \frac{\pi}{\sqrt{3}} + t_k \frac{\operatorname{arctanh}(1/\sqrt{5})}{\sqrt{5}},$$

again for appropriate rationals.

1.7.2 Results in the Negative Case

For $k = -1$ and $k = -2$, the sums $b_2(k)$ and $b_3(k)$ can still be computed explicitly via the integrals (1.25) and (1.26):

$$\sum_{n\geq 1} \frac{1}{n \binom{2n}{n}} = \frac{1}{3}\frac{\pi}{\sqrt{3}},$$

$$\sum_{n\geq 1} \frac{1}{n^2 \binom{2n}{n}} = \frac{1}{18}\pi^2 = \frac{\zeta(2)}{3},$$

$$\sum_{n\geq 1} \frac{1}{n \binom{3n}{n} 2^n} = \frac{1}{10}\pi - \frac{1}{5}\log(2),$$

$$\sum_{n\geq 1} \frac{1}{n^2 \binom{3n}{n} 2^n} = \frac{1}{24}\pi^2 - \frac{1}{2}\log(2)^2.$$

For $k < -2$, however, it seems that the integrals have no accessible antiderivative, so that a direct computation does not appear possible. One might conjecture that the sums $b_3(k)$ then have no explicit reduction to simpler constants, but that conjecture would be wrong. One just has to expand the range of constants among which to hunt for a relation. In fact, it turns out that *multidimensional polylogarithms*

$$L_{a_1,\cdots,a_m}(z) = \sum_{n_1 > \cdots > n_m > 0} \frac{z^{n_1}}{n_1^{a_1} \cdots n_m^{a_m}}$$

(with positive integers a_j) will appear in the evaluations, for suitable z. For example, some of the relations proved in [54], as part of a more comprehensive analysis, are the following (note that $L_n(1) = \zeta(n)$):

$$\sum_{n \geq 1} \frac{1}{n^3 \binom{2n}{n}} = \frac{2}{3} \pi \operatorname{Im}\left(L_2(e^{i\pi/3})\right) - \frac{4}{3}\zeta(3),$$

$$\sum_{n \geq 1} \frac{1}{n^4 \binom{2n}{n}} = \frac{17}{36} \zeta(4),$$

$$\sum_{n \geq 1} \frac{1}{n^5 \binom{2n}{n}} = 2\pi \operatorname{Im}\left(L_4(e^{i\pi/3})\right) - \frac{19}{3}\zeta(5) + \frac{2}{3}\zeta(3)\zeta(2),$$

$$\sum_{n \geq 1} \frac{1}{n^6 \binom{2n}{n}} = -\frac{4}{3} \pi \operatorname{Im}\left(L_{4,1}(e^{i\pi/3})\right) + \frac{3341}{1296}\zeta(6) - \frac{4}{3}\zeta^2(3).$$

In [54, 59], the terms such as $\operatorname{Im}\left(L_{4,1}(e^{i\pi/3})\right)$ were termed *Clausen functions*.

Motivated by these results, one may conjecture that also the sums $b_3(k)$ for negative k can be evaluated in terms of multidimensional polylogarithms, with suitable parameters and at suitable points z. To find the right parameters, we employ integer relation detection schemes between the sums $b_3(k)$ and various polylogarithms. This of course involves a lot of trial and error. But in the end, the evaluations below were found by PSLQ.

None of these evaluations is yet proven. But they all are almost certainly true, since they have been verified to at least 100 decimal digits. Only the evaluation for $b_3(-3)$ can be proved rigorously, by laborious polylog manipulations—the interested reader may feel challenged to try it.

$$\sum_{n=1}^{\infty} \frac{1}{n^3 \binom{3n}{n} 2^n} = -\frac{33}{16} \zeta(3) + \frac{1}{6} \log^3(2) - \frac{1}{24} \pi^2 \log(2) + \pi \operatorname{Im}(L_2(i))$$

$$= -\frac{1}{4} \zeta(3) + \frac{1}{6} \log^3(2) - \frac{1}{24} \pi^2 \log(2) - 4\operatorname{Re}(L_{2,1}(i))$$

$$= -\frac{39}{16} \zeta(3) + \frac{1}{8} \pi^2 \log(2) - 4\operatorname{Re}\left(L_{2,1}(\tfrac{1+i}{2})\right) + 4\operatorname{Re}\left(L_3(\tfrac{1+i}{2})\right),$$

$$\sum_{n=1}^{\infty} \frac{1}{n^4 \binom{3n}{n} 2^n} = -\frac{143}{16} \zeta(3) \log(2) + \frac{91}{640} \pi^4 - \frac{3}{8} \log^4(2) + \frac{3}{8} \pi^2 \log^2(2)$$
$$- 8 L_4(\tfrac{1}{2}) - 8 \operatorname{Re}\left(L_{3,1}(\tfrac{1+i}{2})\right) - 8 \operatorname{Re}\left(L_4(\tfrac{1+i}{2})\right),$$

$$\sum_{n=1}^{\infty} \frac{1}{n^5 \binom{3n}{n} 2^n} = \frac{405}{32} \zeta(5) + \frac{21}{4} \zeta(3) \log^2(2) - \frac{1}{10} \pi^4 \log(2) - \frac{23}{144} \pi^2 \log^3(2)$$
$$- \frac{13}{8} \pi^2 \zeta(3) + \frac{1}{240} \log^5(2) - 13 L_5(\tfrac{1}{2}) - \frac{25}{2} L_{4,1}(\tfrac{1}{2})$$
$$+ 3 \pi \operatorname{Im}\left(L_4(i)\right) - 16 \operatorname{Re}\left(L_{4,1}(\tfrac{1+i}{2})\right) + 16 \operatorname{Re}\left(L_5(\tfrac{1+i}{2})\right).$$

Note that $\operatorname{Im}(L_2(i)) = G$, namely Catalan's constant.

Of course, there are still relations between the various polylogarithmic constants employed here, as is evidenced, for example, by the different evaluations for $b_3(-3)$; note also that $L_2(1/2) = \pi^2/12 - \log^2(2)/2$ and $L_3(1/2) = 7\zeta(3)/8 - \pi^2 \log(2)/12 + \log^3(2)/6$. This means that evaluations other than the ones given above are possible. We have tried to find those evaluations with the smallest rational factors. Again, we must emphasize that the evaluations given here are still conjectural.

1.8   Continued Fractions of Tails of Series

We have already seen several examples of continued fractions in this chapter —see, for example, formulas (1.16) and (1.17). In this section, we observe that the tails of the Taylor series for many standard functions, such as arctan and log, can be expressed as continued fractions in a variety of ways. A surprising side effect is that some of these continued fractions provide dramatic accelerations for the underlying power series. These investigations were motivated by a surprising observation about Gregory's series (see Sections 1.3 and 5.6.4 of [43]).

1.8.1   Gregory's Series Reexamined

As discussed in Section 1.3 of [43], Gregory's series for π,

$$\pi = 4 \sum_{k=1}^{\infty} \frac{(-1)^{k+1}}{2k-1} = 4(1 - 1/3 + 1/5 - 1/7 + \cdots), \quad (1.33)$$

when truncated to 5,000,000 terms, gives a value that differs strangely from the true value of π:

3.141592453589793238464643383279502784197169399387305820974941822307816 40...
3.141592653589793238464264338327950288419716939937510582097494459230781640...
    2   -2   10   -122   2770

The series value differs, as one might expect from a series truncated to 5,000,000 terms, in the seventh decimal place—a "4" where there should be a "6." But the next 13 digits are correct! Then, following another erroneous digit, the sequence is once again correct for an additional 12 digits. This pattern continues as shown. It is explained, *ex post facto*, by substituting $N = 10^7$ in the result below:

Theorem 1.8. *For integer N divisible by 4 the following asymptotic expansion holds:*

$$\frac{\pi}{2} - 2 \sum_{k=1}^{N/2} \frac{(-1)^{k-1}}{2k-1} \sim \sum_{m=0}^{\infty} \frac{E_{2m}}{N^{2m+1}} \tag{1.34}$$

$$= \frac{1}{N} - \frac{1}{N^3} + \frac{5}{N^5} - \frac{61}{N^7} + \cdots,$$

where the coefficients E_{2m} are the even Euler numbers: 1, −1, 5, −61, 1385, −50521, $\cdots$.

The observation on the digits in the Gregory series arrived in the mail from Joseph Roy North in 1987. After verifying its truth numerically (which is much quicker today), it was an easy matter to generate a large number of the "errors" to high precision. The authors of [37] then recognized the sequence of errors above as the Euler numbers—with the help of Sloane's *Handbook of Integer Sequences*. The presumption that this sequence of errors is a form of Euler-Maclaurin summation is now formally verifiable for any fixed N in *Maple*. This allowed them to determine that this phenomenon is equivalent to a set of identities between Bernoulli and Euler numbers, which could with considerable effort have been established. Secure in the knowledge that this observation holds, it is then easier, however, to use the *Boole summation formula*, which applies directly to alternating series and Euler numbers (see [37]). Because N was a power of ten, the asymptotic expansion was obvious on the computer screen.

This is a good example of a phenomenon that really does not become apparent without working to reasonably high precision (who recognizes 2, −2, 10?), and which highlights the role of pattern recognition and hypothesis validation in experimental mathematics.

It was an amusing additional exercise to compute π to 5,000 digits from the Gregory series. Indeed, with $N = 200,000$ and correcting using the first thousand even Euler numbers, Borwein and Limber [62] obtained 5,263 digits of π (plus 12 guard digits). Thus, while the alternating Gregory series is very slowly convergent, the errors are highly predictable.

1.8.2   Euler's Continued Fraction

Identities such as

$$a_0 \quad + \quad a_1 + a_1 a_2 + a_1 a_2 a_3 + a_1 a_2 a_3 a_4 \qquad (1.35)$$
$$= a_0 \quad + \quad \cfrac{a_1}{1 - \cfrac{a_2}{1 + a_2 - \cfrac{a_3}{1 + a_3 - \cfrac{a_4}{1 + a_4}}}}$$

are easily verified symbolically. The general form can then be obtained by substituting $a_N + a_N\, a_{N+1}$ for a_N and checking that the shape of the right-hand side is preserved. This allows many series to be re-expressed as finite continued fractions. For example, with $a_0 = 0, a_1 = x, a_2 = -x^2/3, a_3 = -3x^2/5, \cdots$, we obtain, in the limit, the continued fraction for arctan due to Euler:

$$\arctan(x) = \cfrac{x}{1 + \cfrac{x^2}{3 - x^2 + \cfrac{9x^2}{5 - 3x^2 + \cfrac{25x^2}{7 - 5x^2 + \cdots}}}}. \qquad (1.36)$$

When $x = 1$, this becomes the first continued fraction for $2/\pi$ given by Lord Brouncker (1620–1684):

$$\frac{2}{\pi} = \cfrac{1}{1 + \cfrac{9}{2 + \cfrac{25}{2 + \cfrac{49}{2 + \cdots}}}}.$$

If we let $a_0 = \sum_1^N b_k$ be the initial segment of a similar series, we may use (1.35) to replace the next M terms, say, by a continued fraction. Applied to arctan, this leads to

$$\arctan(z) = \sum_{n=1}^{N} (-1)^{n-1} \frac{z^{2n-1}}{2n-1} + \frac{(-1)^N z^{2N+1}}{2N+1} + \frac{(2N+1)^2 z^2}{(2N+3) - (2N+1)z^2}$$
$$+ \frac{(2N+3)^2 z^2}{(2N+5) - (2N+3)z^2} + \frac{(2N+5)^2 z^2}{(2N+7) - (2N+5)z^2} + \cdots.$$

$$(1.37)$$

1.8.3   Gauss's Continued Fraction

An immediately richer vein lies in Gauss's continued fraction for the ratio of two hypergeometric functions $\dfrac{F(a, b+1; c+1; z)}{F(a, b; c; z)}$ (see [212]). Recall that within its radius of convergence, the Gaussian hypergeometric function is defined by

$$
\begin{aligned}
F(a, b; c; z) \;=\;& 1 + \frac{ab}{c}\,z + \frac{a(a+1)b(b+1)}{c(c+1)}\,z^2 \\
&+ \frac{a(a+1)(a+2)b(b+1)(b+2)}{c(c+1)(c+2)}\,z^3 + \cdots .
\end{aligned}
\tag{1.38}
$$

The general continued fraction is developed by a reworking of the *contiguity relation*

$$
F(a, b; c; z) = F(a, b+1; c+1; z) - \frac{a(c-b)}{c(c+1)}\,z\,F(a, b+1; c+2; z),
$$

and formally, at least, this is quite easy to derive. Convergence and convergence estimates are more delicate. In the limit, for $b = 0$, this process yields

$$
F(a, 1; c; z) \;=\; \cfrac{1}{1 - \cfrac{\frac{a}{c}z}{1 - \cfrac{\frac{(c-a)}{c(c+1)}z}{1 - \cfrac{\frac{c(a+1)}{(c+1)(c+2)}z}{1 - \cfrac{\frac{2(c-a+1)}{(c+2)(c+3)}z}{1 - \cfrac{\frac{(c+1)(a+2)}{(c+3)(c+4)}z}{1 - \cdots}}}}}}
\tag{1.39}
$$

which is the case of present interest.

It is well known and easy to verify that $\log(1+z) = z\,F(1, 1; 2; -z)$. It is then a pleasant surprise to discover that $\log(1+z) - z = \frac{1}{2}z^2\,F(2, 1; 3; -z)$, $\log(1+z) - z + \frac{1}{2}z^2 = \frac{1}{3}z^3\,F(3, 1; 4; -z)$, and to conjecture that

$$
\log(1+z) + \sum_{n=1}^{N-1} \frac{(-1)^n z^n}{n} = \frac{z^N}{N}\,F(N, 1; N+1; -z).
\tag{1.40}
$$

This is easy to first verify for a few cases and then confirm rigorously. As always, a formula for log leads correspondingly to one for arctan:

$$
\arctan(z) - \sum_{n=0}^{N-1} \frac{(-1)^n z^{2n+1}}{2n+1} = \frac{z^{2N+1}}{2N+1}\,F\!\left(N + \frac{1}{2}, 1; N + \frac{3}{2}; -z^2\right)
\tag{1.41}
$$

Happily, in both cases (1.39) is applicable—as it is for a variety of other functions, including for example $\log[(1+z)/(1-z)]$, $(1+z)^k$, and $\int_0^z (1+t^n)^{-1} dt = z\,F(1/n, 1; 1+1/n; -z^n)$. Note that this last function recaptures $\log(1+z)$ and $\arctan(z)$ for $n = 1$ and 2, respectively.

We give the explicit continued fractions for (1.40) and (1.41) in the conventional, more compact form.

Theorem 1.9. *Gauss's continued fractions for* log *and* arctan *are:*

$$\log(1+z) + \sum_{n=1}^{N-1} \frac{(-1)^n z^n}{n} = \tag{1.42}$$

$$\frac{(-1)^{N+1} z^N}{N} + \frac{N^2 z}{N+1} + \frac{1^2 z}{N+2} + \frac{(N+1)^2 z}{N+3} + \frac{2^2 z}{N+4} + \cdots$$

and

$$\arctan(z) - \sum_{n=0}^{N-1} \frac{(-1)^n z^{2n+1}}{2n+1} = \tag{1.43}$$

$$\frac{(-1)^N z^{2N+1}}{2N+1} + \frac{(2N+1)^2 z^2}{2N+3} + \frac{2^2 z^2}{2N+5} + \frac{(2N+3)^2 z^2}{2N+7} + \frac{4^2 z^2}{2N+9} + \cdots .$$

See [56] for details.

Suppose we return to Gregory's series, but add a few terms of the continued fraction for (1.41). One observes numerically that if the results are with $N = 500,000$, adding only five terms of the continued fraction has the effect of increasing the precision by more than 30 digits.

Example 1.10. Hypergeometric functions.

Let

$$E_1(N, M, z) = \log(1+z)$$

$$- \left(-\sum_{n=1}^{N} \frac{(-z)^n}{n} - \frac{(-z)^{N+1}}{N+1} F_M(N+1, 1; N+2; -z) \right)$$

$$E_2(N, M, z) = \arctan(z)$$

$$- \left(-\sum_{n=0}^{N-1} \frac{(-1)^n z^{2n+1}}{2n+1} - \frac{(-1)^N z^{2N+1}}{2N+1} F_M\left(N+\frac{1}{2}, 1; N+\frac{3}{2}; -z^2\right) \right) .$$

$$\tag{1.44}$$

		5×10	5×10^2	5×10^3	5×10^4
	0	0.48×10^{-4}	0.13×10^{-25}	0.15×10^{-232}	0.13×10^{-2292}
	1	0.43×10^{-4}	0.11×10^{-25}	0.14×10^{-232}	0.11×10^{-2292}
	2	0.40×10^{-8}	0.11×10^{-31}	0.14×10^{-240}	0.11×10^{-2302}
M	3	0.34×10^{-8}	1.00×10^{-32}	0.12×10^{-240}	0.10×10^{-2302}
	4	0.12×10^{-11}	0.40×10^{-37}	0.50×10^{-248}	0.41×10^{-2312}
	5	0.10×10^{-11}	0.35×10^{-37}	0.45×10^{-248}	0.37×10^{-2312}
	6	0.78×10^{-15}	0.31×10^{-42}	0.40×10^{-255}	0.33×10^{-2321}

Table 1.1. Error $|E_1(N, M, 0.9)|$ for $N = 5 \times 10^k (1 \le k \le 4)$ and $0 \le M \le 6$.

		5×10	5×10^2	5×10^3	5×10^4	5×10^5	5×10^6
	0	0.99×10^{-2}	1.00×10^{-3}	1.00×10^{-4}	1.00×10^{-5}	1.00×10^{-6}	1.00×10^{-7}
	1	0.97×10^{-2}	1.00×10^{-3}	1.00×10^{-4}	1.00×10^{-5}	1.00×10^{-6}	1.00×10^{-7}
	2	0.91×10^{-6}	1.00×10^{-9}	1.00×10^{-12}	1.00×10^{-15}	1.00×10^{-18}	1.00×10^{-21}
M	3	0.86×10^{-6}	1.00×10^{-9}	1.00×10^{-12}	1.00×10^{-15}	1.00×10^{-18}	1.00×10^{-21}
	4	0.31×10^{-9}	0.39×10^{-14}	0.40×10^{-19}	0.40×10^{-24}	0.40×10^{-29}	0.40×10^{-34}
	5	0.28×10^{-9}	0.39×10^{-14}	0.40×10^{-19}	0.40×10^{-24}	0.40×10^{-29}	0.40×10^{-34}
	6	0.22×10^{-12}	0.34×10^{-19}	0.36×10^{-26}	0.36×10^{-33}	0.36×10^{-40}	0.36×10^{-47}

Table 1.2. Error $|E_1(N, M, 1)|$ for $N = 5 \times 10^k (1 \le k \le 6)$ and $0 \le M \le 6$.

		5×10	5×10^2	5×10^3	5×10^4	5×10^5	5×10^6
	0	0.50×10^{-2}	0.50×10^{-3}	0.50×10^{-4}	0.50×10^{-5}	0.50×10^{-6}	0.50×10^{-7}
	1	0.48×10^{-2}	0.50×10^{-3}	0.50×10^{-4}	0.50×10^{-5}	0.50×10^{-6}	0.50×10^{-7}
	2	0.44×10^{-6}	0.49×10^{-9}	0.50×10^{-12}	0.50×10^{-15}	0.50×10^{-18}	0.50×10^{-21}
M	3	0.42×10^{-6}	0.49×10^{-9}	0.50×10^{-12}	0.50×10^{-15}	0.50×10^{-18}	0.50×10^{-21}
	4	0.15×10^{-9}	0.19×10^{-14}	0.20×10^{-19}	0.20×10^{-24}	0.20×10^{-29}	0.20×10^{-34}
	5	0.14×10^{-9}	0.19×10^{-14}	0.20×10^{-19}	0.20×10^{-24}	0.20×10^{-29}	0.20×10^{-34}
	6	0.10×10^{-12}	0.17×10^{-19}	0.18×10^{-26}	0.18×10^{-33}	0.18×10^{-40}	0.18×10^{-47}

Table 1.3. Error $|E_2(N+1, M, 1)|$ for $N = 5 \times 10^k (1 \le k \le 6)$ and $0 \le M \le 6$.

		5×10	5×10^2
	0	0.31×10^{-32}	0.37×10^{-304}
	1	0.19×10^{-33}	0.23×10^{-305}
	2	0.11×10^{-37}	0.15×10^{-311}
M	3	0.26×10^{-38}	0.37×10^{-312}
	4	0.56×10^{-42}	0.92×10^{-318}
	5	0.13×10^{-42}	0.23×10^{-318}
	6	0.59×10^{-46}	0.13×10^{-323}

Table 1.4. Error $|E^*(N, M)|$ for $N = 5 \times 10^k (1 \le k \le 2)$ and $0 \le M \le 6$.

Then $E_1(N, M, z)$ and $E_2(N, M, z)$ measure the precision of the approximations to $\log(1 + z)$ and $\arctan(x)$ by the first N terms of Taylor series and then adding M terms of their continued fractions respectively. Let

$$E^*(N, M) = E_2(N, M, 1/2) + E_2(N, M, 1/5) + E_2(N, M, 1/8).$$

Tables 1.1, 1.2, 1.3, and 1.4 record the data for the approximations to the constants $\log(1.9), \log(2), \arctan(1)$, and $\arctan(1/2) + \arctan(1/5) + \arctan(1/8)$, respectively. Note that $\arctan(1) = \arctan(1/2) + \arctan(1/5) + \arctan(1/8)$ is a Machin formula, as we saw in Chapter 3 of [43]. □

After some further numerical experimentation, it is clear that for large a and c, the continued fraction $F(a, 1, c; z)$ is rapidly convergent. And, indeed, the rough rate is apparent. This is part of the content of the next theorem:

Theorem 1.11. ([56]) *Suppose* $-1 \leq z < 0$, *with* $a \geq 2$ *and* $a+1 \leq c \leq 2a$. *Then the following error estimate holds for all* $M \geq 2$:

$$|F(a, 1, c; z) - F_M(a, 1; c; z)| \leq$$

$$\frac{\Gamma(n + 1)(n + a)\Gamma(n + c - a)\Gamma(a)\Gamma(c)}{\Gamma(n + a)\Gamma(n + c)a\Gamma(c - a)} \left(\frac{2a}{(c - 2)\left(1 - \frac{2}{z}\right) + (2a - c)} \right)^M,$$

where $n = \lfloor M/2 \rfloor$ *and* $F_M(a, 1; c; z)$ *is the* M-*th convergent of the continued fraction to* $F(a, 1, c; z)$.

We leave it as an exercise to compare the estimates in Theorem 1.11 with the computed errors in Tables 1.1 and 1.2 (using $a = N$ and $c = N+1$) and Table 1.3 (using $a = N + 1/2$ and $c = N + 3/2$). The results are very good.

In [212], one can find listed many explicit continued fractions, which can be derived from Gauss's continued fraction or various of its limiting cases. These include exp, tanh, tan, and various less elementary functions. One especially attractive fraction is that for $J_{n-1}(z)/J_n(z)$ and $I_{n-1}(z)/I_n(z)$, where J and I are *Bessel functions of the first kind*. In particular,

$$\frac{J_{n-1}(2z)}{J_n(2z)} = \frac{n}{z} - \cfrac{\frac{z}{(n+1)}}{1 - \cfrac{\frac{z^2}{(n+1)(n+2)}}{1 - \cfrac{\frac{z^2}{(n+2)(n+3)}}{1 - \cdots}}}. \tag{1.45}$$

Setting $z = i$ and $n = 1$ leads to the very beautiful continued fraction
$I_1(2)/I_0(2) = [1, 2, 3, 4, \cdots]$. In general, arithmetic simple continued fractions correspond to such ratios.

An example of a more complicated situation is

$$\frac{(2\,z')^{2\,N+1}\,\mathrm{F}\left(N + \frac{1}{2}, \frac{1}{2}; N + \frac{3}{2}; z^2\right)}{(N+1)\binom{2\,N+2}{N+1}\mathrm{F}\left(-\frac{1}{2}; \cdot; z^2\right)} = \frac{\arcsin(z)}{\sqrt{1 - z^2}} - \sigma_{2N}(z), \qquad (1.46)$$

where σ_{2N} is the $2N$-th Taylor polynomial for $(\arcsin z)/\sqrt{1 - z^2}$. Only for $N = 0$ is this precisely of the form of Gauss's continued fraction.

1.8.4   Perron's Continued Fraction

Another continued fraction expansion is based on Stieltjes' work on the moment problem (see Perron [179]) and leads to similar acceleration. In volume 2, page 18 of [179], one finds a beautiful continued fraction for

$$\int_0^z \frac{t^\mu}{1+t}\,dt = \cfrac{z}{(\mu+1) + \cfrac{(\mu+1)^2 z}{(\mu+2) - (\mu+1)z + \cfrac{(\mu+2)^2 z}{(\mu+3) - (\mu+2)z + \cdots}}}\,,$$
$$(1.47)$$

valid for $\mu > -1, -1 < z \leq 1$. One can observe that this can be proved by Euler's continued fraction if we write

$$\frac{1}{z^\mu}\int_0^z \frac{t^\mu}{1+t}\,dt = \frac{z}{\mu+1} - \frac{z^2}{\mu+2} + \frac{z^3}{\mu+3} - \frac{z^4}{\mu+4} + \cdots$$

and observe that (1.47) follows from (1.35) in the limit.

Since

$$\frac{z^{\mu+1}}{\mu+1}\,\mathrm{F}\left(\mu+1, 1; \mu+2; -z\right) = \int_0^z \frac{t^\mu}{1+t}\,dt, \qquad (1.48)$$

$$\frac{z^{2\,\mu+1}}{2\,\mu+1}\,\mathrm{F}\left(\mu+\frac{1}{2}, 1; \mu+\frac{3}{2}; -z^2\right) = \int_0^z \frac{t^{2\,\mu}}{1+t^2}\,dt, \qquad (1.49)$$

for $\mu > 0$, on examining (1.40) and (1.41), this is immediately applicable to provide Euler continued fractions for the tail of the log and arctan series. Explicitly, we obtain:

Theorem 1.12. *Perron's continued fractions for (1.40) and (1.41) are*

$$\log(1+z) + \sum_{n=1}^{N-1} \frac{(-1)^n z^n}{n} = \tag{1.50}$$

$$\frac{(-1)^{N+1}z^N}{N} + \cfrac{N^2 z}{(N+1)-Nz} + \cfrac{(N+1)^2 z}{(N+2)-(N+1)z} + \cdots$$

and

$$\arctan z - \sum_{n=0}^{N-1} \frac{(-1)^n z^{2n+1}}{2n+1} = \tag{1.51}$$

$$\frac{(-1)^N z^{2N+1}}{2N+1} + \cfrac{(2N+1)^2 z^2}{(2N+3)-(2N+1)z^2} + \cfrac{(2N+3)^2 z^2}{(2N+5)-(2N+3)z^2} + \cdots .$$

Moreover, while the Gauss and Euler/Perron continued fractions obtained are quite distinct, the convergence behavior is very similar to that of the previous section. Note also the coincidence of (1.51) and (1.37). Indeed, as we have seen, Theorem 1.12 coincides with a special case of (1.35).

1.9 Partial Fractions and Convexity

We consider a network *objective function* p_N given by

$$p_N(q) = \sum_{\sigma \in S_N} \left(\prod_{i=1}^N \frac{q_{\sigma(i)}}{\sum_{j=i}^N q_{\sigma(j)}} \right) \left(\sum_{i=1}^N \frac{1}{\sum_{j=i}^N q_{\sigma(j)}} \right),$$

summed over *all N!* permutations; so a typical term is

$$\left(\prod_{i=1}^N \frac{q_i}{\sum_{j=i}^N q_j} \right) \left(\sum_{i=1}^N \frac{1}{\sum_{j=i}^n q_j} \right).$$

For example, with $N = 3$ this is

$$q_1 q_2 q_3 \left(\frac{1}{q_1 + q_2 + q_3} \right) \left(\frac{1}{q_2 + q_3} \right) \left(\frac{1}{q_3} \right) \left(\frac{1}{q_1 + q_2 + q_3} + \frac{1}{q_2 + q_3} + \frac{1}{q_3} \right).$$

This arose as the objective function in research into coupon collection. The researcher, Ian Affleck, wished to show p_N was *convex* on the positive orthant.

First, we try to simplify the expression for p_N. The *partial fraction decomposition* gives:

$$
\begin{aligned}
p_1(x_1) &= \frac{1}{x_1}, \\
p_2(x_1, x_2) &= \frac{1}{x_1} + \frac{1}{x_2} - \frac{1}{x_1 + x_2}, \\
p_3(x_1, x_2, x_3) &= \frac{1}{x_1} + \frac{1}{x_2} + \frac{1}{x_3} - \frac{1}{x_1 + x_2} - \frac{1}{x_2 + x_3} - \frac{1}{x_1 + x_3} \\
&\quad + \frac{1}{x_1 + x_2 + x_3}.
\end{aligned}
\tag{1.52}
$$

Partial fraction decompositions are another arena in which computer algebra systems are hugely useful. The reader is invited to try performing the third case in (1.52) by hand. It is tempting to predict the "same" pattern will hold for $N = 4$. This is easy to confirm (by computer if not by hand) and so we are led to:

Conjecture 1.13. *For each $N \in \mathbb{N}$, the function*

$$
p_N(x_1, \cdots, x_N) = \int_0^1 \left(1 - \prod_{i=1}^{N} (1 - t^{x_i}) \right) \frac{dt}{t}
\tag{1.53}
$$

is convex; indeed $1/p_N$ is concave.

One may check symbolically that this is true for $N < 5$ via a large Hessian computation. But this is impractical for larger N. That said, it is easy to numerically sample the Hessian for much larger N, and it is always positive definite. Unfortunately, while the integral is convex, the integrand is not, or we would be done. Nonetheless, the process was already a success, as the researcher was able to rederive his objective function in the form of (1.53).

A year later, Omar Hjab suggested re-expressing (1.53) as the *joint expectation* of Poisson distributions. See "Convex," *SIAM Electronic Problems and Solutions* at the URL http://www.siam.org/journals/problems/99-002.htm. Explicitly, this leads to:

Lemma 1.14. *If $x = (x_1, \cdots, x_n)$ is a point in the positive orthant $\mathbb{R}^n_+$, then*

$$
\int_0^\infty \left(1 - \prod_{i=1}^{n} (1 - e^{-tx_i}) \right) dt = \left(\prod_{i=1}^{n} x_i \right) \int_{\mathbb{R}^n_+} e^{-\langle x, y \rangle} \max(y_1, \cdots, y_n) \, dy,
\tag{1.54}
$$

where $\langle x, y \rangle = x_1 y_1 + \cdots + x_n y_n$ is the Euclidean inner product.

Proof. Let us denote the left-hand side of (1.54) by f. Since

$$1 - e^{-tx_i} = x_i \int_0^t e^{-x_i y_i} \, dy_i,$$

it follows that

$$1 - \prod_{i=1}^n (1 - e^{-tx_i}) = \left(\prod_{i=1}^n x_i \right) \left(\int_{\mathrm{R}_+^n} e^{-\langle x,y \rangle} \, dy - \int_{\mathrm{S}_t^n} e^{-\langle x,y \rangle} \, dy \right),$$

where

$$S_t^n = \{ y \in \mathrm{R}_+^n \mid 0 < y_i \le t \text{ for } i = 1, \cdots, n \}.$$

Hence,

$$f(x) = \left(\prod_{i=1}^n x_i \right) \int_0^\infty dt \int_{\mathrm{R}_+^n \setminus \mathrm{S}_t^n} e^{-\langle x,y \rangle} \, dy$$

$$= \left(\prod_{i=1}^n x_i \right) \int_0^\infty dt \int_{\mathrm{R}_+^n} e^{-\langle x,y \rangle} \chi_t(y) \, dy,$$

where

$$\chi_t(y) = \begin{cases} 1 & \text{if } \max(y_1, \cdots, y_n) > t, \\ 0 & \text{otherwise.} \end{cases}$$

Therefore,

$$f(x) = \left(\prod_{i=1}^n x_i \right) \int_{\mathrm{R}_+^n} e^{-\langle x,y \rangle} \, dy \int_0^\infty \chi_t(y) \, dt$$

$$= \left(\prod_{i=1}^n x_i \right) \int_{\mathrm{R}_+^n} e^{-\langle x,y \rangle} \max(y_1, \cdots, y_n) \, dy.$$

$\square$

It follows from the lemma that

$$p_N(x) = \int_{\mathrm{R}_+^N} e^{-(y_1 + \cdots + y_N)} \max\left(\frac{y_1}{x_1}, \cdots, \frac{y_N}{x_N} \right) dy,$$

and hence that p_N is positive, decreasing, and convex, as is the integrand. To derive the stronger result that $1/p_N$ is concave, we proceed as follows. Let

$$h(a, b) = \frac{2ab}{a + b}.$$

Then h is concave and concavity of $1/p_N$ is equivalent to

$$p_N\left(\frac{x+x'}{2}\right) \le h(p_N(x), p_N(x')) \quad \text{for all } x, x' \in \mathbb{R}_+^N. \tag{1.55}$$

To establish this, define

$$m(x, y) = \min\left(\frac{x_1}{y_1}, \cdots, \frac{x_n}{y_n}\right) \quad \text{for } x, y \in \mathbb{R}_+^N.$$

Then, since

$$m(x, y) + m(x', y) \le 2m\left(\frac{x+x'}{2}, y\right),$$

we have

$$p_N\left(\frac{x+x'}{2}\right) = \int_{\mathbb{R}_+^n} \frac{e^{-(y_1+\cdots+y_n)}}{m\left(\frac{x+x'}{2}, y\right)}\, dy \le \int_{\mathbb{R}_+^n} \frac{2e^{-(y_1+\cdots+y_n)}}{m(x, y) + m(x', y)}\, dy$$

$$= \int_{\mathbb{R}_+^n} e^{-(y_1+\cdots+y_n)} h\left(\frac{1}{m(x, y)}, \frac{1}{m(x', y)}\right)\, dy$$

$$\le h(p_N(x), p_N(x')),$$

where we leave it to the reader to confirm that the final assertion follows since h is concave and $\int_{\mathbb{R}_+^N} e^{-(y_1+\cdots+y_N)}\, dy = 1$. This is a form of *Jensen's inequality*. $\square$

Observe that since $h(a, b) \le \sqrt{ab} \le (a+b)/2$, it follows from (1.55) that p_N is log-convex (and convex). A little more analysis of the integrand shows p_N is strictly convex. There is still no truly direct proof of the convexity of p_N.

An amusing related example, which cries out for generalization, is that for $a > b > c > d > 0$, the function

$$f(x) = \frac{a^x - b^x}{c^x - d^x}$$

is convex on the real line, but $\log f(x)$ is convex on the real line only when $ad \ge bc$, and is concave on the real line when $ad < bc$. These assertions are fairly easy to deduce from:

Proposition 1.15. *Let* $g_\mu(x) = (e^{2\mu x} - 1)/(e^{2x} - 1)$, $\ell_\mu(x) = \log g_\mu(x)$, *and* $\ell_{\mu,\nu}(x) = \log\big(g_\mu(x) - g_\nu(x)\big)$. *Then, for* $\mu > 1$, $a > b > 1$, *and all real* x,

1. $g'_\mu(x) \geq 0$, $\ell'_\mu(x) \geq 0$, $\ell''_\mu(x) \geq 0$, $g''_\mu(x) \geq 0$,

2. $\ell''_a(x) - \ell''_b(x) \geq 0$, $g''_a(x) - g''_b(x) \geq 0$,

3. $\ell''_{a,b}(x) \geq 0$ when $a - b \geq 1$, and $\ell''_{a,b}(x) < 0$ when $a - b < 1$.

Note that Item 2 says ℓ_μ becomes more convex as the parameter μ increases, and similarly for g_μ. Note also that

$$\log\left(\frac{\sinh(\mu x)}{\sinh(x)}\right) = \log\left(\frac{e^{2\,\mu x} - 1}{e^{2\,x} - 1}\right) - 2\,x.$$

As an example, with some care, the convex conjugate of the function $f : x \mapsto \log(\sinh(3\,x)\,/\sinh(x))$ can be symbolically nursed to obtain the result g :

$$y \mapsto y/2 \cdot \log\left[(y + \sqrt{-3y^2 + 16})/(-2y + 4)\right] + \log\left[(-2 + \sqrt{-3y^2 + 16})/2\right].$$

Since the conjugate of g is much more easily computed to be f, this produces a symbolic computational proof that f and g are convex and are mutually conjugate.

1.10   Log-Concavity of Poisson Moments

Recall that a sequence $\{a_n\}$ is *log-convex* if $a_{n+1}a_{n-1} \geq a_n^2$, for $n \geq 1$ and is log concave when the sign is reversed. Consider the *unsolved* Problem 10738 posed by Radu Theodorescu in the 1999 *American Mathematical Monthly* [205]:

Problem: For $t > 0$, let

$$m_n(t) = \sum_{k=0}^{\infty} k^n \exp(-t)\frac{t^k}{k!}$$

be the n-th moment of a *Poisson distribution* with parameter t. Let $c_n(t) = m_n(t)/n!$. Show

(a) $\{m_n(t)\}_{n=0}^{\infty}$ is log-convex for all $t > 0$.

(b) $\{c_n(t)\}_{n=0}^{\infty}$ is not log-concave for $t < 1$.

(c*) $\{c_n(t)\}_{n=0}^{\infty}$ is log-concave for $t \geq 1$.

Solution:

(a) Neglecting the factor of $\exp(-t)$ as we may, this reduces to

$$\sum_{k,j\geq0}\frac{(jk)^{n+1}t^{k+j}}{k!j!}\leq\sum_{k,j\geq0}\frac{(jk)^{n}t^{k+j}}{k!\,j!}k^2=\sum_{k,j\geq0}\frac{(jk)^{n}t^{k+j}}{k!j!}\frac{k^2+j^2}{2},$$

and this now follows from $2jk\leq k^2+j^2$.

(b) As

$$m_{n+1}(t)=t\sum_{k=0}^{\infty}(k+1)^n\exp(-t)\frac{t^k}{k!},$$

on applying the binomial theorem to $(k+1)^n$, we see that $m_n(t)$ satisfies the recurrence

$$m_{n+1}(t)=t\sum_{k=0}^{n}\binom{n}{k}m_k(t),\qquad m_0(t)=1.$$

In particular for $t=1$, we obtain the sequence

$$1,1,2,5,15,52,203,877,4140\ldots.$$

These are the *Bell numbers*, which again can be discovered by consulting Sloane's on-line integer sequence recognition tool at http://www.research. att.com/~njas/sequences. This tool can also tell us that for $t=2$, we have obtained generalized Bell numbers, and can give us the exponential generating functions. The Bell numbers were known earlier to Ramanujan.

Now an explicit computation shows that

$$t\frac{1+t}{2}=c_0(t)\,c_2(t)\leq c_1(t)^2=t^2$$

exactly if $t\geq1$. Also, preparatory to the next part, a simple calculation shows that

$$\sum_{n\geq0}c_nu^n=\exp\left(t(e^u-1)\right). \tag{1.56}$$

(c*) (The * indicates this was the unsolved component.) We appeal to a recent theorem due to E. Rodney Canfield [80]. A search in 2001 on *MathSciNet* for "Bell numbers" since 1995 turned up 18 items. This article showed up as paper #10. Later, *Google* found the paper immediately! Canfield proves the next lovely and quite difficult result.

Theorem 1.16. *If a sequence* $1, b_1, b_2, \cdots$ *is non-negative and log-concave, then so is the sequence* $1, c_1, c_2, \cdots$ *determined by the generating function equation*

$$\sum_{n \geq 0} c_n u^n = \exp\left(\sum_{j \geq 1} b_j \frac{u^j}{j}\right).$$

Using Equation (1.56) above, we apply this to the sequence $b_j = t/(j-1)!$ which is log-concave exactly for $t \geq 1$.   □

Indeed, symbolic computation—facilitated by the recursion above—strongly suggests the *only* violation of log-concavity of the sequence $\{c_n(t)\}_{n=0}^{\infty}$ for $t > 0$, occurs as illustrated in (b). We have not been able to prove this conjecture. It seems to require a significant strengthening of Theorem 1.16 to cover the case when the first term of $\{1, b_n\}$ is replaced by $b_0 \neq 1$.

It transpired that the given solution to (c) was the only one received by the *Monthly*. This is quite unusual. The reason might well be that it relied on the following sequence of steps:

Question ⇒ Computer Algebra System ⇒ Web Interface

⇒ Search Engine ⇒ Digital Library

⇒ Hard New Paper ⇒ Answer

Now if only we could automate this!

1.11   Commentary and Additional Examples

1. **Dictionaries are like timepieces.** Samuel Johnson observed that dictionaries are like clocks: The best do not run true, and the worst are better than none. The same is true of tables and databases. We quoted Michael Berry as saying "Compounding the impiety, I would give up Shakespeare in favor of Prudnikov, Brychkov and Marichev" [23, 182]. That excellent compendium contains

$$\sum_{k=1}^{\infty} \sum_{l=1}^{\infty} \frac{1}{k^2 (k^2 - kl + l^2)} = \frac{\pi^\alpha \sqrt{3}}{30}, \tag{1.57}$$

where the "$\propto$" is probably "4" [182, volume 1, entry 9, page 750]. Integer relation methods suggest that no reasonable value of $\propto$ works. What is intended in (1.57)? Note that

(a)

$$\sum_{n=1}^{\infty}\sum_{m=1}^{\infty}\frac{2}{n^2\left(n^2-mn+m^2\right)}+\sum_{n=1}^{\infty}\sum_{m=1}^{\infty}\frac{2}{n^2\left(n^2+mn+m^2\right)}$$
$$=\sum_{m,n\in\mathbb{Z}\ mn\neq0}\frac{1}{n^2\left(n^2+mn+m^2\right)}$$
$$=6\zeta(4).$$

(b)

$$\sum_{n=1}^{\infty}\sum_{m=1}^{n-1}\frac{1}{nm\left(n^2+mn+m^2\right)}+\sum_{n=1}^{\infty}\sum_{m=1}^{\infty}\frac{1}{n^2\left(n^2+mn+m^2\right)}$$
$$=\frac{13}{12}\zeta(4).$$

(c)

$$\sum_{n=1}^{\infty}\sum_{m=1}^{\infty}\frac{1}{n^2\left(n^2+mn+m^2\right)}7=\frac{2}{\sqrt{3}}\operatorname{Im}\sum_{n=1}^{\infty}\frac{\Psi\left(1+n\frac{-1+i\sqrt{3}}{2}\right)}{n^3}$$
$$=\frac{2}{\sqrt{3}}\operatorname{Im}\int_{0}^{\infty}\frac{\operatorname{Li}_3\left(e^{\left(\frac{-1+i\sqrt{3}}{2}\right)t}\right)}{e^t-1}\,dt$$
$$=1.00445719820157402755414025\ldots$$

2. **A series for π with first term 22/7.** An estimate for $\pi-355/113$ is given in [103]:

(a) Let $P(t)=4-4t^2+5t^4-4t^5+t^6$ and $Q(t)=t(1-t)$. Show that
$$\frac{4}{1+t^2}=\frac{4-4t^2+5t^4-4t^5+t^6}{1+t^4\left(1-t\right)^4/4},$$

and so
$$\pi=\int_{0}^{1}\frac{P(t)}{1+Q^4(t)/4}\,dt=\frac{1}{2}\int_{0}^{1}\frac{P(t)+P(1-t)}{1+Q^4(t)/4}\,dt.$$

(b) Observe that $P(t) + P(1-t) = 6 + 2Q(t) - Q^2(t) - 2Q^3(t)$ and deduce that

$$\pi = \frac{1}{2} \sum_{n=0}^{\infty} \left(-\frac{1}{4}\right)^n \int_0^1 \frac{6 + 2Q(t) - Q^2(t) - 2Q^3(t)}{1 + Q^4(t)/4}\, dt = \sum_{n=0}^{\infty} a_n,$$

where

$$a_n = \left(-\frac{1}{4}\right)^n \left\{ \frac{3\,(4\,n)!^2}{(8\,n+1)!} + \frac{(4\,n+1)!^2}{(8\,n+3)!} - \frac{(4\,n+2)!^2}{2\,(8\,n+5)!} - \frac{(4\,n+3)!^2}{(8\,n+7)!} \right\}.$$

(c) This series, which can be neatly written as a sum of four $_5F_4$ functions, gains roughly three digits per term. Check that the series has constant term $22/7$ and continues

$$\frac{22}{7} + \frac{76}{15015}\left(-\frac{1}{4}\right) + \frac{543}{37182145}\left(-\frac{1}{4}\right)^2 + \frac{308}{6511704225}\left(-\frac{1}{4}\right)^3 + \cdots .$$

(d) Find an economical way to show

$$\frac{355}{113} - \frac{33}{10^8} < \pi < \frac{355}{113} + \frac{24}{10^8}.$$

3. **A sophomore's dream.** Show that

$$\int_0^1 x^x\, dx = \sum_{n=1}^{\infty} \frac{(-1)^{n+1}}{n^n},$$

and that

$$\int_0^1 x^{-x}\, dx = \sum_{n=1}^{\infty} \frac{1}{n^n}.$$

Give meaning to and evaluate $\int_0^1 x^{x^x}\, dx$. In each case, it helps to start by writing the integrand as a series and to justify integrating term by term.

4. **Some binary digit algorithms.** Whenever a function satisfies a suitable addition formula, its inverse admits algorithms to compute binary (or other base) representations [57]. We begin with an introductory example. Let $x \geq 0$. Set

$$a_0 = x \quad \text{and} \quad a_{n+1} = \frac{2a_n}{1 - a_n^2} \quad \text{(with } a_{n+1} = -\infty \text{ if } a_n = \pm 1\text{)}.$$

Then

$$\sum_{\substack{a_n < 0 \\ n \geq 0}} \frac{1}{2^{n+1}} = \frac{\arctan x}{\pi}.$$

In general, let an interval $I \subseteq \mathbb{R}$ and subsets $D_0, D_1 \subseteq I$ with $D_0 \cup D_1 = I$ and $D_0 \cap D_1 = \emptyset$ be given, as well as functions $r_0 : D_0 \to I$, $r_1 : D_1 \to I$. Then consider the system (S) of the following two functional equations for an unknown function $f : I \to [0, 1]$.

$$2f(x) = f(r_0(x)) \quad \text{if } x \in D_0, \tag{S_0}$$
$$2f(x) - 1 = f(r_1(x)) \quad \text{if } x \in D_1. \tag{S_1}$$

Such a system always leads to an iteration:

$$a_0 = x \quad \text{and} \quad a_{n+1} = \begin{cases} r_0(a_n) & a_n \in D_0, \\ r_1(a_n) & a_n \in D_1. \end{cases} \tag{1.58}$$

Then:

$$f(a_0) = f(x) = \sum_{\substack{a_n \in D_1 \\ n \geq 0}} \frac{1}{2^{n+1}}. \tag{1.59}$$

Several examples. Here are some elementary transcendental functions f that satisfy a system of type (S) with algebraic r_0, r_1.

(a) $f(x) = \log x / \log 2$, $\quad I = [1, 2]$, $\quad D_0 = [1, \sqrt{2})$, $\quad D_1 = [\sqrt{2}, 2]$, $r_0(x) = x^2$, $\quad r_1(x) = x^2/2$.

   Of course, the recursion (1.58) and (1.59) can be used to compute the binary expansion of the function f. Take for example $x = a_0 = 3/2$. Then $a_1 = 9/8$, $a_2 = 81/64$, $a_3 = 6561/4096$, $a_4 = 3^{16}/2^{25}$, and so on. The first 20 binary digits of $\log 3 / \log 2$ are 1.10010101110000000001_2.

(b) $f(x) = \arccos(x)/\pi$, $\quad I = [-1, 1]$, $\quad D_0 = (0, 1]$, $\quad D_1 = [-1, 0]$, $\quad r_0(x) = 2x^2 - 1$, $\quad r_1(x) = 1 - 2x^2$. Note that $r_0^k = r_0 \circ \cdots \circ r_0$ is the Chebyshev polynomial of the first kind T_{2^k} for $[-1, 1]$.

(c) $f(x) = 2\arcsin(x)/\pi$, $\quad I = [0, 1]$, $\quad D_0 = [0, 1/\sqrt{2})$, $\quad D_1 = [1/\sqrt{2}, 1]$, $\quad r_0(x) = 2x\sqrt{1 - x^2}$, $\quad r_1(x) = 2x^2 - 1$.

(d) $f(x) = \begin{cases} \arctan(x)/\pi & x \in [0, \infty) \\ 1 + \arctan(x)/\pi & x \in [-\infty, 0), \end{cases}$ $\quad D_0 = [0, \infty)$, $\quad D_1 = [-\infty, 0)$, $\quad r_0(x) = \frac{2x}{1-x^2}$, $r_0(1) = -\infty$, $\quad r_1(x) = \frac{2x}{1-x^2}$, $r_1(-1) = -\infty$.

(e) $f(x) = \text{arccot}(x)/\pi$, $\quad I = \mathbb{R} \cup \{-\infty\}$, $\quad D_0 = [0, \infty)$, $\quad D_1 = [-\infty, 0)$, $\quad r_0(x) = \frac{x^2-1}{2x}$, $\quad r_0(0) = -\infty$, $\quad r_1(x) = \frac{x^2-1}{2x}$.

(f) $f(x) = \text{arsinh}(x)/\log 2$, $\quad I = [0, 3/4]$, $\quad D_0 = [0, 1/2\sqrt{2})$, $\quad D_1 = [1/2\sqrt{2}, 3/4]$, $\quad r_0(x) = 2x\sqrt{1+x^2}$, $\quad r_1(x) = 5/2x\sqrt{1+x^2} - 3/2x^2 - 3/4$.

(g) $f(x) = \arccos(x)/\pi$ satisfies

$$
\begin{array}{rcll}
3f(x) &=& f(4x^3 - 3x) & \text{if } x \in (1/2, 1], \\
3f(x) - 1 &=& f(-4x^3 + 3x) & \text{if } x \in (-1/2, 1/2], \\
3f(x) - 2 &=& f(4x^3 - 3x) & \text{if } x \in [-1, -1/2].
\end{array}
$$

That means that ternary representations of f can be computed by the following recursion:

$$
\text{Set } a_0 = x, \ a_{n+1} = \begin{cases} 4a_n^3 - 3a_n & a_n \in (1/2, 1] \cup [-1, -1/2] \\ -4a_n^3 + 3a_n & a_n \in (-1/2, 1/2]. \end{cases}
$$

Then

$$
\frac{\arccos(x)}{\pi} = \sum_{a_n \in (-1/2, 1/2]} \frac{1}{3^{n+1}} + \sum_{a_n \in [-1, -1/2]} \frac{2}{3^{n+1}}.
$$

Full details are given in [57].

5. **A two-term recursion.** Determine the behavior of the iteration $u_{n+1} = |u_n| - u_{n-1}$ for arbitrary real starting points $u_0 = x$ and $u_1 = y$. Then attempt to generalize this behavior. (Taken from [132].)

Solution: Numerical testing shows the iteration has period nine. One proof is to explicitly compute (preferably using a symbolic math program) the messy looking algebraic function this determines. It can be seen that each case returns $[x, y]$. More explicitly

$$
A = \begin{bmatrix} 0 & 1 \\ -1 & 1 \end{bmatrix} \qquad B = \begin{bmatrix} 0 & 1 \\ -1 & -1 \end{bmatrix}
$$

can be used to represent the iteration. Then $B^3 = I = -A^3$ and analysis of the cases above shows that it always devolves to $A^3 B A^3 B^2$ which is the identity. A nice generalization is that for $M, K > 1$ and integer, the iteration

$$
u_{k+1} = \cos\left(\frac{\pi}{K}\right)(|u_k| + u_k) + \cos\left(\frac{\pi}{M}\right)(|u_k| - u_k) - u_{k-1}
$$

has period $KM + K - M$. This can again be checked symbolically for many small M and K. One may prove this by considering

$$A = \begin{bmatrix} 0 & 1 \\ -1 & 2\cos(\pi/K) \end{bmatrix} \qquad B = \begin{bmatrix} 0 & 1 \\ -1 & 2\cos(\pi/M) \end{bmatrix},$$

and reducing the iteration to $(A^K B)^{M-2} A^K B^2$.

6. **A rational function recursion.** Solve the recursion $u_0 = 2$ and

$$u_{n+1} = \frac{2\,u_n + 1}{u_n + 2}.$$

(Taken from [132]).

Solution: The first few numerators are

$$2, 5, 14, 41, 122, 365, 1094, 3281, 9842, 29525, 88574$$

and clearly satisfy $n_{k+1} = 3\,n_k - 1$. Sloane's online sequence recognition tool announces that the numerators are $(3^n + 1)/2$. The denominators are one smaller, and so

$$u_n = \frac{3^{n+1} + 1}{3^{n+1} - 1}.$$

This can now easily proved by induction.

7. **A sequence with nines.** Determine the number of nines in the tail of the sequence

$$u_{n+1} = 3\,u_n^4 + 4\,u_n^3,$$

with $u_0 = 9$. (Taken from [132]).

Solution: The first six or so cases show there is a 5 followed by 2^k occurrences of 9. Define $u_k = 6 \cdot 10^{2^k} - 1$ and show inductively that this is so.

8. **Continued fraction of Champernowne's number.** Compute the first ten or twenty terms of the continued fraction for Champernowne's number, namely the decimal constant $0.123456789101112131415\ldots$ Explain the phenomenon observed.

9. **A sequence involving square roots.** Consider the iteration

$$c_{n+1} = c_n + r - \frac{c_n}{\sqrt{1 + c_n^2}}, \qquad c_0 \geq 1,$$

where r is a positive constant. For which r does the sequence $\{c_n\}$ converge? In case of convergence to $c \neq c_0$, prove that $\lim(c_{n+1} - c)/(c_n - c)$ exists and determine its value. In case of divergence, find a precise asymptotic expression for c_n.

Solution: Justify the following assertions: When $0 < r < 1$, the sequence converges to $c = r/\sqrt{1-r^2}$ and either $c_n = c$ for every integer $n \geq 0$, or $\lim(c_{n+1}-c)/(c_n-c) = 1 - (1-r^2)^{3/2}$. When $r = 1$, $c_n \sim (3n/2)^{1/3}$; and when $r > 1$, $c_n \sim (r-1)n$.

10. **A sequence involving exponentials.** Define a sequence $\{t_k\}$ by setting

$$t_1 = 1, \qquad t_{k+1} = t_k \exp(-t_k), \quad k = 1, 2, \cdots .$$

Determine the behavior of the sequence.

Solution: Note that t_k tends monotonically to a limit ℓ which must necessarily be zero. Hence $t_{k+1}^{-1} - t_k^{-1} = t_k^{-1}(\exp t_k - 1)$, which tends to $\exp'(0) = 1$ as k tends to infinity. Whence, since Cesàro averaging preserves limits,

$$\frac{1}{mt_m} = \frac{1}{m} \sum_{k=1}^{m-1} \frac{e^{t_k} - 1}{t_k} + \frac{1}{mt_1}$$

also tends to 1, and

$$\lim_{m \to \infty} mt_m = 1.$$

The reader is invited to perform a similar analysis for a more general $g : [0, 1] \mapsto [0, 1]$.

11. **Some double integrals.** Evaluate the following integrals:

(a)
$$\int_0^1 \int_0^1 \frac{1}{(1 - x^2 y^2)} \, dx \, dy \qquad \left(= \frac{\pi^2}{8} \right)$$

(b)
$$\int_0^1 \int_0^1 \frac{1}{(1 - xy)} \, dx \, dy \qquad \left(= \frac{\pi^2}{6} \right)$$

(c)
$$\int_{-1}^1 \int_{-1}^1 \frac{1}{\sqrt{1 + x^2 + y^2}} \, dx \, dy \qquad \left(= 4\log(2 + \sqrt{3}) - \frac{2\pi}{3} \right)$$

(d)

$$\frac{1}{4} \int_0^1 \int_0^1 \frac{1}{(x+y)\sqrt{(1-x)(1-y)}} \, dx \, dy \qquad (= G)$$

(e)

$$\int_0^1 \int_0^1 \frac{1-x}{(1-xy)|\log(xy)|} \, dx \, dy \qquad (= \gamma).$$

12. **A double summation.** Evaluate

$$\sum_{n=1}^{\infty} \sum_{m=1}^{\infty} \frac{1}{(m^2 n + mn^2 + 2\,mn)} \qquad (= 7/4).$$

13. **Infinite series and dilogarithms.** Prove that for positive integers a and b,

$$\sum_{m=0}^{\infty} \frac{z^m}{(am+b)^2} = \frac{z^{-b/a}}{a} \sum_{k=0}^{a-1} e^{-2\pi ibk/a} \mathrm{Li}_2(z^{1/a} e^{2\pi ik/a}),$$

and note the finitude of the sum. If one believes in Li evaluations as "fundamental," then this leads to a finite expression for the Broadhurst V constant (mentioned in Chapter 2 of [43]), as well as some other BBP sums.

14. **An infinite product evaluation.** Evaluate

$$\frac{z-1}{w-1} \prod_{n=1}^{\infty} \frac{w^{1/2^n}+1}{z^{1/2^n}+1}.$$

for $w, z > 0$.

Hint: Examine the case $w = 2$.

15. **Quasi-elliptic integrals.** Show that

$$\int \frac{6\,x}{\sqrt{x^4 + 4\,x^3 - 6\,x^2 + 4\,x + 1}} \, dx = \log\left(a(x) + b(x)\sqrt{D(x)}\right) \quad (1.60)$$

where

$$\begin{aligned}
a(x) &= x^6 + 12\,x^5 + 45\,x^4 + 44\,x^3 - 33\,x^2 + 43 \\
b(x) &= x^4 + 10\,x^3 + 30\,x^2 + 22\,x - 11 \\
D(x) &= x^4 + 4\,x^3 - 6\,x^2 + 4\,x + 1.
\end{aligned}$$

Hint: Confirm that in (1.60) $a' = 6x, /b$ and hence that (1.60) is a pseudo-elliptic integral in the sense described in [211]. The following is taken from the abstract of [211]:

In this report we detail the following story. Several centuries ago, Abel noticed that the well-known elementary integral

$$\int \frac{dx}{\sqrt{x^2 + 2bx + c}} = \log\left(x + b + \sqrt{x^2 + 2bx + c}\right)$$

is just a presage of more surprising integrals of the form

$$\int \frac{f(x)dx}{\sqrt{D(x)}} = \log\left(p(x) + q(x)\sqrt{D(x)}\right).$$

Here f is a polynomial of degree g and the D are certain polynomials of degree $\deg D(x) = 2g + 2$. Specifically, $f(x) = p'(x)/q(x)$ (so q divides p'). Note that, morally, one expects such integrals to produce inverse elliptic functions and worse, rather than an innocent logarithm of an algebraic function.

Abel went on to study abelian integrals, and it was Chebyshev who explained—using continued fractions—what is going on with these "quasi-elliptic" integrals. Recently, the second author computed all the polynomials D over the rationals of degree 4 that have an f as above. We explain various contexts in which the present issues arise. These contexts include symbolic integration of algebraic functions, the study of units in function fields and, given a suitable polynomial g, the consideration of the period length of the continued fraction expansion of the numbers $\sqrt{g(n)}$ as n varies over the integers. But the major content of this survey is an introduction to period continued fractions in hyperelliptic—thus quadratic—function fields.

16. **Clausen's product.** Prove that

$$\begin{aligned} {}_2F_1{}^2\left(a, b, a + b + 1/2, z\right) = \\ {}_3F_2\left(2\,a, a + b, 2\,b, a + b + 1/2, 2\,a + 2\,b, z\right). \end{aligned} \tag{1.61}$$

Hint: Show both sides satisfy the following differential equation

$$\begin{aligned} 0 = {}&x^2\left(x - 1\right)y^{(3)}(x) - 3\,x\left(a + b + 1/2 - (a + b + 1)x\right)y^{(2)}(x) \\ &+ \left[\left(2(a^2 + b^2 + 4\,ab) + 3(a + b) + 1\right)x \right. \\ &\qquad\qquad \left. - (a + b)\left(2(a + b) + 1\right)\right]y'(x) \\ &+ 4\,ab\left(a + b\right)y(x), \end{aligned}$$

are analytic at zero and have appropriate initial values. Use Clausen's product to deduce that

$$\arcsin^2(x) = \frac{1}{2} \sum_{n=1}^{\infty} \frac{(2x)^{2n}}{n^2 \binom{2n}{n}}.$$

Hint: $\arcsin(x) = x \cdot {}_2F_1\left(1/2, 1/2, 3/2; x^2\right)$.

17. **An application of Clausen's product.** Consider the hypergeometric function

$$G_a : x \mapsto F\left(a, a + \frac{1}{2}, 2a + 1; x\right),$$

and show that

$$G_{ab} = G_a^b \qquad \text{for all } a, b \in \mathbb{R}.$$

By considering $G_{-1/2}$, determine the closed form for G_a.

18. **Putnam problem 1987–A6.** Let n be a positive integer and let $a_3(n)$ be the number of zeroes in the ternary expansion of n. Determine for which positive x the series $\sum_{n=1}^{\infty} x^{a_3(n)}/n^3$ converges.

Solution: For $x < 25$. In the b-ary analogue $\sum_{n=1}^{\infty} x^{a_b(n)}/n^b$ converges if and only if $x < b^b - b + 1$.

19. **Putnam problem 1987–B2.** For r, s nonnegative with $r + s \leq t$, evaluate

$$\sum_{k=0}^{s} \frac{\binom{s}{k}}{\binom{t}{r+k}} \qquad \left(= \frac{t+1}{(t+1-s)\binom{t-s}{r}}\right).$$

20. **Putnam problem 1987-B4.** Let $x_0 = 4/5$ and $y_0 = 3/5$ and consider the dynamical system

$$x_{n+1} \leftarrow x_n \cos(y_n) - y_n \sin(y_n) \qquad y_{n+1} \leftarrow x_n \sin(y_n) + y_n \cos(y_n).$$

Does the system converge and if so, to what?

Hint: It may help to consider $z = x + i\,y$.

21. **Putnam problem 1989–A2.** Evaluate

$$\int_0^a \int_0^b e^{\max\left(a^2 y^2, b^2 x^2\right)} dy\, dx.$$

Hint:

$$\frac{2}{ab} \int_0^{ab} \int_0^z e^{z^2} dw\, dz = \frac{e^{a^2 b^2} - 1}{ab}.$$

22. **Putnam problem 1990–A1.** Let $T_0 = 2$, $T_1 = 3$, and $T_2 = 6$. Find a simple formula for T_n where

$$T_n = (n+4)\, T_{n-1} - 4n\, T_{n-2} + (4n-8)\, T_{n-3} \qquad (= n! + 2^n).$$

23. **Putnam problem 1990–B5.** Is there an infinite sequence of nonzero reals $\{a_n\}$ so that

$$p(x) = \sum_{k=0}^{n} a_k x^k$$

has n distinct real roots for all n?

Hint: Let $a_n = (-10)^{-n^2}$ and evaluate p at 10^{2k}.

24. **Putnam problem 1993–A2.** Suppose that a sequence $\{x_n\}$ of nonzero real numbers satisfies $x_n^2 = 1 + x_{n+1}\, x_{n-1}$ for all n. Show that for some real a, the sequence satisfies $x_{n+1} = a\, x_n - x_{n-1}$. *Hint*: Examine values of $(x_{n+1} + x_{n-1})/x_n$ for various n.

25. **Putnam problem 1997–A3.** Evaluate

$$\mathcal{E} = \int_0^\infty \sum_{k=0}^{\infty} \frac{(-1)^k x^{2k+1}}{2^k k!} \sum_{k=0}^{\infty} \frac{x^{2k}}{4^k \, (k!)^2} \, dx.$$

Answer: $1.6487212707001281\ldots = \sqrt{e}$.

Solution: The integrand is the expression $x\, \exp(-x^2/2)\, I_0(x)$, which *Maple* can identify and integrate. Alternately, one can interchange the integral and the second sum to obtain

$$\mathcal{E} = \sum_{k=0}^{\infty} \frac{\int_0^\infty x^{2k+1} e^{-1/2\, x^2} \, dx}{4^k \, (k!)^2} = \sum_{k=0}^{\infty} \frac{1}{2^k k!}.$$

26. **Putnam problem 1999–A3.** Consider the power series expansion

$$\frac{1}{1 - 2x - x^2} = \sum_{n \geq 0} a_n x^n.$$

Prove that for each integer $n \geq 0$, there is an integer m such that

$$a_n^2 + a_{n+1}^2 = a_m.$$

Answer: $a_n^2 + a_{n+1}^2 = a_{2n+1},$ (1.62)

which remains to be proven.

Hint: The first 15 coefficients are

$$1, 2, 5, 12, 29, 70, 169, 408, 985, 2378, 5741, 13860, 33461, 80782, 195025,$$

and the desired squares are

$$5, 29, 169, 985, 5741, 33461, 195025,$$

which is more than enough to spot the pattern. To prove this either explicitly use the closed form for

$$a_n = \frac{1}{2\sqrt{2}} \left(\left(1 + \sqrt{2}\right)^{n+1} - \left(-\sqrt{2} + 1\right)^{n+1} \right),$$

or show that both sides of (1.62) satisfy the same recursion (and initial conditions).

27. **Log-concavity.** An easy criterion for log-concavity of a sequence is Newton's lemma that *if* $\sum_{k=0}^{n} a_k x^k$ *is a real polynomial with only real roots then its coefficients are log-concave, as are* $a_k / \binom{n}{k}$ $(0 < k < n)$.

   (a) Use Rolle's theorem to prove Newton's lemma.
   (b) Deduce that the binomial coefficients, row by row, as the Stirling numbers of the first and second kind are log-concave. In particular, they are unimodal.
   (c) The Catalan numbers are log-convex.
   (d) The Motzkin numbers, M_n (see Item 41), are log concave. This was first shown in 1998 and may be established analytically, as in [111], starting with the recursion

   $$(n + 2)M_n = (2n + 1)M_{n-1} + 3(n - 1)M_{n-2},$$

   with ordinary generating function

   $$\frac{(1 - x) - \sqrt{1 - 2x - 3x^2}}{2x^2}.$$

   (e) Prove that the sequence $x_n = M_n / M_{n-1}$ is increasing (equivalent to log-concavity) by considering that the function $f : [2, \infty) \to \mathbb{R}$ by $f(x) = 2$ on $[2, 3]$ and by

   $$f(x) = \frac{2x + 3}{x + 2} + \frac{3(x - 1)}{x + 2} \frac{1}{f(x - 1)}$$

   thereafter. Thus $f(n) = x_n$. Show that f is continuous, increasing and piece-wise smooth with $\lim_{x \to \infty} f(x) = 3$.

28. **Putnam problem 1999–A4.** Sum the series

$$S = \sum_{m=1}^{\infty} \sum_{n=1}^{\infty} \frac{m^2 n}{3^m (n \, 3^m + m \, 3^n)} \qquad \left(= \frac{9}{32} \right). \qquad (1.63)$$

Hint: Interchange m and n and average to obtain $S = \left(\sum_{n \geq 1} n/3^n \right)^2$.

29. **Putnam problem 1999–A6.** Consider the sequence defined by $a_1 = 1, a_2 = 2, a_3 = 24$, and, for $n > 3$,

$$a_n = \frac{6a_{n-1}^2 a_{n-3} - 8a_{n-1}a_{n-2}^2}{a_{n-2} \, a_{n-3}}. \qquad (1.64)$$

Show that n divides a_n for all n.

Hint: Consider the much simpler linear recursion satisfied by $b_n = a_n/a_{n-1}$, which is solved by $2^{n-1}(2^{n-1} - 1)$ so that

$$a_n = 2^{n(n-1)/2} \prod_{k=1}^{n-1} \left(2^k - 1 \right).$$

Write n as $2^a b$ where b is odd and observe that $a \leq n(n-1)/2$ and b divides $2^{\phi(b)} - 1$, where ϕ is Euler's totient function.

30. **Berkeley problem 1.3.3.** Establish the limit of the recursion $x_0 = 1$ and

$$x_{n+1} = \frac{3 + 2x_n}{3 + x_n},$$

for $n > 0$. Does the initial value matter?

Answer: $\ell = (\sqrt{13} - 1)/2$. Note: The *Maple* code

```
iter1:=proc(y,n) local x,k; x:=y; for k to n do
x:=(3+2*x)/(3+x);od;solve(Minpoly(x,2))[1];end:
```

answers this symbolically.

31. **Berkeley problem 1.3.4.** Similarly for

$$x_{n+1} = \frac{1}{2 + x_n}.$$

Answer: $\ell = \sqrt{2} - 1$.

32. **Continued fraction errors.** Compare the estimates in Theorem 1.11 with the computed errors in Table 1.1 using $a = N$ and $c = N+1$, and in Table 1.3 using $a = N + 1/2$ and $c = N + 3/2$.

33. **Berkeley problems 5.7.1, 5.7.2, and 5.7.3.** Evaluate

$$\int_0^{2\pi} e^{e^{it}} \, dt \qquad (= 2\pi),$$

$$\int_0^{2\pi} e^{e^{it} - it} \, dt \qquad (= 2\pi),$$

and, for $a > b > 0$, show

$$\frac{1}{2\pi} \int_0^{2\pi} \frac{1}{|ae^{it} - b|^4} \, dt = \frac{b^2 + a^2}{(a^2 - b^2)^3}.$$

Hint: Proofs use the Cauchy integral formula for derivatives.

34. **A double integral-series equivalence.** For nonnegative integer m, (a) show that

$$-i \int_0^\infty \int_0^\infty \frac{t^m e^{(ix-1)t}}{1 + x^2} \, dt \, dx = \frac{m!}{2^{m+1}} \left(\sum_{n=0}^m \frac{2^n}{n+1} - i\frac{\pi}{2} \right). \quad (1.65)$$

(b) Hence, establish the moment evaluations:

$$\int_0^\infty t^m \int_0^\infty \frac{\cos(tx)}{1 + x^2} \, dx \, dt = \frac{m!}{2^{m+1}} \frac{\pi}{2} \quad (1.66)$$

and

$$\int_0^\infty t^m \int_0^\infty \frac{\sin(tx)}{1 + x^2} \, dx \, dt = \frac{m!}{2^{m+1}} \sum_{n=0}^m \frac{2^n}{n+1}. \quad (1.67)$$

(c) Show that (1.67) also equals

$$\frac{m!}{2(m+1)} \sum_{k=0}^m \frac{1}{\binom{m}{k}}.$$

Hint: The left-hand side is

$$-i \, m! \int_0^\infty \frac{(1 - ix)^{-1-m}}{1 + x^2} \, dx.$$

Both sides then of (1.65) satisfy the recursion

$$2(m+1)r(m) - m(m+1)r(m-1) = m!,$$

with initial value $r(0) = \frac{1}{2} - i\frac{\pi}{4}$. Now justify exchanging the order of integration.

35. **An open summation problem.** Does

$$\sum_{n=1}^{\infty} \frac{\left(\frac{2}{3} + \frac{1}{3}\sin(n)\right)^n}{n}$$

converge? This is an open problem.

36. **Hadamard inequality.** Let $A = \{a_{ij}\}$ be a real $n \times n$ positive semidefinite Hermitian matrix.

(a) Show

$$\det(A) \le \prod_{i=1}^{n} a_{ii},$$

with equality if and only if A is a diagonal matrix or if some a_{ii} is zero.

(b) By applying the result of (a) to AA^*, obtain the inequality

$$|\det(A)| \le \left(\prod_{i=1}^{n}\sum_{j=1}^{n}|a_{ij}|^2\right)^{1/2},$$

for arbitrary square matrices. *Hint*: Apply the arithmetic-geometric mean inequality to the diagonalization of A.

37. **A tangent series.** (From [140, pg. 83]). For each positive n, evaluate

$$T_n = \sum_{k=0}^{n-1} \tan^2\left(\frac{\pi(2k+1)}{4n}\right),$$

with proof.

Answer: $T_n = n(2n-1)$.

38. **Putnam problem 1989–B3.** Let $f : [0, \infty) \to [0, \infty)$ be differentiable and satisfy $f'(x) = 6f(2x) - 3f(x)$. Assume that $|f(x)| \le \exp(-\sqrt{x})$. Determine a formula for the moments

$$\mu_n = \int_0^{\infty} x^n f(x)\, dx$$

of f for $n = 1, 2, 3, \cdots$. Deduce that $\{3^n \, \mu_n/n!\}$ converges and that the limit is only zero if μ_0 is 0.

Hint: Use integration by parts in the formula for μ_n, which then involves $f'(x)$; a recursion for μ_n will then be apparent.

39. **Putnam problem 1992–A4.** Consider an infinitely differentiable real function f with

$$f\left(\frac{1}{n}\right) = \frac{n^2}{1 + n^2}.$$

for positive integers n. Determine the Maclaurin series of f.

Hint: Consider the series of $f(x) - 1/(1 + x^2)$, and recall that if a C^∞ function is zero on a sequence converging to zero, then all of its derivatives are zero (although this doesn't necessarily imply that the function is analytic in a neighborhood of zero).

40. **Putnam problem 2000–A4.** Show that the improper integral

$$\mathcal{I} = \lim_{M \to \infty} \int_0^M \sin(x) \sin(x^2) \, dx \qquad (1.68)$$

exists.

Hint: Numerical experimentation shows that a limit of approximately 0.4917 is reached. The existence of the limit can be rigorously established in two ways:

(a) Since the integrand equals $\cos(x^2 - x) - \cos(x^2 + x))/2$, it suffices to show that $\lim_{M \to \infty} \int_0^M \cos(x + x^2) \, dx$ exists. After a change of variables, it suffices to consider

$$\sum_{k=0}^{n-1} \int_{(k-1/2)\pi}^{(k+1/2)\pi} \frac{\cos(u)}{\sqrt{1 + 4u}} \, du.$$

This converges by the alternating series test.

(b) Use Cauchy's theorem to integrate the entire functions $\exp(ix^2 \pm ix)$ over a triangular path with vertices at $0, M$ and $(1 + i)M$. Easy estimates show that the integrals over the vertical and the diagonal edges converge.

41. **Several classic sequences.** In each case, try to find a generating function or rule for the sequence given:

(a) $6, 28, 496, 8128, 33550336, 8589869056, 137438691328,$
    $2305843008139952128, \cdots$

(b) $1, 1, 2, 5, 15, 52, 203, 877, 4140, 21147, 115975, 678570, 4213597,$
    $27644437, \cdots$

(c) $1, 1, 2, 4, 9, 21, 51, 127, 323, 835, 2188, 5798, 15511, 41835, \cdots$

(d) $1, 2, 6, 22, 94, 454, 2430, 14214, 89918, 610182, 4412798, \cdots$

(e) $1, 4, 11, 16, 24, 29, 33, 35, 39, 45, 47, 51, 56, 58, 62, 64, \cdots$

(f) $1, 20, 400, 8902, 197281, 4865617, \cdots$

Answers:

(a) The first few perfect numbers.

(b) The *Motzkin numbers*. Among other interpretations they count
    the number of ways to join n points on a circle by nonintersecting
    chords, and the number of length n paths from $(0, 0)$ to $(n, 0)$
    that do not go below the horizontal axis and are made up of steps
    $(1, 1), (1, -1)$ and $(1, 0)$. The generating function (see Item 27)
    is the expression $(1 - x - \sqrt{1 - 2x - 3x^2})/(2x^2)$.

(c) The *Bell numbers*, whose exponential generating function is
    $\exp(e^x - 1)$.

(d) Values of *Bell polynomials*, in this case counting ways of placing
    n labeled balls into n unlabeled (but two-colored) boxes (see
    Section 1.10).

(e) *Aronson's sequence*, whose definition is: "t is the first, fourth,
    eleventh, $\cdots$ letter of this sentence."

(f) The number of possible chess games after n moves.

42. **Duality for Mahler's generating function.** Let $\alpha > 0$ be irra-
    tional and consider $G_\alpha(z, w) = \sum_{n=1}^{\infty} z^n w^{\lfloor n\alpha \rfloor}$ as in (1.18). Define
    $F_\alpha(z, w) = \sum_{n=1}^{\infty} z^n \sum_{m=1}^{\lfloor n\alpha \rfloor} w^m$. Show

(a)
$$F_\alpha(z, w) = \frac{z}{1 - z} G_{\alpha^{-1}}(w, z)$$

(b)
$$F_\alpha(z, w) + F_{\alpha^{-1}}(w, z) = \frac{z}{1 - z} \frac{w}{1 - w}.$$

43. **A continued fraction form of Mahler's generating function.**
Let $\alpha > 0$ be irrational. Show that

$$\frac{1-w}{w} \sum_{n=1}^{\infty} z^n w^{\lfloor n\alpha \rfloor} = \cfrac{1}{c_0 + \cfrac{1}{c_1 + \cfrac{1}{c_2 + \cdots}}}, \qquad (1.69)$$

where

$$c_0 = \frac{z^{-1} w^{-a_0} - 1}{w^{-1} - 1}$$

and for $n \geq 1$

$$c_n = \frac{z^{-q_{n-2}} w^{-p_{n-2}} \left[z^{-a_n q_{n-1}} w^{-a_n p_{n-1}} - 1 \right]}{z^{-q_{n-1}} w^{-p_{n-1}} - 1} = \frac{z^{q_n} w^{p_n} - z^{-q_{n-2}} w^{-p_{n-2}}}{z^{-q_{n-1}} w^{-p_{n-1}} - 1}.$$

Here $\{a_n\}$ are the convergents, and p_n/q_n are the partial quotients of the continued fraction for α. In particular, sums like

$$\sum_{n=1}^{\infty} \frac{1}{3^n \, 2^{\lfloor n\alpha \rfloor}} \qquad \text{and} \qquad \sum_{n=1}^{\infty} \frac{1}{2^n \, 3^{\lfloor n\alpha \rfloor}}$$

are nonquadratic irrationals since their continued fractions are clearly unbounded.

(a) Apply (1.69) for $\alpha = (1 + \sqrt{5})/2$ so that $a_n = 1$ and p_n and q_n are Fibonacci numbers. Apply this also to $\sqrt{2} \pm 1$.

(b) Using Exercise 42, deduce a continued fraction for $\sum_{n=1}^{\infty} \lfloor n\alpha \rfloor z^n$.

More details and quite broad extensions—found experimentally—may be found in [48].

44. **Beatty's theorem.** Let irrational numbers $\sigma, \tau > 0$ be given. Use the Mahler continued fraction to show that the sets of integer parts

$$\mathcal{S} = \{\lfloor n\sigma \rfloor : 0 < n \in \mathbb{N}\} \quad , \quad \mathcal{T} = \{\lfloor n\tau \rfloor : 0 < n \in \mathbb{N}\}$$

partition $\mathbb{N} \setminus \{0\}$ if and only if $\sigma > 1$ and

$$\frac{1}{\sigma} + \frac{1}{\tau} = 1.$$

What happens in the case that σ is rational?

Beatty's Theorem, as is often the case, was rediscovered by Beatty. It was known to Lord Raleigh and others earlier. This is a good example of Stigler's Law of Eponymy, namely that "No scientific law is named after its original discoverer." [141, page 60]. Stigler's law is named after Stephen Stigler, the son of the 1982 Nobel-prize-winning economist George Stigler. Neither Stigler was the "discoverer" of this principle.

Hint: Observe first that $\sigma, \tau > 1$ is necessary. Set $\alpha = \sigma - 1$ and $\beta = \tau - 1$. Note that $\mathcal{S}$ and $\mathcal{T}$ partition the positive integers if and only if

$$G_\alpha(z, z) + G_\beta(z, z) = \sum_{n=1}^{\infty} z^n = \frac{z}{1 - z}.$$

Use Exercise 42 to show this happens if and only if

$$G_\beta(z, z) = G_{\alpha^{-1}}(z, z),$$

which in turn happens exactly when $\lfloor n\beta \rfloor = \lfloor n(\alpha^{-1}) \rfloor$ for all $n \geq 1$. This last equivalence holds if and only if $\beta\,\alpha = 1$, and this is the same as

$$\frac{1}{\sigma} + \frac{1}{\tau} = 1.$$

A direct proof of the "if" is quite easy.

45. **Wilker's inequalities.** Show that for $0 < x < \pi/2$, one has

$$2 + \frac{2}{45}x^3 \tan(x) > \frac{\sin^2(x)}{x^2} + \frac{\tan(x)}{x} > 2 + \frac{16}{\pi^4}x^3 \tan(x),$$

and that the constants $2/45$ and $16/\pi^4$ are the best possible.

46. **A Gamma integral.** Show that

$$\int_0^\infty e^{iy} y^{a-1}\, dy = i^a \, \Gamma(a)$$

for $0 < a < 1$. Hence, evaluate $\int_0^\infty \cos\left(x^b/b\right)\, dx$ for $b > 1$.

Hint: Use Cauchy's theorem on a contour that goes from 0 to R and then on a quarter circle to iR and back on the vertical axis to 0.

47. **The Airy integral.** For real x, the Airy integral is defined by

$$\mathrm{Ai}(x) = \frac{1}{\pi} \int_0^\infty \cos\left(\frac{1}{3}t^3 + xt\right)\, dt.$$

Integrate by parts to show that the integral is well defined. Then obtain the value of Ai(0). Finally, show that $\text{Ai}^2(z)$ satisfies $w'' - 4zw' - 2w = 0$. See also [143]. Some difficult Airy-related integrals are derived and discussed in connection with the quantum bouncer in [95].

48. **Zeroes of the Airy function.** Let $N(\text{Ai})$ denote the real zeroes of the Airy function Ai. Then, for $n = 2, 3, \cdots$, we define the Airy zeta function

$$Z(n) = \sum_{a \in N(\text{Ai})} \frac{1}{a^n} = \frac{\pi T_{n-1}(0)}{\Gamma(n)},$$

where

$$
\begin{aligned}
T_n(z) &= C^{(n)}(z) \int_0^\infty \text{Ai}^2(u)\, du \\
&\quad - \sum_{j=1}^n \binom{n}{j} C^{(n-j)}(z) \frac{d^{j-1}}{dz^{j-1}} \text{Ai}^2(z) + \frac{d^{n-1}}{dz^{n-1}} (\text{Ai}(z)\text{Bi}(z)),
\end{aligned}
$$

with $C(z) = \text{Bi}(z)/\text{Ai}(z)$, [95]. Here, Ai and Bi are the fundamental solutions to $w'' = zw$ with $\text{Ai}(0) = \sqrt[3]{3}/(3\Gamma(2/3))$ and $\text{Bi}(0) = 3^{5/6}/(3\Gamma(2/3))$, and Ai is the Airy integral.

49. **Fun with Airy functions.** Let $\alpha_n = \text{Ai}^{(n)}(0)$ and $\beta_n = \text{Bi}^{(n)}(0)$.

   (a) Show that $\alpha_{n+3} = (n+1)\alpha_n$, with a similar recursion for β_n.

   (b) Show that $\alpha_n^* = (\text{Ai}^2)^{(n)}(0)$ and $\beta_n^* = (\text{Bi}^2)^{(n)}(0)$ both satisfy $\delta_{n+3} = (4n+2)\delta_n$.

   (c) Consider $\gamma_n = (\text{Ai}/\text{Bi})^{(n)}(0)$ and $\delta_n = (\text{Ai}\,\text{Bi})^{(n)}(0)$. Show that $\delta_{n+3} = (4n+2)\delta_n$, and obtain a recursion for γ_n via the convolution

$$\sum_{j=0}^n \binom{n}{j} \delta_j\, \gamma_{n-j} = \beta_n^*.$$

   Note the analogy with the Bernoulli numbers.

50. **More on the Airy zeta function.** Express $Z(n)$ of Exercise 48 explicitly as a polynomial in

$$X = \frac{3^{5/6}}{2\pi} \Gamma^2(2/3) = \frac{1}{2\pi\,\text{Ai}(0)\text{Bi}(0)}.$$

For example,

$$Z(2) = X^2 = 3\,\frac{3^{2/3}\Gamma^4\,(2/3)}{4\,\pi^2},$$

$$Z(7) = X^7 - \frac{7}{12}X^4 + \frac{13}{180}X,$$

and

$$Z(10) = X^{10} - \frac{5}{6}X^7 + \frac{209}{1008}X^4 - \frac{17}{1296}X.$$

Note that

$$\int_0^\infty \mathrm{Ai}^2\,(t)\;dt = \mathrm{Ai}'\,(0)^2 = \frac{\sqrt[3]{3}\,\Gamma^2\,(2/3)}{4\,\pi^2}.$$

Exercise 48 may help find the desired polynomials.

51. **Gosper's continued fraction for π.** During his record 1985 computation of the simple continued fraction for π, Gosper found some
large convergents, but no strong evidence to suggest that π has a
bounded or unbounded continued fraction. Gosper describes how
continued fractions allow you to "see" what a number is. "[I]t's completely astounding ... it looks like you are cheating God somehow"
[4, page 112]. He goes on to talk about how this sense of surprise has
driven him to extensive work with continued fractions.

52. **A positivity problem.** Show that f defined by

$$f(x) = \sum_{j=0}^\infty \frac{(-2)^j\,x^{2^j}}{\prod_{i=1}^j\,(2^i - 1)}$$

is strictly positive for $0 < x < 1$. *Hint:* $g = x \mapsto f(x)/x$ satisfies
$g'(x) + 2\,g(x^2) = 0$ and $g(0) = 1, g(1) = 0$.

53. **Ramanujan's AGM continued fraction.** The assertions below
are extensions of those in Entry 12 of Chapter 18 of Ramanujan's
Second Notebook. Let

$$\mathcal{R}_\eta(a,b) = \cfrac{a}{\eta + \cfrac{b^2}{\eta + \cfrac{4a^2}{\eta + \cfrac{9b^2}{\eta + \ddots}}}}$$

for $a, b, \eta > 0$.

(a) Then show the marvelous fact that

$$\mathcal{R}_\eta\left(\frac{a+b}{2}, \sqrt{ab}\right) = \frac{\mathcal{R}_\eta(a,b) + \mathcal{R}_\eta(b,a)}{2}.$$

This relies on knowing for $y > 0$ that

$$\frac{2}{\eta}\sum_{k\geq 0}\frac{\mathrm{sech}((2k+1)\pi y/2)}{1+\{(2k+1)/\eta\}^2} = \mathcal{R}_\eta(\theta_2^2(q), \theta_3^2(q))$$

$$\frac{1}{\eta}\sum_{k\in Z}\frac{\mathrm{sech}(k\pi y)}{1+\{(2k)/\eta\}^2} = \mathcal{R}_\eta(\theta_3^2(q), \theta_2^2(q)),$$

where $q = \exp(-\pi y)$, and observing how $y \mapsto \frac{y}{2}$ interacts with these two identities. While these two series are hard to derive, they are easy to verify numerically.

(b) The continued fraction is hard to compute directly for $a = b$. We shall find a way to use the last hyperbolic series to compute it.

(c) Determine the relationship between $\mathcal{R}_\eta$ and $\mathcal{R}_1$. Hence, for $0 < b < a$, show that

$$\mathcal{R}_1(a,b) = \frac{\pi}{2}\sum_{n\in Z}\frac{a\,\mathrm{K}(k)}{\mathrm{K}^2(k) + a^2\,n^2\pi^2}\,\mathrm{sech}\left(n\pi\frac{\mathrm{K}(k')}{\mathrm{K}(k)}\right), \quad (1.70)$$

where $k = b/a = \theta_2^2/\theta_3^2$.

(d) For $y = 1$, this evaluates

$$\mathcal{R}_1\left(1, \frac{1}{\sqrt{2}}\right) = \frac{\pi}{2}\sum_{n\in Z}\frac{\mathrm{K}(1/\sqrt{2})\,\mathrm{sech}(n\pi)}{\mathrm{K}^2(1/\sqrt{2}) + n^2\pi^2},$$

with similar evaluations for $\mathcal{R}_1(1, k_N)$ at the N-th singular value discussed in Section 4.2.

(e) Denote $\mathcal{R}(a) = \mathcal{R}_1(a,a)$. Write (1.70) as a Riemann sum to deduce that

$$\mathcal{R}(a) = \int_0^\infty\frac{\mathrm{sech}\left(\frac{\pi x}{2a}\right)}{1+x^2}\,dx = 2\,a\sum_{k=1}^\infty\frac{(-1)^{k+1}}{1+(2k-1)\,a}, \quad (1.71)$$

where the final equality comes from the Cauchy-Lindelöf Theorem (Theorem 6.10). The final sum provides an analytic continuation of $\mathcal{R}$, and can be written in the computationally efficient form $\mathcal{R}(a) = 1/2\,\Psi\left(3/4 + 1/4\,a^{-1}\right) - 1/2\,\Psi\left(1/4 + 1/4\,a^{-1}\right)$. (Here, $\Psi = \Gamma'/\Gamma$ is the digamma function.) Observe that

$$
\begin{aligned}
\mathcal{R}(a) &= \frac{2a}{1+a}\,\mathrm{F}\left(\frac{1}{2a} + \frac{1}{2}, 1; \frac{1}{2a} + \frac{3}{2}; -1\right) \\
&= 2\int_0^1 t^{1/a}(1+t^2)^{-1}\,dt = a\int_0^\infty \mathrm{sech}(ax)\,e^{-x}\,dx
\end{aligned}
$$

is now of the form for Gauss's continued fraction of Section 7.8 of [43].

(f) Conclude that $\mathcal{R}(1) = \log 2$ and $\mathcal{R}(1/2) = 2 - \pi/2$ with similar evaluations for all Egyptian fractions $(1/n)$.

(g) Prove that
$$
\mathcal{R}(2) = \sqrt{2}\left\{\frac{\pi}{2} - \log(1 + \sqrt{2})\right\}.
$$

Evaluate also $\mathcal{R}(3/2)$ and $\mathcal{R}(5)$.

(h) One can derive the rapidly convergent ζ-series

$$
\mathcal{R}(a) = \frac{\pi}{2}\sec\left(\frac{\pi}{2a}\right) + \frac{2\,a^2(1 + 8a - 106a^2 + 280a^3 + 9a^4)}{(a^2 - 1)\,(9\,a^2 - 1)\,(5\,a - 1)\,(7\,a - 1)} + \mathcal{C}(a)
$$

where

$$
\mathcal{C}(a) = \frac{1}{2}\sum_{n\geq 1}\{\zeta(2n+1) - 1\}\frac{(3a - 1)^{2n} - (a - 1)^{2n}}{(4a)^{2n}}.
$$

54. **A fast series.** It is possible to deduce from the previous exercise, using Poisson summation (see Theorem 2.12), that

(a) For $0 < b \leq a, k = b/a, K = \mathrm{K}(k)$, and $K' = \mathrm{K}(k')$ we have

$$
\begin{aligned}
\mathcal{R}_1(a, b) &= \mathcal{R}\left(\frac{\pi a}{2K'}\right) + \frac{\pi}{\cos(K'/a)}\frac{1}{e^{2K/a} - 1} \\
&\quad + \frac{2\pi a}{K'}\sum_{d\in O^+}\frac{(-1)^{(d-1)/2}}{1 - \pi^2 d^2 a^2/(4K'^2)}\frac{1}{e^{\pi dK/K'} - 1}.
\end{aligned}
$$

This allows a highly effective computation of the continued fraction when a is close to b, and so the series (1.70) is not effective. Determine the proper interpretation when $a = b$.

(b) The AGM relationship, $\mathcal{R}_1(b, a) = 2\,\mathcal{R}_1((a+b)/2, \sqrt{ab}) - \mathcal{R}_1(a, b)$, then allows one to compute $\mathcal{R}_1(a, b)$ for $a < b$ equally effectively. (See [39] for details.)

(c) Denoting by O^+ the positive odd integers, one may establish

$$\int_{-\infty}^{\infty} \frac{\operatorname{sech}(bx)}{1+x^2} e^{iax}\, dx = \frac{\pi}{\cos b} e^{-a} + \frac{2\pi}{b} \sum_{d \in O^+} \frac{(-1)^{(d-1)/2} e^{-\pi da/(2b)}}{1 - \pi^2 d^2/(4b^2)}.$$

Interpret this relation for the possibility that b is an odd integer times $\pi/2$. Compare the next exercise.

55. A cosh integral.

(a) Show that, for $|a| \le b$, one has

$$\int_0^\infty \frac{\cosh(at)}{\cosh(bt)(1+t^2)}\, dt =$$

$$\frac{\pi}{2}\frac{1}{b\sin(b)}\left(\sin(b)\int_0^a \frac{\sin(a-t)}{\cos\left(\frac{1}{2}\frac{\pi t}{b}\right)}\, dt - \sin(a)\int_0^b \frac{\sin(b-t)}{\cos\left(\frac{1}{2}\frac{\pi t}{b}\right)}\, dt\right)$$

$$+ \frac{\pi}{2}\frac{\sin(a)}{\sin(b)} - \frac{\sin(a-b)}{\sin(b)}\int_0^\infty \frac{1}{\cosh(bt)(1+t^2)}\, dt.$$

Hint: Show that both sides satisfy the same second-order differential equation in a with the same values for $a = 0$ (namely the value $\int_0^\infty \operatorname{sech}(bx)/(1+x^2)\, dx = \mathcal{R}(\pi/2/b)$) and for $a = b$ (namely $\pi/2$).

(b) Deduce that

$$\int_0^\infty \frac{\operatorname{sech}(bt)}{1+t^2}\cos(at)\, dt =$$

$$\frac{\cosh(a)}{2}\left\{\Psi\left(\frac{3}{4}+\frac{1}{2}\frac{b}{\pi}\right) - \Psi\left(\frac{1}{4}+\frac{1}{2}\frac{b}{\pi}\right)\right\}$$

$$- \frac{\pi}{2b}\int_0^a \sinh(a-t)\operatorname{sech}\left(\frac{\pi t}{2b}\right)\, dt$$

for all $a, b \ge 0$.

(c) Note that as $b \to 0$, the final integrand is convergent to zero, but not uniformly, and the prior evaluation becomes

$$\int_0^\infty \frac{\cos(at)}{1+t^2}\, dt = \frac{\pi}{2}e^{-a}.$$

The final integral in (b) equals $\int_0^a \cosh(a-t)\arctan\left(\sinh\left(\frac{\pi t}{2b}\right)\right)\, dt$.

(d) Show for $|a| < b$ that

$$\int_0^\infty \frac{\operatorname{cosech}(bt)}{1+t^2} \sinh(at)\,dt = \pi \sum_{k=1}^\infty \frac{\sin(k\pi(1-a/b))}{b+k\pi}.$$

56. **Infinitely differentiable functions.** A lovely result of E. Borel [203, pg. 191] shows that for every real sequence $\{a_n\}$, there is an infinitely differentiable function on R such that $f^{(n)}(0) = a_n$. A remarkable explicit example occurs via Ramanujan's continued fraction. Use (1.71) to show that the function $a \mapsto \mathcal{R}(a)$ has Maclaurin series $\sum_{n\geq 0} E_{2n}\, a^{2n+1}$, with zero radius of convergence. *Hint:* Show that $\int_0^\infty \operatorname{sech}(\pi x/2)x^{2n}\,dx = E_{2n}$ for each positive integer n. Then estimate

$$\left| \mathcal{R}(a) - \sum_{n=0}^{N-1} a^{2n+1} E_{2n} \right| \leq a^{2N+1} \int_0^\infty \operatorname{sech}\left(\frac{\pi x}{2}\right) x^{2N}\,dx$$

$$= a^{2N+1} |E_{2N}|,$$

where E_{2N} denote the Euler numbers.

57. **Two expected distances.** These results originate with James D. Klein.

(a) The expected distance between two random points on different sides of the unit square:

$$\frac{2}{3} \int_0^1 \int_0^1 \sqrt{x^2+y^2}\,dx\,dy + \frac{1}{3} \int_0^1 \int_0^1 \sqrt{1+(y-u)^2}\,du\,dy$$

$$= \quad 0.8690090552745344638849705943454064856719\ldots$$

$$= \quad \frac{2}{9} + \frac{1}{9}\sqrt{2} + \frac{5}{9}\log\left(1+\sqrt{2}\right).$$

(b) The expected distance between two random points on different faces of the unit cube:

$$\frac{4}{5} \int_0^1 \int_0^1 \int_0^1 \int_0^1 \sqrt{x^2+y^2+(z-w)^2}\,dw\,dx\,dy\,dz$$

$$+ \quad \frac{1}{5} \int_0^1 \int_0^1 \int_0^1 \int_0^1 \sqrt{1+(y-u)^2+(z-w)^2}\,du\,dw\,dy\,dz$$

$$= 0.926390055174046729218163586547779014444496019010734\ldots$$

$$= \frac{4}{75} + \frac{17}{75}\sqrt{2} - \frac{2}{25}\sqrt{3} - \frac{7}{75}\pi$$
$$+ \frac{7}{25}\log\left(1 + \sqrt{2}\right) + \frac{7}{25}\log\left(7 + 4\sqrt{3}\right).$$

(c) Show that the first term in (b) is

$$\frac{\sqrt{2\pi}}{5}\sum_{n=2}^{\infty}\frac{F\left(1/2, -n+2; 3/2; 1/2\right)}{(2n+1)\,\Gamma\left(n+2\right)\Gamma\left(5/2 - n\right)}$$

$$+ \frac{4}{15}\sqrt{2} + \frac{2}{5}\log\left(\sqrt{2} + 1\right) - \frac{1}{75}\pi$$

and the second term is

$$\frac{\sqrt{\pi}}{10}\sum_{n=0}^{\infty}\frac{F\left(1, 1/2, -1/2 - n, -n-1; 2, 1/2 - n, 3/2; -1\right)}{(2n+1)\,\Gamma\left(n+2\right)\Gamma\left(3/2 - n\right)}$$

$$-\frac{2}{25} + \frac{\sqrt{2}}{50} + \frac{1}{10}\log\left(\sqrt{2} + 1\right).$$

This allows one to numerically compute the expectation to high precision and to express both of the individual integrals in terms of the same set of constants. These expectations have actually been checked by computer simulations.

Hint: Reduce the first integral to a three-dimensional one, and use the binomial theorem on both.

58. **Euler: The Master of Us All.** Laplace once wrote, "Read Euler, read Euler. He is the master of us all." This quote is used for the title of Dunham's lovely book about Euler [113]. It contains a concise biography and emphasizes his breadth and mastery of all the topics of his period. Dunham's discussions on Euler and logarithms, and on Euler and infinite series, are especially worth reading in conjunction with this chapter. The former emphasizes the centrality of logarithms and their generalizations to Euler's work, and the latter discusses in illuminating detail Euler's conquest of the *Basel problem* of evaluating $\zeta(2)$.

2 | Fourier Series and Integrals

Having contested the various results [Biot and Poisson] now recognise that they are exact but they protest that they have invented another method of expounding them and that this method is excellent and the true one. If they had illuminated this branch of physics by important and general views and had greatly perfected the analysis of partial differential equations, if they had established a principal element of the theory of heat by fine experiments ... they would have the right to judge my work and to correct it. I would submit with much pleasure ... But one does not extend the bounds of science by presenting, in a form said to be different, results which one has not found oneself and, above all, by forestalling the true author in publication.

John Herivel, *Joseph Fourier: The Man and the Physicist*, 1992

It is often useful to decompose a given function into components, analyze them, and then reassemble the function again, possibly in a different way. One classical and mathematically very interesting method is to use trigonometric functions. This is the basis for the theory of Fourier analysis.

One can think of a sound (a certain tone played on the violin, say) as consisting of countably many oscillations with different discrete frequencies, which together define the pitch and the specific timbre of the tone. These component frequencies can be identified via Fourier analysis, in particular by computing the Fourier series of a periodic function. Of course, in reality no oscillation is precisely periodic, and a sound will consist of a continuum of frequencies. Mathematically, this is analyzed by taking the continuous Fourier transform. Thus, the Fourier transform arises from Fourier series by taking more and more frequencies into account, a process described by the Poisson summation formula. Finally, the question arises as to whether a function thus analyzed can be reconstructed from its Fourier series. It turns out that even for a continuous function, the Fourier series may not be everywhere convergent. Thus one is led to consider special summation methods, or "kernels."

These are among the topics covered in the next few sections. The interested reader can find more details than we have room to give here in Chapter 8 of Stromberg's book [203], in the classical treatment available in Zygmund [216], or in the more modern books by Katznelson [149] and Butzer-Nessel [75].

One reason for studying Fourier methods is that they are very useful for evaluation of sums and integrals. Once such an object is identified as a Fourier series or integral, that knowledge can be used for the evaluation, or methods such as the Parseval equation or Poisson summation can be brought to bear. We will see numerous examples of this in the present chapter and elsewhere in the book.

2.1   The Development of Fourier Analysis

We start with some historical background here, which we have adapted in part from the MacTutor web site at http://www-gap.dcs.st-and.ac.uk/ ˜history, and also from R. Bhatia's monograph on Fourier series [25].

Joseph Fourier was one of the more colorful figures of mathematical history. Originally intending to be a Catholic priest, Fourier declined to take his vows when he realized that he could not extinguish his interest in mathematics. Shortly afterwards, he became involved in the movement that led to the French Revolution in 1793, but, fortunately for modern mathematics, he was spared the guillotine, and was able to study mathematics at the Ecole Normale in Paris under the tutelage of Lagrange. A few years later, he was appointed as a scientific adviser for Napoleon's expedition to Egypt. When Napoleon's army was defeated by Nelson at the battle of the Nile, Fourier and the other French advisers insisted that they be able to retain some of the artifacts they had found there. The British refused, but at least permitted the French to make a catalog of what they felt were the more important items. Fourier was given this task by the French commanders. The eighth item of his catalog was the Rosetta stone, which had been recognized by the French scientists on the expedition as a possible key to understanding the Egyptian language. Later in Europe, when published copies of the inscriptions were made available, Champollion, a student who had been inspired by Fourier himself to study Egyptology, succeeded in the first translation.

Fourier's principal contributions to mathematics, namely Fourier analysis and Fourier series, paralleled and even stimulated the development of the entire field of real analysis. Fourier analysis had its origin in the 1700s, when d'Alembert derived the wave equation that describes the motion of a vibrating string, starting with an initial "function," which at the time was restricted to an analytic expression. In 1755, Daniel Bernoulli gave another solution for the problem in terms of *standing waves*, namely waves associated with the $n + 1$ points $0, 1/n, 2/n, \cdots, (n - 1)/n, 1$ on the string that remain fixed. The motion for $n = 1, 2, \cdots$ is the *first harmonic*, the *second harmonic*, and so on. Bernoulli asserted that every solution to the problem of the plucked string is merely a sum of these harmonics.

Beginning in 1804, Joseph Fourier began to analyze the conduction of heat in solids. He not only discovered the basic equations governing heat conduction, but he developed methods to solve them, and, in the process, developed and extended Fourier analysis to a much broader range of scientific problems. He described this work in his book *The Analytical Theory of Heat*, which is regarded as one of the most important books in the history of physics.

Like Bernoulli, Fourier asserted that any continuous function can be written as

$$f(x) = \sum_{n=-\infty}^{\infty} A_n e^{int}. \tag{2.1}$$

But Fourier claimed that his representation is valid not only for f given by a single analytical formula, but for f given by any *graph*, which at the time was a more general object, encompassing, for example, a piecewise combination of different analytic expressions. Fourier was not able prove his assertions, at least not to the satisfaction of the mathematical community at the time (and certainly not to the standards required today). But other mathematicians were intrigued, and they pursued these questions with renewed determination.

Dirichlet was the first to find a rigorous proof. He defined a real function as we now understand the term, namely as a general mapping from one set of reals to another, thus decoupling analysis from geometry. He then was able to prove that for every "piecewise smooth" function f, the Fourier series of f converges to $f(x)$ at any point x where f is continuous, and to the average value $(f(x+) + f(x-))/2$ if f has a jump discontinuity at x. This was the first major convergence result for Fourier series.

Mathematicians realized that to handle functions that have infinitely many discontinuities, it was necessary to generalize the notion of an integral beyond the intuitive idea of the area under a curve. Riemann succeeded in developing a theory of integration that could handle such functions, and using this theory, he was able to exhibit an example of a function that did not satisfy Dirichlet's piecewise continuous condition, yet still possessed a pointwise convergent Fourier series. Cantor observed that changing a function f at a few points does not change its Fourier coefficients. In the course of asking how many points can be changed while preserving Fourier coefficients, he was led to the notion of countably infinite and uncountably infinite sets. Ultimately, Lebesgue extended Dirichlet's, Riemann's, and Cantor's results into what we now know as measure theory, where sets of measure zero, almost everywhere equality of functions, and almost everywhere convergence of functions supersede the simple concepts that prevailed in the 1700s.

In summary, it is not an exaggeration to say that all of modern real and complex analysis has its roots in Fourier series and Fourier analysis.

2.2   Basic Theorems of Fourier Analysis

2.2.1   Fourier Series

We will consider 2π-periodic functions $f : R \to C$. For $p > 0$, we write $f \in L^p(T)$ if such an f is Lebesgue measurable and satisfies

$$\|f\|_p = \left[\int_{-\pi}^{\pi} |f(t)|^p \, dt \right]^{1/p} < \infty$$

(T stands for *Torus*). Note that $\| \cdot \|_p$ is not a norm on L^p, since any function f which is 0 almost everywhere will have $\|f\|_p = 0$. (Later we will identify functions which are equal a.e.) In what follows, we will be mainly interested in the spaces $L^1(T)$ and $L^2(T)$. Note that $\| \cdot \|_1 \le \| \cdot \|_2$ and $L^2(T) \subsetneq L^1(T)$. If f is 2π-periodic and continuous on R, then we write $f \in C(T)$ and equip this space with the uniform norm.

For a function $f \in L^1(T)$, define the n-th Fourier coefficient ($n \in Z$) by

$$\widehat{f}_n = \int_{-\pi}^{\pi} f(t) \, e^{-int} \, dt.$$

This is motivated by the insight that if we write

$$s_n(t) = \frac{1}{2\pi} \sum_{k=-n}^{n} c_k \, e^{ikt}, \tag{2.2}$$

for some sequence c_k and assume $L^1(T)$-convergence of (s_n) to some $f \in L^1(T)$, then $\widehat{f}_n = c_n$. Thus, for arbitrary $f \in L^1(T)$, we will write (formally and suggestively)

$$f(t) \sim \frac{1}{2\pi} \sum_{n=-\infty}^{\infty} \widehat{f}_n \, e^{int},$$

where in general no assertion about convergence of this series is implied. Any convergence statement is to be read in the sense of symmetric limits, i.e., of

$$s_n(f,t) = \frac{1}{2\pi} \sum_{k=-n}^{n} \widehat{f}_k \, e^{ikt}.$$

Fourier coefficients usually are complex numbers, even when f is a real-valued function. Sometimes it is desirable to have a real-valued series for f. Then the Fourier series can be equivalently written as

$$f(t) \sim \frac{1}{2\pi}a_0 + \frac{1}{\pi} \sum_{n=1}^{\infty} (a_n \cos(nt) + b_n \sin(nt)),$$

where

$$a_n = \int_{-\pi}^{\pi} f(t) \, \cos(nt) \, dt$$

$$b_n = \int_{-\pi}^{\pi} f(t) \, \sin(nt) \, dt.$$

Note that

$$s_n(f,t) = \frac{1}{2\pi}a_0 + \frac{1}{\pi} \sum_{k=1}^{n} (a_k \cos(kt) + b_k \sin(kt)).$$

Example 2.1. Fourier series of a symmetric function.

Define $f \in L^1(\mathrm{T})$ by $f(t) = (\pi - t)/2$ for $t \in [0, 2\pi)$. Then $\widehat{f}_n = -i\pi/n$ for $n \neq 0$ and $\widehat{f}_0 = 0$, and its Fourier series is given by

$$f(t) \sim \frac{-i}{2} \sum_{\substack{n=-\infty \\ n \neq 0}}^{\infty} \frac{1}{n} e^{int}, \tag{2.3}$$

or, equivalently,

$$f(t) \sim \sum_{n=1}^{\infty} \frac{\sin(nt)}{n}. \tag{2.4}$$

The right-hand side of (2.4) equals 0 at $t = 0$, while the left-hand side by definition equals $\pi/2$. Thus, equality cannot hold pointwise here. The situation would improve if we were to define $f(0) = 0$. In fact, it follows from

$$\sum_{n=1}^{\infty} \frac{z^n}{n} = -\mathrm{Ln}(1-z) \quad \text{for } |z| \leq 1, \ z \neq 1, \tag{2.5}$$

by setting $z = e^{it}$ and taking real and imaginary parts, that

$$\sum_{n=1}^{\infty} \frac{\sin(nt)}{n} = \frac{\pi - t}{2} \quad \text{and} \tag{2.6}$$

$$\sum_{n=1}^{\infty} \frac{\cos(nt)}{n} = -\ln|2\sin(t/2)|, \tag{2.7}$$

for all $t \in (0, 2\pi)$, and even with uniform convergence on every closed subinterval of $(0, 2\pi)$. □

The question now is under what condition the Fourier series of a function converges to that function. The answer depends on the definition of "convergence," and is most interesting in the cases of pointwise, $L^1(\mathrm{T})$- and $L^2(\mathrm{T})$-convergence.

Pointwise and uniform convergence. As we have seen in the above example (and as is clear from the computation of Fourier series), $L^1(\mathrm{T})$-functions which are equal a.e. will have the same Fourier series. By the *uniqueness theorem for Fourier series*, the converse is also true: Functions with the same Fourier series are equal a.e. It is *not* true that the Fourier series of any continuous function is pointwise convergent to that function. An example due to Lebesgue is given in Item 6 at the end this chapter. Such functions must have a complicated structure, because they cannot be of bounded variation. A function $f : \mathrm{R} \to \mathrm{C}$ is of bounded variation on (a, b) if there is an $M > 0$ with

$$\sum_{k=0}^{n-1} |f(t_{k+1}) - f(t_k)| \leq M \quad \text{for all } a < t_0 < t_1 < \cdots < t_n < b.$$

The infimum of all such M is the total variation of f, thus

$$V_a^b(f) = \sup \left\{ \sum_{k=0}^{n-1} |f(t_{k+1}) - f(t_k)| : a < t_0 < t_1 < \cdots < t_n < b, \ n \in \mathrm{N} \right\}.$$

The set of all functions of bounded variation on (a, b) is denoted $BV(a, b)$. An equivalent characterization (due to Lebesgue) is that f can be written as the difference of two bounded increasing functions. Thus, any BV-function f is differentiable a.e., and the one-sided limits $f(t+)$ and $f(t-)$ exist in (a, b).

Theorem 2.2. (Jordan test). *For $f \in L^1(\mathrm{T}) \cap BV(a, b)$, we have*

$$\lim_{n \to \infty} s_n(f, t) = \frac{f(t-) + f(t+)}{2} \quad \text{for every } t \in (a, b)$$

and uniformly on every compact subinterval of (a, b) where f is continuous. If $f \in C(\mathrm{T})$, then the convergence is uniform on R.

Note that this theorem is proved via Cesàro summation, and thus via the Fejér kernel, which we will discuss in Section 2.3.3.

The Jordan test is another explanation of the convergence properties of the Fourier series for the function f from the previous example. If another function $f \in C(\mathrm{T})$ is defined by $f(t) = t^2$ on $[-\pi, \pi]$, then

$$f(t) \sim \frac{1}{3}\pi^2 + 4 \sum_{n=1}^{\infty} \frac{(-1)^n}{n^2} \cos(nt).$$

Since f is continuous on R, the Fourier series converges uniformly to f, by the Jordan test. However, since $s_n(f, t)$ is uniformly convergent, this also follows directly from the uniqueness theorem for Fourier series.

For $t = 0$, we get from this, the identity

$$\sum_{n=1}^{\infty} \frac{(-1)^{n-1}}{n^2} = \frac{\pi^2}{12}.$$

By separating even and odd parts, this proves that

$$\zeta(2) = \sum_{n=1}^{\infty} \frac{1}{n^2} = \frac{\pi^2}{6}.$$

To conclude the section on pointwise convergence, we note that the Fourier series of an $f \in L^1(\mathrm{T})$ may be divergent everywhere (by an example due to Kolmogorov in 1926). If, however, $f \in L^p(\mathrm{T})$ for $p > 1$, then $s_n(f, t)$ converges to $f(t)$ a.e., a result proved by Carleson and Hunt in 1966 [81, 144].

Convergence in $L^1(\mathrm{T})$. From now on it makes sense to identify functions that are equal almost everywhere, so that $\| \cdot \|_1$ and $\| \cdot \|_2$ are now norms in the respective spaces. Thus we will deal, strictly speaking, with equivalence classes of functions instead of pointwise defined functions, although this will not be denoted explicitly. For example, the statement that such a function is continuous will mean that in its equivalence class we can find a continuous function, then denoted by the same symbol. This notation is unusual but convenient, and it corresponds to how one deals with these objects informally. It can lead to dangerous pitfalls, though, as we will see later. In general, the Fourier series of an $f \in L^1(\mathrm{T})$ need not be convergent to f with respect to the $L^1(\mathrm{T})$-norm. Of the many restrictions on f which imply convergence, we mention here only one which is particularly simple (and has a convincing analog in the case of Fourier transforms).

Theorem 2.3. *Let* $f \in L^1(\mathrm{T})$. *If* $\sum_{n=-\infty}^{\infty} |\widehat{f}_n| < \infty$, *then* $f \in C(\mathrm{T})$ *and*

$$f(t) = \frac{1}{2\pi} \sum_{n=-\infty}^{\infty} \widehat{f}_n \, e^{int},$$

with convergence in the $L^1(\mathrm{T})$*-norm as well as uniformly on* T.

There are functions in $L^1(\mathrm{T})$ (even continuous ones) for which the Fourier coefficients are not absolutely summable. It is a difficult and in general open problem to characterize those sequences which are Fourier sequences of an $L^1(\mathrm{T})$-function. A weak necessary condition follows from the Riemann-Lebesgue lemma (see next subsection and Exercise 14): The Fourier coefficients of any $f \in L^1(\mathrm{T})$ satisfy $\lim_{|n| \to \infty} \widehat{f}_n = 0$.

Convergence in $L^2(\mathrm{T})$. In contrast to the L^1-case, for $p > 1$ the Fourier series of any $f \in L^p(\mathrm{T})$ converges to f in the $L^p(\mathrm{T})$-norm. For $p = 2$, this follows directly from the usual Hilbert space theory: The trigonometric functions constitute an orthogonal basis for $L^2(\mathrm{T})$. This implies that the Fourier coefficients of an $f \in L^2(\mathrm{T})$ are square-summable and that every square-summable sequence is the sequence of Fourier coefficients of an $f \in L^2(\mathrm{T})$. (That is the Riesz-Fischer theorem.) Another consequence of Hilbert space theory is the Parseval equation.

Theorem 2.4. (Parseval's formula). *For any* $f, g \in L^2(\mathrm{T})$, *the identity*

$$\frac{1}{2\pi} \sum_{n=-\infty}^{\infty} \widehat{f}_n \, \overline{\widehat{g}_n} = \int_{-\pi}^{\pi} f(t) \, \overline{g(t)} \, dt \tag{2.8}$$

holds. In particular, for $f = g$ *we get* $\frac{1}{2\pi} \sum_{n=-\infty}^{\infty} \left| \widehat{f}_n \right|^2 = \int_{-\pi}^{\pi} |f(t)|^2 \, dt.$

Example 2.5. Parseval's formula and the zeta function.

Applying the Parseval equation to the function $f(t) = (\pi - t)/2$ on $(0, 2\pi)$ from the first example in this section again gives the identity, after simplifying,

$$\zeta(2) = \sum_{n=1}^{\infty} \frac{1}{n^2} = \frac{\pi^2}{6}.$$

This is the same result as in the previous example, but the Parseval equation is conceptually simpler than the Jordan test. Applying the Parseval equation to $f(t) = t^2/4 - \pi t/2 + \pi^2/6$ gives

$$\zeta(4) = \sum_{n=1}^{\infty} \frac{1}{n^4} = \frac{\pi^4}{90},$$

and this can be continued to give $\zeta(2n)$ as a rational multiple of π^{2n} for any $n \in \mathbb{N}$. The general formula is given in the next chapter, by a different method. Similar formulas for $\zeta(2n+1)$ are unknown (and highly unlikely to exist; see the next chapter for more information). □

Example 2.6. Parseval's formula and Euler sums.

Multiplying (2.6) and (2.7), using the Cauchy product, simplifying, and performing a partial fraction decomposition gives

$$-\ln|2\sin(t/2)| \cdot \frac{\pi - t}{2} = \sum_{n=1}^{\infty} \frac{1}{n} \sum_{k=1}^{n-1} \frac{1}{k} \sin(nt) \quad \text{on } (0, 2\pi). \qquad (2.9)$$

Now Parseval proves the Euler sum formula that we first mentioned in [43]:

$$\frac{1}{4} \int_0^{2\pi} (\pi - t)^2 \ln^2(2\sin(t/2)) \, dt = \pi \sum_{n=1}^{\infty} \frac{\left(\sum_{k=1}^{n-1} k^{-1}\right)^2}{n^2}.$$

□

2.2.2 Fourier Transforms

We now consider functions $f : \mathbb{R} \to \mathbb{C}$. For $p > 0$, we write $f \in L^p(\mathbb{R})$ if such an f is Lebesgue-measurable and satisfies

$$\|f\|_p = \left[\int_{-\infty}^{\infty} |f(t)|^p \, dt\right]^{1/p} < \infty.$$

As before, functions which are equal a.e. will be identified, so that $\|\cdot\|_p$ is a norm. In what follows, we will be mainly interested in the spaces $L^1(\mathbb{R})$ and $L^2(\mathbb{R})$. In contrast to the periodic case, there is now no inclusion relation between these spaces. If $f : \mathbb{R} \to \mathbb{C}$ is continuous, we write $f \in C(\mathbb{R})$ and equip this space with the uniform norm.

The Fourier transform on $L^1(\mathbb{R})$ is now a direct analog of the Fourier coefficients on $L^1(\mathbb{T})$. By further analogy to the previous subsection, a Fourier transform on $L^2(\mathbb{R})$ would also be of interest. However, the definition of such an $L^2(\mathbb{R})$-transform is not as straightforward as before, since these spaces are not contained in each other. There is, however, a meaningful transform on $L^2(\mathbb{R})$, and we will discuss this after giving the properties of the $L^1(\mathbb{R})$-transform.

Fourier Transform on $L^1(\mathbb{R})$. For a function $f \in L^1(\mathbb{R})$, define the Fourier transform of f to be the function $\widehat{f} : \mathbb{R} \to \mathbb{C}$ given by

$$\widehat{f}(x) = \int_{-\infty}^{\infty} f(t)\, e^{-ixt}\, dt.$$

As the example, $f = I_{(-\pi,\pi)}$ (characteristic function) with $\widehat{f}(x) = 2\sin(\pi x)/\pi$ shows, the Fourier transform of an $f \in L^1(\mathbb{R})$ need not be in $L^1(\mathbb{R})$. It is not difficult to show, however, that such an $\widehat{f}$ is always continuous, with $\|\widehat{f}\|_\infty \le \|f\|_1$. The *Riemann-Lebesgue lemma* says that, additionally, $\lim_{|x|\to\infty} \widehat{f}(x) = 0$. It is proved by approximating f by step functions.

Under what conditions can an $f \in L^1(\mathbb{R})$ be reconstructed from its Fourier transform? In principle, this is always possible: By the *uniqueness theorem for Fourier transforms*, functions with the same Fourier transform must be equal a.e. In practice, one would like a simple formula for this inversion. Such a formula is given in the inversion theorem below, whose proof is not easy: It depends on constructing and investigating a suitable summation kernel for the Fourier transform (often the Gauss or the Fejér kernel is used).

Theorem 2.7. *If $f \in L^1(\mathbb{R})$ is such that $\widehat{f} \in L^1(\mathbb{R})$, then $f, \widehat{f} \in C(\mathbb{R})$ and*

$$f(t) = \frac{1}{2\pi} \int_{-\infty}^{\infty} \widehat{f}(x)\, e^{ixt}\, dx \tag{2.10}$$

for all $t \in \mathbb{R}$.

Conditions which are good for $\widehat{f} \in L^1(\mathbb{R})$ are given in Item 10 at the end of this chapter.

Example 2.8. Sinc integrals.

For $f(t) = \max\{1 - |t|, 0\}$, we compute $\widehat{f}(x) = \mathrm{sinc}^2(x/2) \in L^1(\mathbb{R})$, where $\mathrm{sinc}\, x = (\sin x)/x$. Thus by Theorem 2.7, we immediately get

$$\int_{-\infty}^{\infty} \text{sinc}^2(x/2)\, e^{ixt}\, dx \;=\; 2\pi \max\{1 - |t|, 0\} \qquad (2.11)$$

for all t, and especially $\int_{-\infty}^{\infty} \text{sinc}^2(x/2)\, dx = 2\pi$. The integral transform package of *Maple* knows this integral:

```
inttrans[fourier]((sin(x/2)/(x/2))^2,x,t);
```

   returns

```
2 Pi (-2 t Heaviside(t) + t Heaviside(t - 1) - Heaviside(t - 1)
+ t Heaviside(t + 1) + Heaviside(t + 1))
```

This is an inelegant form of the answer we gave. Note that newer versions of *Maple* can evaluate the integral directly as well, while in previous versions this was only possible with the use of the Fourier transform, so that researchers required more knowledge about what they were doing. Thus this is another example of being able to watch technology in progress!   □

Fourier Transform on $L^2(\mathrm{R})$. Since $L^2(\mathrm{R})$ is not a subset of $L^1(\mathrm{R})$, in contrast to the periodic case, the definition of the Fourier transform cannot be directly transferred onto this space: The function $f(t)\, e^{-ixt}$ may not be integrable! For this reason, the Fourier transform on $L^2(\mathrm{R})$ is usually defined as the continuation of the Fourier transform on $L^1(\mathrm{R}) \cap L^2(\mathrm{R})$ which is dense in $L^2(\mathrm{R})$. An equivalent (and more practical) definition is the following. First of all, note that for $f \in L^2(\mathrm{R})$, $f_A = f \cdot \chi_{(-A,A)} \in L^1(\mathrm{R})$ for any $A > 0$. It can be proved, with some effort, that then $\widehat{f_A} \in L^2(\mathrm{R})$ and that the $L^2(\mathrm{R})$-limit of the functions $\widehat{f_A}$ for $A \to \infty$ exists in $L^2(\mathrm{R})$.

Definition 2.9. *Let $f \in L^2(\mathrm{R})$. The $L^2(\mathrm{R})$-limit of the functions $\widehat{f_A}$ for $A \to \infty$ is called the Fourier (or Plancherel) transform of f and is again denoted by $\widehat{f}$.*

By definition, the Fourier transform of an $f \in L^2(\mathrm{R})$ is always in $L^2(\mathrm{R})$. It need not be continuous, nor does the Riemann/Lebesgue lemma hold. It can be proved that $\|f\|_2 = \|\widehat{f}\|_2$ *(Parseval equation)*, and that every function in $L^2(\mathrm{R})$ is the Fourier transform of an $f \in L^2(\mathrm{R})$. An $f \in L^2(\mathrm{R})$ is reconstructible from its Fourier transform by the same process as in Theorem 2.7.

Theorem 2.10. *For any $f \in L^2(\mathrm{R})$,*

$$f(t) \;=\; \frac{1}{2\pi} \int_{-\infty}^{\infty} \widehat{f}(x)\, e^{ixt}\, dx \;=\; \lim_{A \to \infty} \frac{1}{2\pi} \int_{-A}^{A} \widehat{f}(x)\, e^{ixt}\, dx \qquad (2.12)$$

in $L^2(\mathrm{R})$.

Example 2.11. Fourier transform of sine-exponential.

We have already computed $\widehat{f}(x) = 2\sin(\pi x)/x$ for $f(t) = \chi_{(-\pi,\pi)}(t)$. This $\widehat{f}$ is in $L^2(\mathrm{R})$, but not in $L^1(\mathrm{R})$. Theorem 2.10 now says that

$$\int_{-\infty}^{\infty} \frac{\sin(\pi x)}{\pi x} e^{ixt}\, dx \;=\; \chi_{(-\pi,\pi)}(t) \quad \text{a.e.} \qquad\qquad \square$$

It would be interesting to replace this a.e. statement with a pointwise statement, since a.e. statements are more difficult to handle when the goal is exact evaluation of a series or integral, and so they are less useful for experimental mathematics. In fact, a standard theorem of Fourier analysis (Jordan's theorem, see [75, page 205]) is: *If $f \in L^1(\mathrm{R})$ is of bounded variation in an interval including the point t, then*

$$\lim_{a \to \infty} \frac{1}{2\pi} \int_{-a}^{a} \widehat{f}(x)\, e^{ixt}\, dx \;=\; \frac{1}{2}\left(f(t+) + f(t-)\right).$$

Applied to the above example, this gives

$$\lim_{a \to \infty} \int_{-a}^{a} \frac{\sin(\pi x)}{\pi x} e^{ixt}\, dx \;=\; \begin{cases} 0 & \text{for } |t| > \pi, \\ 1 & \text{for } |t| < \pi, \\ \frac{1}{2} & \text{for } |t| = \pi. \end{cases} \qquad (2.13)$$

2.3 More Advanced Fourier Analysis

2.3.1 The Poisson Summation Formula

There are obvious similarities between Fourier series and Fourier transforms. Thus one would think that there are connections between the two concepts. That is indeed so, and the link is provided by the Poisson summation formula.

As a first example, take a function $F \in L^1(\mathrm{T})$, and note that $\widehat{F}_n = \widehat{f}(n)$ if $f \in L^1(\mathrm{R})$ is defined to equal F on $(-\pi, \pi)$ and to vanish everywhere else. This already is a simple special case of the Poisson summation formula.

A second approach to Poisson summation can be motivated by the following question. Take a function $g \in C(\mathrm{R})$, and assume that the sequence $(g(\frac{n}{w}))$ is absolutely summable for each $w > 0$. Consider $F_w(t) = \sum_{n=-\infty}^{\infty} g\left(\frac{n}{w}\right) e^{int}$ for $w > 0$. Such functions F_w are often investigated in connection with summation procedures for Fourier series, and we will do later. A first graphical observation is that the restrictions to $(-\pi, \pi)$ of these functions F_w tend to be concentrated more and more around 0 for increasing w. If, however, the functions $(1/w) \cdot F_w(t/w)$ are plotted, then there seems to be convergence to a limit function that depends on g. What is this limit function, and how can we prove convergence? The answer is again given by the Poisson summation formula.

In general, the Poisson formula links the finite and the infinite transforms via the so-called *periodization*, an operation which associates $L^1(\mathrm{R})$-functions in a natural way with 2π-periodic functions: For $f \in L^1(\mathrm{R})$, set $F(t) = \sum_{j \in \mathrm{Z}} f(t + 2\pi j)$ for all t for which the limit exists. The next theorem is (one version of) the classical Poisson formula, linking the Fourier series of F with the Fourier transform of f.

Theorem 2.12. (Poisson summation formula). *Let $f \in L^1(\mathrm{R})$.*

(a) The periodization F exists for almost every $t \in \mathrm{T}$, and we have $F \in L^1(\mathrm{T})$ and $\|F\|_1 \le \|f\|_1$.

(b) The Fourier series of F is

$$\sum_{j=-\infty}^{\infty} f(t + 2\pi j) \ \sim \ \frac{1}{2\pi} \sum_{n=-\infty}^{\infty} \widehat{f}(n)\, e^{int}, \qquad (2.14)$$

or in other words, we have $\widehat{F}_n = \widehat{f}(n)$ for all $n \in \mathrm{Z}$.

Proof.
(a) Obviously $f(t + 2\pi j)$ is integrable over $[-\pi, \pi]$, and we have

$$\sum_{j=-\infty}^{\infty} \int_{-\pi}^{\pi} |f(t + 2\pi j)|\, dt \ = \ \sum_{j=-\infty}^{\infty} \int_{-\pi+2\pi j}^{\pi+2\pi j} |f(t)|\, dt$$

$$= \ \int_{-\infty}^{\infty} |f(t)|\, dt = \|f\|_{L^1(\mathrm{R})} < \infty.$$

By B. Levi's theorem, the series F is absolutely convergent a.e., we have $F \in L^1(\mathrm{T})$, and summation and integration can be exchanged. Now we also get $\|F\|_{L^1(\mathrm{T})} \le \|f\|_{L^1(\mathrm{R})}$ by using the triangle inequality, doing the exchange and then using the above computation.

(b) Now applying B. Levi's theorem to the functions $f(t + 2\pi j) \cdot e^{-nt}$ for fixed n, and using the fact that the function e^{-int} is 2π-periodic, we get with the same summation trick as before,

$$\widehat{F}_n \ = \ \int_{-\pi}^{\pi} F(t)\, e^{-int}\, dt \ = \ \int_{-\infty}^{\infty} f(t)\, e^{-int}\, dt \ = \ \widehat{f}(n). \qquad \square$$

Of course, it is interesting to ask when the identity holds pointwise instead of just in the sense of Fourier series. From the Jordan test, it can be deduced that if $f \in L^1(\mathbb{R}) \cap BV(\mathbb{R})$ and $f(t) = \frac{1}{2}(f(t+) + f(t-))$ everywhere, then equality holds for all t.

Example 2.13. Fourier series of hyperbolic functions.

Choose $y > 0$ and define

$$f(t) = \begin{cases} e^{-yt} & \text{for } t > 0, \\ 0 & \text{for } t < 0, \end{cases}$$

and $f(0) = 1/2$. Then $\widehat{f}(x) = (y + ix)^{-1}$, and Poisson says that

$$\sum_{j=-\infty}^{\infty} f(t + 2\pi j) = \frac{1}{2\pi} \sum_{n=-\infty}^{\infty} \frac{1}{y + in} e^{int},$$

which is equivalent to

$$\frac{1}{2\pi} \left(\frac{1}{y} + 2 \sum_{n=1}^{\infty} \frac{y\cos(nt) + n\sin(nt)}{y^2 + n^2} \right) = \sum_{j > t/(2\pi)} e^{-y(t+2\pi j)} + \begin{cases} \frac{1}{2}, & t \in 2\pi\mathbb{Z}, \\ 0, & \text{otherwise.} \end{cases}$$

Setting $t = 0$, we get

$$\pi \coth(\pi y) = \frac{1}{y} + 2y \sum_{n=1}^{\infty} \frac{1}{y^2 + n^2}.$$

Setting $t = 1/2$, we get

$$\pi \operatorname{cosech}(\pi y) = \frac{1}{y} + 2y \sum_{n=1}^{\infty} \frac{(-1)^n}{x^2 + n^2}. \qquad \square$$

Example 2.14. A quadratic exponential identity.

Now choose $s > 0$ and set $f(t) = e^{-st^2}$. Then $\widehat{f}(x) = \sqrt{\pi/s}\, e^{-x^2/(4s)}$ (see Section 5.6 of [43]), and Poisson says that

$$\sum_{j=-\infty}^{\infty} e^{-s(t+2\pi j)^2} = \frac{1}{2}\sqrt{\frac{1}{\pi s}} \sum_{n=-\infty}^{\infty} e^{-n^2/(4s)} e^{int}. \qquad (2.15)$$

By analyticity, this can be extended to hold for all $\operatorname{Re}(s) > 0$. We will meet this formula again in Section 3.1. $\qquad \square$

Example 2.15. An experimental version of Poisson's formula.

Take an even real-valued continuous $g \in L^1(\mathrm{R})$ such that $(g(\frac{n}{w}))$ is absolutely summable for each $w > 0$ and assume that $\widehat{g} \in L^1(\mathrm{R})$. Then using Poisson and the inversion theorem 2.7 it follows that

$$\sum_{n=-\infty}^{\infty} g\left(\frac{n}{w}\right) e^{int} = w \sum_{j=-\infty}^{\infty} \widehat{g}\left(w(t + 2\pi j)\right), \qquad (2.16)$$

where equality is meant in $L^1(\mathrm{T})$. This equality does not hold pointwise! In Katznelson's book [149], an example is given of a function $f \in L^1(\mathrm{R})$ with $\widehat{f} \in L^1(\mathrm{R})$ for which equality does not hold pointwise in Poisson's formula (2.14), because the periodization does not converge uniformly. This is the "dangerous pitfall" that we mentioned earlier: Even though the left-hand side of (2.16) is a continuous function, the right-hand side is not necessarily continuous; it is only equal to a continuous function a.e. To deduce pointwise equality from this would be wrong! This is a noteworthy difference to the situation in Theorem 2.7.

In any case, this gives an answer to the question at the beginning of the present subsection: The limit function (in the $L^1(\mathrm{R})$ sense) of the F_w, restricted to $(-\pi, \pi)$ and then suitably rescaled, is the Fourier transform of g. This follows from (2.16), since

$$\|\tfrac{1}{w} F_w\left(\tfrac{t}{w}\right) \cdot \chi_{(-\pi w, \pi w)} - \widehat{g}(t)\|_1 = \int_{-\pi w}^{\pi w} \Big| \sum_{\substack{j=-\infty \\ j \neq 0}}^{\infty} \widehat{g}(t + 2\pi j w) \Big| \, dt$$

$$+ \int_{|t| > \pi w} |\widehat{g}(t)| \, dt$$

$$\leq 2 \int_{|t| > w} |\widehat{g}(t)| \, dt \;\to\; 0 \quad (w \to \infty)$$

if $\widehat{g} \in L^1(\mathrm{R})$. Often enough (but not always) this convergence is even uniform on R.

This is a very visual theorem; it can be discovered and explored on the computer. In this sense, Formula (2.16) is the experimental version of Poisson's summation formula! $\qquad \square$

2.3.2 Convolution Theorems

As we have seen, the Fourier series even of a continuous function need not converge back to the function, neither in the L^1-sense nor pointwise.

Often, convergence properties of such a series can be improved by putting additional factors into the series to "force convergence." For example, it can be proved (and will be in the next subsection) that for $f \in C(\mathrm{T})$, the series

$$\sigma_n(f,t) \;=\; \frac{1}{2\pi} \sum_{k=-n}^{n} \left(1 - \frac{|k|}{n+1} \right) \widehat{f_k}\, e^{ikt}$$

converges to f uniformly as well as in the L^1-sense; this is Fejér's famous theorem.

How do these convergence factors work? If we set

$$F_n(t) \;=\; \sum_{k=-n}^{n} \left(1 - \frac{|k|}{n+1} \right) e^{ikt},$$

then the Fourier coefficients of $\sigma_n(f,t)$ are the product of those of f and those of F_n. Thus it seems reasonable that convergence properties of σ_n can be deduced from suitable properties of F_n. An important relation between these objects is given by the following theorem.

Theorem 2.16. (Convolution theorem for the torus). *For $f, g \in L^1(\mathrm{T})$, define*

$$h(t) \;=\; (f * g)(t) \;=\; \frac{1}{2\pi} \int_{-\pi}^{\pi} f(t-u)\, g(u)\, du.$$

(a) Then the integral exists for a.e. $t \in \mathrm{T}$, and we have $h \in L^1(\mathrm{T})$ and $\|h\|_1 \leq \frac{1}{2\pi} \|f\|_1 \|g\|_1$.

(b) Moreover, $\widehat{h}_n = \frac{1}{2\pi} \widehat{f}_n \cdot \widehat{g}_n$ holds for the Fourier coefficients.

Applied to the question above, this convolution theorem says that $\sigma_n(f,t) = (F_n * f)(t)$. In the next section, this will lead to a proof of Fejér's theorem. We will in fact be able to treat much more general summation kernels.

In the following sections, we will also need a convolution theorem in $L^1(\mathrm{R})$.

Theorem 2.17. (Convolution theorem for the real line). *For $f, g \in L^1(\mathrm{R})$, define*

$$h(t) \;=\; (f * g)(t) \;=\; \int_{-\infty}^{\infty} f(t-u)\, g(u)\, du.$$

(a) Then the integral exists for a.e. $t \in R$, and we have $h \in L^1(R)$ and $\|h\|_1 \leq \|f\|_1 \|g\|_1$.

(b) Moreover, $\widehat{h}(x) = \widehat{f}(x) \cdot \widehat{g}(x)$ holds for the Fourier transforms.

The convolution in $L^1(R)$ tends to make functions smoother but less localized. If g is, for example, the characteristic function of an interval, then $h = f * g$ will be absolutely continuous for every L^1-function f. If, on the other hand, both f and g are L^1-functions with bounded supports, equal to, say, $[-a, a]$ and $[-b, b]$, then the support of $h = f * g$ will be equal to $[-(a + b), a + b]$.

Convolution theorems are quite important in computational mathematics! To compute a convolution, one usually has to perform many multiplications, so that it is expensive in terms of time and memory. On the other hand, a single multiplication is often cheap. The convolution theorems (which have many analogs for different types of convolutions) say that convolutions can be transformed into multiplications and thus may be much easier to compute than appears on first glance. Techniques for computing convolutions using the fast Fourier transform (FFT) are presented in Chapter 6 of [43]. The speed of convolutions computed using FFTs is a principal reason that Fourier theory in general, and the FFT in particular, are so important in computational science.

2.3.3 Summation Kernels

With regard to Fejér's sum σ_n, the convolution theorem says that $\sigma_n(f, t) = (F_n * f)(t)$, and we are interested in the question of whether $F_n * f \to f$ for $n \to \infty$ in a suitable norm (L^1 or uniformly). This question can be generalized: Under what properties on a family of functions $(K_w) \subseteq L^1(T)$ is there convergence $\|K_w * f - f\|_1 \to 0$ ($w \to \infty$) for all $f \in L^1(T)$? Under what conditions is there uniform convergence for continuous f? Any family of type (K_w) is called a *kernel*. If there is suitable convergence, then (K_w) is also called an *approximate identity*, and the question is now open to systematic experimentation. Which conditions of a kernel make it an approximate identity?

It is proved in Katznelson's book [149] that the following conditions imply that a kernel $(K_w) \subseteq L^1(T)$ is an approximate identity for $L^p(T)$ ($1 \leq p < \infty$) and for $C(T)$:

(S1) $\dfrac{1}{2\pi} \displaystyle\int_{-\pi}^{\pi} K_w(t)\, dt \;=\; 1 \quad$ for all w,

(S2) $\displaystyle\int_{-\pi}^{\pi} |K_w(t)|\, dt \leq M$ uniformly in w,

(S3) $\displaystyle\lim_{w\to\infty} \int_{\delta\leq|t|\leq\pi} |K_w(t)|\, dt = 0$ for all $0 < \delta < \pi$.

Of particular interest here are kernels of the form

$$K_w(t) = \sum_{k=-\infty}^{\infty} g\Big(\frac{k}{w}\Big) e^{ikt}$$

for suitable functions g, since many classical kernels are of this form. Now our direction should be clear: The "experimental" version (2.16) of the Poisson formula will be of use.

Theorem 2.18. *Let $K_w(t) = \sum_{k=-\infty}^{\infty} g\big(\frac{k}{w}\big) e^{ikt}$ and assume that $g \in C(\mathrm{R}) \cap L^1(\mathrm{R})$, that $(g(\frac{n}{w}))$ is absolutely summable for each $w > 0$, and that $g(0) = 1$ and $\widehat{g} \in L^1(\mathrm{R})$. Then (K_w) is an approximate identity.*

Proof. We have to check (S1)–(S3) above. Condition (S1) is a condition on the middle Fourier coefficient of K_w and follows from $g(0) = 1$. Regarding (S2), we have, using (2.16) and the theorem of B. Levi,

$$\int_{-\pi}^{\pi} |K_w(t)|\, dt = \frac{1}{2\pi} \int_{-\pi}^{\pi} \left| w \sum_{j=-\infty}^{\infty} \overline{\widehat{g}}(w(t + 2\pi j)) \right| dt$$

$$\leq w \sum_{j=-\infty}^{\infty} \int_{-\pi}^{\pi} |\widehat{g}(w(t + 2\pi j))|\, dt$$

$$= w \int_{-\infty}^{\infty} |\widehat{g}(wt)|\, dt = \|\widehat{g}\|_1,$$

which is a uniform bound. Finally, we get by similar computations as before that

$$\int_{\delta\leq|t|\leq\pi} |K_w(t)|\, dt \leq \int_{|t|>w\delta} |\widehat{g}(t)|\, dt,$$

which tends to 0 for $w \to \infty$ since $\widehat{g} \in L^1(\mathrm{R})$. $\square$

All of the conditions in Theorem 2.18 are easily checked symbolically. Moreover, the methods employed here allow more detailed investigations of the approximation properties of the kernels by direct computation of the Fourier transform. For example, often the *Lebesgue constants*, defined as

$\|K_w\|_1$, determine the rate of convergence of $K_w * f$ to f, or, if K_w is not an approximate identity, the growth rate of $K_w * f$. The Lebesgue constants, as computed in the proof of (S2), satisfy $\|K_w\|_1 \leq \|\widehat{g}\|_1$. This bound is precise, since the same Poisson methods also give

$$\|K_w\|_1 \geq \|\widehat{g}\|_1 - 2 \int_{|t| > \pi w} |\widehat{g}(t)| \, dt.$$

Example 2.19. The Fejér kernel.

The kernel F_n as defined above comes from the Cesàro summation method applied to Fourier series; thus $\sigma_n(f, t) = (F_n * f)(t)$. By geometric summation, F_n can also be written as

$$F_n(t) = \frac{\sin^2(\frac{n+1}{2} t)}{(n+1) \sin^2(\frac{t}{2})}.$$

If we set $g(x) = \max\{1 - |x|, 0\}$, then $F_n = K_{n+1}$. Since $\widehat{g}(t) = \text{sinc}^2(t/2) \in L^1(\mathbb{R})$, the conditions of Theorem 2.18 are satisfied. Thus we deduce directly that F_n is an approximate identity in $L^p(\mathbb{T})$ and in $C(\mathbb{T})$. Moreover, since $\widehat{g}$ is non-negative, for the Lebesgue constants we obtain $\|F_n\|_1 = \|\widehat{g}\|_1 = 2\pi$ (compare with the example in Section 2.2.2). □

Example 2.20. The Poisson kernel.

Another important summation method is Abel summation. Applied to Fourier series, it leads to the Poisson kernel, defined as

$$P_r(t) = \sum_{k=-\infty}^{\infty} r^{|k|} e^{ikt} = \frac{1 - r^2}{1 - 2r \cos(t) + r^2} \quad \text{for } 0 \leq r < 1.$$

The question is whether $P_r * f \to f$ in $L^p(\mathbb{T})$ or in $C(\mathbb{T})$ for $r \to 1$. By setting $w = -1/\ln(r)$ and $g(x) = e^{-|x|}$, we have $P_r(t) = K_w(t)$. Since $\widehat{g}(t) = 2/(1 + t^2) \in L^1(\mathbb{R})$, Theorem 2.18 produces convergence of $P_r * f$ to f. Similarly to the Fejér kernel, P_r as well as g and $\widehat{g}$ are nonnegative, so that we again have $\|P_r\|_1 = \|K_w\|_1 = \|\widehat{g}\|_1 = 2\pi$. □

Example 2.21. The Dirichlet kernel.

The same methods also explain why the usual summation of Fourier series does not always give convergence. This summation corresponds to the

Dirichlet kernel

$$D_n(t) \;=\; \sum_{k=-n}^{n} e^{ikt} \;=\; \frac{\sin\left(\left(n+\frac{1}{2}\right)t\right)}{\sin\left(\frac{t}{2}\right)}$$

via $s_n(f,t) = (D_n * f)(t)$. The Dirichlet kernel is of the form K_n with $g = \chi_{[-1,1]}$. This g is neither continuous, nor does it have a Fourier transform in $L^1(\mathbb{R})$ (its Fourier transform is $\widehat{g}(t) = 2\,\mathrm{sinc}(t)$.) Thus Theorem 2.18 is not applicable. But the experimental Poisson formula (2.16) can still be used to estimate the Lebesgue constants and gives $\|D_n\|_1 = \int_{-n\pi}^{n\pi} |2\,\mathrm{sinc}(t)|\, dt + O(1) = (8/\pi)\ln n + O(1)$. These are unbounded, and so the Dirichlet kernel can be expected to have worse summation properties than the Fejér and Poisson kernels above.　　　　　　　　　　　　　　　　　　　　□

In summary, the methods described here are very useful to gauge norm-convergence properties of kernels of the special form $K_w(t) = \sum g(\frac{k}{w})e^{ikt}$ in a direct, computational way, and they open the door to further experimentation. Our description has been adapted from [123], but reportedly these methods go back at least to Korovkin.

These methods deal with norm- and uniform convergence. But what about pointwise convergence? As described in [149], instead of (S1)–(S3), the following properties of a kernel can be used to prove pointwise convergence of $K_w * f$ to f : K_w satisfies (S1), is non-negative and even, and satisfies

$$\lim_{w \to \infty} \left(\sup_{\delta \le |t| \le \pi} K_w(t) \right) \;=\; 0 \quad \text{for all } 0 < \delta < \pi.$$

This allows one to decide, for given $f \in L^1(\mathbb{T})$ and $t_0 \in \mathbb{R}$, whether $(K_w * f)(t_0)$ converges to $f(t_0)$.

For the Fejér kernel, it leads to Lebesgue's condition: *If there exists a value $\check{f}(t_0)$ such that*

$$\lim_{h \to 0} \frac{1}{h} \int_0^h \left| \frac{f(t_0+t)+f(t_0-t)}{2} - \check{f}(t_0) \right| dt \;=\; 0,$$

then $\sigma_n(f,t) \to \check{f}(t_0)$ for $n \to \infty$. In particular, $\sigma_n(f,t) \to f(t)$ a.e.

For the Poisson kernel, it leads to Fatou's condition: *If there exists a value $\check{f}(t_0)$ such that*

$$\lim_{h \to 0} \frac{1}{h} \int_0^h \left(\frac{f(t_0+t)+f(t_0-t)}{2} - \check{f}(t_0) \right) dt \;=\; 0,$$

then $(P_r * f)(t_0) \to \check{f}(t_0)$ for $r \to 1-$. In particular, $(P_r * f)(t) \to f(t)$ a.e. The convergence is uniform on closed subintervals where f is continuous.

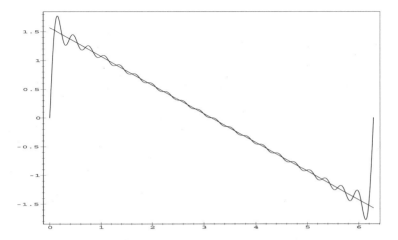

Figure 2.1. The Gibbs phenomenon for Fourier series.

2.4   Examples and Applications

2.4.1   The Gibbs Phenomenon

If a function $f \in L^1(\mathrm{T})$ is of bounded variation, it may have jump discontinuities. The Jordan test says that the Fourier series of f converges to the center of the gap at such a point. Directly to the left and right of the jump, the series converges pointwise, but not uniformly on any interval containing the discontinuity, to the function. The function $f(t) = (\pi - t)/2 = \sum n^{-1}\sin(nt)$ on $[0, 2\pi]$ is a good example for this behavior (see Figure 2.1), where the series for $(\pi - t)/2$ is evaluated to 20 terms.

One notices that the truncated Fourier series "overshoots" the function at the discontinuity. These oscillations do not diminish when more terms are added; they just move closer to the discontinuity. When the experimental physicist A. Michelson (famous for the Michelson-Morley experiment, which led to Einstein's special relativity) had built a machine to calculate Fourier series and fed it a discontinuous function, he noticed this phenomenon. It was unexpected for him, but after hand calculations confirmed this behavior, he wrote a letter to *Nature* in 1898, expressing his doubts that "a real discontinuity can replace a sum of continuous curves" (cited after Bhatia [25]). J. Willard Gibbs, one of the founders of modern thermodynamics, replied to this letter and clarified the matter. Thus here we have another example of a mathematical theorem that was experimentally discovered (by an experimental physicist)!

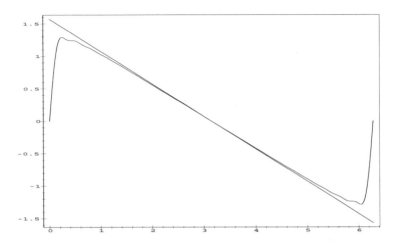

Figure 2.2. The Gibbs phenomenon for Fejér series.

Now what is the explanation for the Gibbs phenomenon? Inspection of the picture shows that the largest overshoot seems to occur around the point π/N if N terms of the Fourier series are added. Thus we compute

$$s_N\left(f, \frac{\pi}{N}\right) = \sum_{n=1}^{N} \frac{\sin\left(\frac{n\pi}{N}\right)}{n} = \frac{\pi}{N} \sum_{n=1}^{N} \frac{\sin\left(\frac{n\pi}{N}\right)}{\frac{n\pi}{N}},$$

where the last sum is a Riemann sum for the integral

$$I = \int_0^\pi \frac{\sin(t)}{t}\, dt.$$

Therefore, $s_N(f, \pi/N) \to I$ for $N \to \infty$. Since $I/f(0+) = I/(\pi/2) \approx 1.178979744$, this explains why the overshoot does not go away for large N. This overshoot of roughly 18% is not dependent on the function f used here as an example, but can be observed (and proved) for any jump discontinuity.

Does the Gibbs phenomenon vanish when we use Fejér's series instead of the Fourier series? Figure 2.2 shows the Fejér approximation to f, again to 20 terms. The oscillation is now replaced by a pronounced "undershoot" to the right of 0 (this can be explained by the positivity of the Fejér kernel); again, the undershoot can be observed to move closer to the discontinuity, but not vanish altogether when more terms are added. In fact, we have to pay for the increased smoothness of the approximation by its reduced willingness to snuggle up to the limit function.

2.4.2   A Function with Given Integer Moments

The k-th moment of a function $f \in L^1(\mathrm{R})$ is defined as

$$\mu_k(f) \;=\; \int_{-\infty}^{\infty} f(t)\, t^k \, dt,$$

provided that $t \mapsto f(t)\, t^k \in L^1(\mathrm{R})$. The *Hamburger moment problem* is the problem to find a function f with a given sequence of moments (μ_k). This problem is underdetermined: There can be nonvanishing functions whose every moment is 0. This is easily seen by the following argument. Assume that f is k times differentiable with every derivative in $L^1(\mathrm{R})$. Then by partial integration,

$$\widehat{f^{(k)}}(x) \;=\; \int_{-\infty}^{\infty} f^{(k)}(t)\, e^{-ixt}\, dt \;=\; (ix)^k\, \widehat{f}(x),$$

and by the inversion theorem, $\int_{-\infty}^{\infty} \widehat{f}(x)\, x^k \, dx = 2\pi f^{(k)}(0)$. Thus if $f \in L^1(\mathrm{R})$ is infinitely differentiable with every derivative in $L^1(\mathrm{R})$ and satisfies $f^{(k)}(0) = 0$ for all k, then all moments of $\widehat{f}$ vanish. Of course, such nontrivial functions f exist, even with compact support.

Obviously, for an even function every odd moment vanishes. To generalize this, it is quite easy to find, for given $n \in \mathrm{N}$, a function whose k-th moment is nonzero precisely when $k \bmod n = 0$. Just choose an infinitely differentiable function $f \in L^1(\mathrm{R})$ with all derivatives in $L^1(\mathrm{R})$, and such that all its derivatives are nonzero at 0. Then set $g(t) = f(t^n)$, and by the chain and product rule of differentiation, $g^{(k)}(0)$ is nonzero precisely for $k \bmod n = 0$. Thus, $\widehat{g}$ satisfies the moment condition. If f is analytic, say $f(t) = \sum_{j=0}^{\infty} a_j\, t^j$, then $g^{(k)}(0) = (qn)!\, a_q$ if $k = qn$ and $g^{(k)}(0) = 0$ otherwise. An example for such a function f is $f(t) = (1+t)\, e^{-t^2/2}$.

When this was first investigated some years ago, numerical fast Fourier transforms were used—as a test—to calculate the moments for $g(t) = f(t^n)$, where $f(t) = (1+t)\, e^{-t^2/2}$ as above. The scheme for doing this is presented in Section 6.1 of [43]. When this was done, it was noticed that the resulting moment values were extremely accurate, far more than one would expect based on what amounts to a simple step-function approximation to the Fourier integrals. Readers may recall that we also encountered this phenomenon in Section 5.2 of [43]. This phenomenon, which is rooted in the Euler-Maclaurin summation formula, is the foundation of some extremely efficient and highly accurate numerical quadrature schemes, which we shall see in Sections 7.4.2 and 7.4.3.

2.4.3   Bernoulli Convolutions

Consider the discrete probability density on the real line with measure $1/2$ at each of the two points ± 1. The corresponding measure is the so-called Bernoulli measure, denoted $b(X)$. For every $0 < q < 1$, the infinite convolution of measures

$$\mu_q(X) \;=\; b(X) * b(X/q) * b(X/q^2) * \cdots \tag{2.17}$$

exists as a weak limit of the finite convolutions. The most basic theorem about these infinite Bernoulli convolutions is due to Jessen and Wintner ([145]). They proved that μ_q is always continuous, and that it is either absolutely continuous or purely singular. This statement follows from a more general theorem on infinite convolutions of purely discontinuous measures (Theorem 35 in [145]); however, it is not difficult to prove the statement directly with the use of Kolmogoroff's 0-1-law (which can be found, e.g., in [26]). The question about these measures is to decide for which values of the parameter q they are singular, and for which q they are absolutely continuous.

This question can be recast in a more real-analytic way by defining the distribution function F_q of μ_q as

$$F_q(t) \;=\; \mu_q(-\infty, t], \tag{2.18}$$

and to ask for which q this continuous, increasing function F_q is singular, and for which it is absolutely continuous. Note that F_q satisfies $F_q(t) = 0$ for $t < -1/(1-q)$ and $F_q(t) = 1$ for $t > 1/(1-q)$.

Another way to define the distribution function F_q is by functional equations: F_q is the only bounded solution of the functional equation

$$F(t) \;=\; \frac{1}{2}F\left(\frac{t-1}{q}\right) + \frac{1}{2}F\left(\frac{t+1}{q}\right) \tag{2.19}$$

with the above restrictions. Moreover, if F_q is absolutely continuous and thus has a density $f_q \in L^1(\mathbf{R})$, then f_q satisfies the functional equation

$$2q\,f(t) \;=\; f\left(\frac{t-1}{q}\right) + f\left(\frac{t+1}{q}\right) \tag{2.20}$$

almost everywhere. This is a special case of a much more general class of equations, namely two-scale difference equations. Those are functional equations of the type

$$f(t) \;=\; \sum_{n=0}^{N} c_n\, f(\alpha t - \beta_n) \qquad (t \in \mathbf{R}), \tag{2.21}$$

with $c_n \in \mathbb{C}$, $\beta_n \in \mathbb{R}$ and $\alpha > 1$. They were first discussed by Ingrid Daubechies and Jeffrey C. Lagarias, who proved existence and uniqueness theorems and derived some properties of L^1-solutions [105, 106]. One of their theorems, which we state here in part for the general equation (2.21) and in part for the specific case (2.20), is the following:

Theorem 2.22.

(a) If $\alpha^{-1}(c_0 + \cdots + c_N) = 1$, then the vector space of $L^1(\mathbb{R})$-solutions of (2.21) is at most one-dimensional.

(b) If, for given $q \in (0,1)$, equation (2.20) has a nontrivial L^1-solution f_q, then its Fourier transform satisfies $\widehat{f_q}(0) \neq 0$, and is given by

$$\widehat{f_q}(x) = \widehat{f_q}(0) \prod_{n=0}^{\infty} \cos(q^n x). \tag{2.22}$$

In particular, for normalization we can assume $\widehat{f_q}(0) = 1$.

(c) On the other hand, if the right-hand side of (2.22) is the Fourier transform of an L^1-function f_q, then f_q is a solution of (2.20).

(d) Any nontrivial L^1-solution of (2.21) is finitely supported. In the case of (2.20), the support of f_q is contained in $[-1/(1-q), 1/(1-q)]$.

This implies in particular that the question of whether the infinite Bernoulli convolution (2.17) is absolutely continuous is equivalent to the question of whether (2.20) has a nontrivial L^1-solution. Now what is known about these questions?

It is relatively easy to see that in the case $0 < q < 1/2$, the solution of (2.19) is singular; it is in fact a Cantor function, meaning that it is constant on a dense set of intervals. This was first proved by R. Kershner and A. Wintner [151]. (An example of a Cantor function is depicted in Figure 6.1 of [43].)

It is also easy to see that in the case $q = 1/2$, there is an L^1-solution of (2.20), namely $f_{1/2}(t) = \frac{1}{4}\chi_{[-2,2]}(t)$. Moreover, this can be used to construct a solution for every $q = 2^{-1/p}$ where p is an integer, namely

$$f_{2^{-1/p}}(t) = 2^{(p-1)/2} \cdot \left[f_{1/2}(t) * f_{1/2}(2^{1/p}t) * \cdots * f_{1/2}(2^{(p-1)/p}t) \right]. \tag{2.23}$$

This was first noted by Wintner via the Fourier transform [214]. Explicitly, we have

$$\widehat{f_{2^{-1/p}}}(x) = \prod_{n=0}^{\infty} \cos(2^{-n/p}x) = \prod_{m=0}^{\infty} \prod_{k=0}^{p-1} \cos(2^{-(m+k/p)}x)$$
$$= \widehat{f_{1/2}}(x) \cdot \widehat{f_{1/2}}(2^{-1/p}x) \cdots \widehat{f_{1/2}}(2^{-(p-1)/p}x),$$

which is equivalent to (2.23) by the convolution theorem.

Note that the regularity of these solutions $f_{2^{-1/p}}$ increases when p and thus $q = 2^{-1/p}$ increases: $f_{2^{-1/p}} \in C^{p-2}(\mathbf{R})$. From the results given so far, one might therefore surmise that (2.20) would have a nontrivial L^1-solution for every $q \geq 1/2$ with increasing regularity when q increases. This supposition, however, would be wrong, and it came as a surprise when Erdős proved in 1939 [116] that there are some values of $1/2 < q < 1$ for which (2.20) does not have an L^1-solution, namely, the inverses of Pisot numbers. A *Pisot number* (discussed further in Exercise 13 of Chapter 7) is defined to be an algebraic integer greater than 1 all of whose algebraic conjugates lie inside the unit disk. The best known example of a Pisot number is the golden mean $\varphi = (\sqrt{5}+1)/2$. The characteristic property of Pisot numbers is that their powers quickly approach integers: If a is a Pisot number, then there exists a θ, $0 < \theta < 1$, such that

$$\text{dist}(a^n, \mathbf{Z}) \leq \theta^n \quad \text{for all } n \in \mathbf{N}. \tag{2.24}$$

Erdős used this property to prove that if $q = 1/a$ for a Pisot number a, then $\limsup_{x \to \infty} \left| \widehat{f_q}(x) \right| > 0$. Thus in these cases, f_q cannot be in $L^1(\mathbf{R})$, since that would contradict the Riemann-Lebesgue lemma. Erdős's proof uses the Fourier transform $\widehat{f_q}$: Consider, for $N \in \mathbf{N}$,

$$\left| \widehat{f_q}(q^{-N}\pi) \right| = \prod_{n=1}^{\infty} |\cos(q^n \pi)| \cdot \prod_{n=0}^{N-1} \left| \cos(q^{-n}\pi) \right| =: C \cdot p_N,$$

where $C > 0$. Moreover, choose $\theta \neq 1/2$ according to (2.24) and note that

$$p_N = \prod_{\substack{n=0 \\ \theta^n \leq 1/2}}^{N-1} \left| \cos(q^{-n}\pi) \right| \cdot \prod_{\substack{n=0 \\ \theta^n > 1/2}}^{N-1} \left| \cos(q^{-n}\pi) \right|$$
$$\geq \prod_{\substack{n=0 \\ \theta^n \leq 1/2}}^{N-1} \cos(\theta^n \pi) \cdot \prod_{\substack{n=0 \\ \theta^n > 1/2}}^{N-1} \left| \cos(q^{-n}\pi) \right|$$

$$\geq \prod_{\substack{n=0 \\ \theta^n \leq 1/2}}^{\infty} \cos(\theta^n \pi) \cdot \prod_{\substack{n=0 \\ \theta^n > 1/2}}^{\infty} \left| \cos(q^{-n} \pi) \right| \; = \; C' \; > \; 0,$$

independently of N.

In 1944, Raphaël Salem [191] showed that the reciprocals of Pisot numbers are the only values of q where $\widehat{f_q}(x)$ does not tend to 0 for $x \to \infty$. In fact, no other $q > 1/2$ are known at all where F_q is singular. Moreover, the set of explicitly given q with absolutely continuous F_q is also not very big: The largest such set known to date was found by Adriano Garsia in 1962 [120]. It contains reciprocals of certain algebraic numbers (such as roots of the polynomials $x^{n+p} - x^n - 2$ for $\max\{p, n\} \geq 2$) besides the roots of $1/2$.

Matters remained in this state for more than 30 years; the question remained settled only for countably many $q \in [1/2, 1)$. The most recent significant progress then was made in 1995 by Boris Solomyak [201], who developed exciting new methods in geometric measure theory to prove that F_q is in fact absolutely continuous for almost every $q \in [1/2, 1)$. (See also [177] for a simplified proof and [176] for a survey and some newer results.)

This, however, yields no explicit result; the set of q for which the behavior of F_q is known explicitly is the same as before. Here we suggest an experimental approach to at least identify q-values for which the behavior of F_q can be guessed. In fact, define a map T_q, mapping the set of L^1-functions with support in $[-1/(1-q), 1/(1-q)]$ and with $\widehat{f_q}(0) = 1$ into itself, by

$$(T_q f)(t) \; = \; \frac{1}{2q} \left(f\left(\frac{t-1}{q}\right) + f\left(\frac{t+1}{q}\right) \right) \quad \text{for } t \in \mathrm{R}.$$

Then note that the fixed points of T_q are the solutions of (2.20) and that T_q is nonexpansive (see Chapter 6). Therefore, one may have hope that by iterating the operator, it may be possible to approximate the fixed point. In fact, if a sequence of iterates $T_q^n f$ converges in $L^1(\mathrm{R})$ for some initial function f, then the limit will be a fixed point of T_q, since T_q is continuous. It is, however, not easy to prove convergence; no convergence proof is known. It is, on the other hand, possible to prove a weaker result, namely convergence in the mean, provided that a fixed point exists: *If a solution $f_q \in L^1(\mathrm{R})$ with $\widehat{f_q}(0) = 1$ of (2.20) exists, then for every initial function $f \in L^1(\mathrm{R})$ with support in $[-1/(1-q), 1/(1-q)]$ and with $\widehat{f}(0) = 1$, we have*

$$\lim_{n \to \infty} \left\| \frac{1}{n} \sum_{k=0}^{n-1} T_q^k f - f_q \right\|_1 \; = \; 0.$$

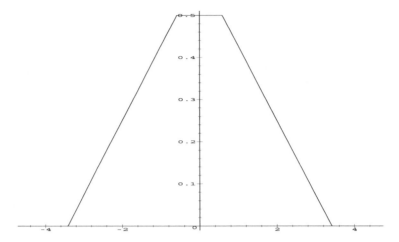

Figure 2.3. 25th iterate for $q = 1/\sqrt{2}$.

This theorem follows from properties of Markov operators [155] and from a result by Mauldin and Simon [164], showing that if an L^1-density f_q exists, then it must be positive a.e. on its support.

In practice, we observe that the iterates usually seem to converge directly, even without the means. Plotting them, we hope to infer existence and regularity of L^1-solutions by visual inspection. Figures 2.3-2.6 show the 25th iterate for $f = (1 - q)/2\, \chi_{[-1/(1-q),1/(1-q)]}$ as initial function.

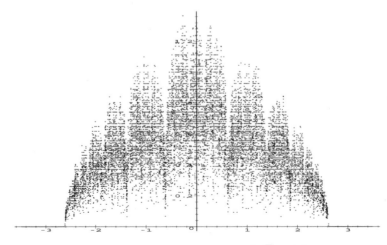

Figure 2.4. 25th iterate for $q = (\sqrt{5} - 1)/2$.

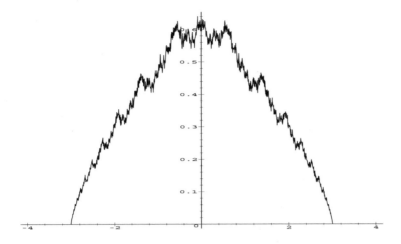

Figure 2.5. 25th iterate for $q = 2/3$.

Figure 2.3 shows convergence to $2^{1/2} \cdot \left[f_{1/2}(t) * f_{1/2}(2^{1/2}t) \right]$ for $q = 1/\sqrt{2}$; Figure 2.4 shows that for $q = (\sqrt{5} - 1)/2$, the iterates do not converge to a meaningful function. It is not known if there is a density for any rational $q \in (1/2, 1)$. Figures 2.5 and 2.6 show that there seems to be a continuous limit in both cases shown; moreover, regularity seems to increase when q increases.

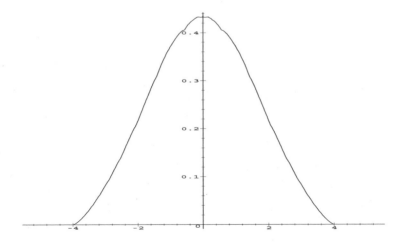

Figure 2.6. 25th iterate for $q = 3/4$.

2.5   Some Curious Sinc Integrals

Define

$$I_n = \int_0^\infty \operatorname{sinc} x \cdot \operatorname{sinc}\left(\frac{x}{3}\right) \cdots \operatorname{sinc}\left(\frac{x}{2n+1}\right) dx.$$

Then *Maple* and *Mathematica* evaluate

$$I_0 = \int_0^\infty \operatorname{sinc} x \, dx = \frac{\pi}{2},$$

$$I_1 = \int_0^\infty \operatorname{sinc} x \cdot \operatorname{sinc}\left(\frac{x}{3}\right) dx = \frac{\pi}{2},$$

$$\vdots$$

$$I_6 = \int_0^\infty \operatorname{sinc} x \cdot \operatorname{sinc}\left(\frac{x}{3}\right) \cdots \operatorname{sinc}\left(\frac{x}{13}\right) dx = \frac{\pi}{2}, \quad \text{but}$$

$$I_7 = \int_0^\infty \operatorname{sinc} x \cdot \operatorname{sinc}\left(\frac{x}{3}\right) \cdots \operatorname{sinc}\left(\frac{x}{15}\right) dx$$

$$= \frac{467807924713440738696537864469}{935615849440640907310521750000}\pi,$$

where the fraction is approximately $0.499999999992646\ldots$

When this fact was recently verified by a researcher using a computer algebra package, he concluded that there must be a "bug" in the software. This conclusion may be too hasty, but it does raise the question: How far can we trust our computer algebra system? Or as computer scientists often ask, "Is it a bug or a feature?"

In this section, we will derive general formulas for this type of sinc integrals, thereby proving that all of the above evaluations are in fact correct. Thus, this is a somewhat cautionary example for too enthusiastically inferring patterns from symbolic or numerical computations. The material comes from [33]; additional information can be found in [34].

2.5.1   The Basic Sinc Integral

It will turn out that the general multi-fold sinc integral can be reduced to the integral I_0, so that it makes sense to first evaluate this integral. Note that the function $\operatorname{sinc} x$ is not an element of $L^1(\mathbb{R})$! Thus, Lebesgue theory cannot be applied here directly, and in fact the integral has to be interpreted correctly. Here we use the usual interpretation

$$I_0 = \int_0^\infty \operatorname{sinc} x \, dx = \lim_{a \to \infty} \int_0^a \operatorname{sinc} x \, dx.$$

Thus we interpret it as an improper Riemann integral, or at best as the limit of Lebesgue integrals.

Of course, *Maple* and *Mathematica* directly evaluate $I_0 = \pi/2$, but this is not helpful for those who demand understanding or a proof. Where does this evaluation come from? Peering behind the covers, we find that *Maple* knows

$$\int_0^a \operatorname{sinc} x \, dx \;=\; \operatorname{Si}(a) \;\to\; \frac{\pi}{2} \quad \text{for } a \to \infty.$$

However, this just shifts the problem to another level, since Si equals the integral by definition. We will now give several proofs of this identity: One will be short (and incomplete), one will be wrong, and one will be constructive!

For the first proof we just remember the Jordan theorem in Section 2.2.2, which directly implies that

$$\lim_{a \to \infty} \int_{-a}^a \operatorname{sinc}(\pi x) \, e^{ixt} \, dx \;=\; \frac{1}{2} \left(\chi_{(-\pi,\pi)}(t+) + \chi_{(-\pi,\pi)}(t-) \right),$$

so that $t = 0$ gives the desired evaluation. However, this is only a proof modulo the Jordan theorem. A direct proof would still be preferable.

The second "proof" is not a proof, just an idea: Write the sinc function as an inner integral and then use Fubini. Writing $1/x = \int_0^\infty e^{-tx} \, dt$, we would have to use Fubini on the function $g(t, x) = e^{-tx} \sin x$ on $R \times R$. The double integral that results from exchanging the integration order does, in fact, give

$$\int_0^\infty \int_0^\infty e^{-tx} \sin x \, dx \, dt \;=\; \int_0^\infty \frac{1}{1 + t^2} \, dt \;=\; \frac{\pi}{2}.$$

However, this exchange is not allowed, since the integrand is not in $L^1(R^2)$.

But this idea can now be made into a proof which is valid and constructive. If g is not L^1 on R^2, then we just have to restrict the domain of g at first. Now Fubini is applicable in

$$\begin{aligned}
\int_0^a \operatorname{sinc} x \, dx &= \int_0^a \int_0^\infty e^{-xt} \sin x \, dt \, dx \\
&= \int_0^\infty \int_0^a e^{-xt} \sin x \, dx \, dt \\
&= \int_0^\infty \frac{1}{1 + t^2} \left[1 - e^{-at}(t \sin a + \cos a) \right] dt, \\
&= \frac{\pi}{2} - \int_0^\infty \frac{e^{-at}}{1 + t^2} (t \sin a + \cos a) \, dt,
\end{aligned}$$

and the final integral goes to 0 for $a \to \infty$ as follows by elementary estimates (see [26]).

Another constructive method to evaluate the sinc integral is given in Item 26 in the Exercises at the end of this chapter.

2.5.2   Iterated Sinc Integrals

Now let $n \geq 1$ and $a_0, a_1, \cdots, a_n$ be positive reals. Our goal is to find inequalities (explicit formulas will be given in the exercises) for the integral

$$\tau = \int_0^\infty \prod_{k=0}^n \mathrm{sinc}(a_k\, x)\, dx,$$

which in particular explain the behavior of the integrals I_n. For simplicity and without loss, we can assume that $a_0 = 1$.

Theorem 2.23. *Let $s = \sum_{k=1}^n a_k$. If $s \leq 1$, then $\tau = \pi/2$; if $s > 1$, then $\tau < \pi/2$.*

Proof. Let $\tau(x) = \prod_{k=0}^n \mathrm{sinc}(a_k\, x)$. Note that $\mathrm{sinc}(a_k\, x) = \widehat{f_k}(x)$ with $f_k = (1/(2a_k))\, \chi_{[-a_k, a_k]}$. Thus by the convolution theorem, $\tau(x) = (f_0 * \cdots * f_n)\widehat{}(x)$, and by the inversion theorem,

$$\int_{-\infty}^\infty \tau(x)\, dx \;=\; 2\pi\, (f_0 * \cdots * f_n)(0) \;=\; 2\pi\, \frac{1}{2} \int_{-1}^1 (f_1 * \cdots * f_n)(u)\, du. \quad (2.25)$$

Now since the support of f_k equals $[-a_k, a_k]$, the support of $(f_1 * \cdots * f_n)$ equals $[-s, s]$. If $s \leq 1$, then

$$\int_{-1}^1 (f_1 * \cdots * f_n)(u)\, du \;=\; \int_{-\infty}^\infty (f_1 * \cdots * f_n)(u)\, du$$

$$=\; (f_1 * \cdots * f_n)\widehat{}(0) \;=\; \prod_{k=1}^n \mathrm{sinc}(a_k\, 0) \;=\; 1,$$

and $\int_{-\infty}^\infty \tau(x)\, dx = \pi$ follows. If, on the other hand, $s > 1$, then the interval $[-1, 1]$ is strictly inside the support of $(f_1 * \cdots * f_n)$. Since $(f_1 * \cdots * f_n)$ is strictly positive in the interior of its support, we get

$$\int_{-1}^1 (f_1 * \cdots * f_n)(x)\, dx \;<\; \int_{-\infty}^\infty (f_1 * \cdots * f_n)(x)\, dx \;=\; 1,$$

and $\int_{-\infty}^\infty \tau(x)\, dx < \pi$ follows. $\square$

This theorem explains why the values of I_n suddenly drop below $\pi/2$ at $n = 7$ and not before: We have $1/3 + 1/5 + \cdots + 1/13 < 1$; however, $1/3 + 1/5 + \cdots + 1/13 + 1/15 > 1$.

A geometric interpretation of this behavior can also be given. Consider the polyhedra

$$P = \{(x_1, \cdots, x_n) : -1 \le \sum_{k=1}^{n} x_k \le 1, \; -a_k \le x_k \le a_k \text{ for } k = 1, \cdots, n\},$$

$$Q = \{(x_1, \cdots, x_n) : -1 \le \sum_{k=1}^{n} a_k x_k \le 1, \; -1 \le x_k \le 1 \text{ for } k = 1, \cdots, n\},$$

$$H = \{(x_1, \cdots, x_n) : -1 \le x_k \le 1 \text{ for } k = 1, \cdots, n\}.$$

Then by formula (2.25),

$$
\begin{aligned}
\tau &= \frac{\pi}{2^n \, a_1 \cdots a_n} \int_0^{\min(1,s)} \left(\chi_{[-a_1, a_1]} * \cdots * \chi_{[-a_n, a_n]} \right)(x)\, dx \\
&= \frac{\pi}{2} \frac{\mathrm{Vol}(P)}{2^n \, a_1 \cdots a_n} = \frac{\pi}{2} \frac{\mathrm{Vol}(Q)}{\mathrm{Vol}(H)}.
\end{aligned}
$$

Thus the value of τ drops below $\pi/2$ precisely when the constraint $-1 \le \sum a_k x_k \le 1$ becomes active and "bites" into the hypercube H.

Of course, the same methods will also work for infinite products. Consider the function

$$C(x) = \prod_{n=1}^{\infty} \cos\left(\frac{x}{n}\right)$$

which is continuous since the product is absolutely convergent. We are interested in the integral $\mu = \int_0^{\infty} C(x)\, dx$. High precision numerical evaluation of this highly oscillatory integral is possible, but by no means straightforward. We get

$$\int_0^{\infty} C(x)\, dx \approx 0.7853805572986328734925830114673325247.61,$$

while $\pi/4 \approx 0.785398$ only differs in the fifth significant place. Can this numerical evaluation $\mu < \pi/4$ be confirmed symbolically? Indeed it can, by reduction to a sinc integral of the above type, only this time with an infinite product. Recall the sine product (1.11) and note that a corresponding product for the cosine can be derived:

$$\sin(x) = x \prod_{n=1}^{\infty} \left(1 - \frac{x^2}{\pi^2 \, n^2}\right), \qquad \cos(x) = \prod_{k=0}^{\infty} \left(1 - \frac{4x^2}{\pi^2 \, (2k+1)^2}\right).$$

Using this result, and exchanging the order of multiplication, we obtain

$$C(x) = \prod_{n=1}^{\infty} \prod_{k=0}^{\infty} \left(1 - \frac{4x^2}{\pi^2 \, n^2 \, (2k+1)^2}\right) = \prod_{k=0}^{\infty} \operatorname{sinc}\left(\frac{2x}{2k+1}\right).$$

Now apply the theorem to get that

$$\mu = \int_0^{\infty} C(x)\, dx = \lim_{N \to \infty} \int_0^{\infty} \prod_{k=0}^{N} \operatorname{sinc}\left(\frac{2x}{2k+1}\right) dx < \frac{\pi}{4}.$$

This remarkable observation was made by Bernard Mares, then 17, and led to the entire development that we have given of the iterated sinc integrals. More examples are given in the Exercises. There is an interesting connection with random harmonic series in [192].

2.6 Korovkin's Three Function Theorems

In 1953, Pavel Korovkin [153] provided an approach to uniform approximation results that is especially well suited to computational assistance and discovery. While the result can be given much more generally, we limit ourselves to the two most basic cases.

In what follows, by ι we denote the identity function $t \mapsto t$; by $C[0,1]$ we denote continuous functions on the unit interval; and by $\rightrightarrows$ we denote uniform convergence (i.e., in the supremum norm). The interval $[0,1]$ can easily be replaced by any finite interval $[a, b]$.

Recall that an operator between continuous function spaces is positive if it maps nonnegative functions to nonnegative functions (when linear, this is necessarily a monotone and bounded linear operator). The motivating example is

Example 2.24. Bernstein operators.

For n in N, let $\mathcal{B}_n(f)$ be defined by

$$\mathcal{B}_n(f)(t) = \sum_{k=0}^{n} f\left(\frac{k}{n}\right) \binom{n}{k} t^k \, (1-t)^{n-k}. \tag{2.26}$$

It is clear that the Bernstein operators are linear and positive on $C[0,1]$ and indeed take values which are polynomials.   □

Theorem 2.25. (First Korovkin three function theorem). *Let L_n be a sequence of positive linear operators from $C[0,1]$ to $C[0,1]$. Suppose that*

$$L_n(1) \rightrightarrows_n 1, \quad L_n(\iota) \rightrightarrows_n \iota, \quad L_n(\iota^2) \rightrightarrows_n \iota^2.$$

Then

$$L_n(f) \rightrightarrows_n f$$

as $n \to \infty$ for all f in $C[0,1]$.

Proof. The hypotheses imply that $L_n(q) \rightrightarrows_n q$ for all quadratic q. Fix f in $C[0,1]$, x in $[0,1]$, and $\varepsilon > 0$. We claim that one can find a quadratic q_x^ε with $f \leq q_x^\varepsilon$ and $f(x) + \varepsilon \geq q_x^\varepsilon$. Thus

$$L_n(f) \ \leq L_n\left(q_x^\varepsilon\right) \rightrightarrows_n q_x^\varepsilon \leq f(x) + \varepsilon.$$

A compactness argument completes the proof. The details are left for the reader as Exercise 33. □

Corollary 2.26. (Stone-Weierstrass). *The Bernstein polynomials are uniformly dense in $C[0,1]$.*

Proof. We check by hand or in a computer algebra system that $\mathcal{B}_n(1) = 1$, $\mathcal{B}_n(\iota) = \iota$, and slightly more elaborately, $\mathcal{B}_n(\iota^2) = \iota^2 + \frac{1}{n}\left(\iota - \iota^2\right) \rightrightarrows_n \iota^2$. □

In the periodic case, the role of t and t^2 is taken by sin and cos, as the second Korovkin theorem shows.

Theorem 2.27. (Second Korovkin three function theorem). *Let L_n be a sequence of positive linear operators from $C(\mathrm{T})$ to $C(\mathrm{T})$. Suppose that*

$$L_n(1) \rightrightarrows_n 1, \quad L_n(\sin) \rightrightarrows_n \sin, \quad L_n(\cos) \rightrightarrows_n \cos.$$

Then

$$L_n(f) \rightrightarrows_n f$$

as $n \to \infty$ for all f in $C(\mathrm{T})$.

The great virtue of the Korovkin approach is that it provides us with a well-formed program. We illustrate with the second theorem. For any kernel (K_n), we may induce a sequence of linear operators $\mathcal{K}_n(f) = K_n * f$ and must answer two questions: (i) Is each $\mathcal{K}_n$ positive? (ii) Does $\mathcal{K}_n(f) \rightrightarrows_n f$

for the three functions $f = 1, \sin, \cos$? The first is usually easy to answer; the second is frequently a direct computation.

Example 2.28. Dirichlet and Fejér Operators.

We revisit the uniform convergence properties of the Dirichlet and Fejér kernels from Section 2.3.3.

1. The Dirichlet kernel induces the operator

$$\mathcal{D}_n(f) = D_n * f$$

   where $D_n = \sin\left((n + 1/2)t\right) / \sin\left(t/2\right)$. This is quite easily seen not to be positive. This is a good thing, since we know that $\mathcal{D}_n(f)$ might not converge uniformly to f, for f in $C(\mathrm{T})$.

2. The Fejér kernel induces the operator

$$\mathcal{F}_n(f) = F_n * f$$

   where $F_n = \sin^2\left((n + 1)/2)t\right) / [(n + 1)\sin^2\left(t/2\right)] \geq 0$. Thus, to recover Fejér's theorem on the uniform convergence of the Cesàro averages, it suffices to compute

$$\mathcal{F}_n(1) = 1, \quad \mathcal{F}_n(\sin) = \frac{n}{n+1}\sin, \quad \mathcal{F}_n(\cos) = \frac{n}{n+1}\cos. \qquad \square$$

2.7   Commentary and Additional Examples

1. **An error function evaluation.** (*Monthly* Problem 11000, Mar. 2003) [154].

   (a) Work out the ordinary generating function of $\binom{2n}{n}$, and so evaluate

   $$\sum_{n+m=N} \binom{2n}{n}\binom{2m}{m}.$$

   (b) Evaluate

   $$\int_0^{\frac{\pi}{2}} \cos^{2n}(x)\sin^{2m}(x)\,dx.$$

(c) Recall the *error function*, $\operatorname{erf}(x) = (2/\sqrt{\pi}) \int_0^x \exp\left(-t^2\right) \, dt$, and show for $a > 0$ that

$$a \int_0^{\frac{\pi}{2}} \operatorname{erf}\left(\sqrt{a}\cos x\right) \operatorname{erf}\left(\sqrt{a}\sin x\right) \sin(2x) \, dx = e^{-a} + a - 1.$$

(d) The previous evaluation can be viewed as an inner product of the functions $\operatorname{erf}\left(\sqrt{a}\sin x\right)\sin x$ and $\operatorname{erf}\left(\sqrt{a}\cos x\right)\cos x$. Determine that

$$\int_0^{\pi/2} \operatorname{erf}^2\left(\sqrt{a}\cos x\right)\cos^2(x) \, dx$$

$$= \sum_{N=0}^{\infty} \frac{\left(\frac{-a}{4}\right)^{N+1}(8N+12)\binom{2N}{N}}{(N+2)!} F\left(\frac{1}{2}, -N, -N - \frac{1}{2}; \frac{3}{2}, -N + \frac{1}{2}; -1\right)$$

$$= \frac{1}{2\pi} - 2\int_0^1 \frac{e^{-1/2\,a\left(1+x^2\right)}}{1+x^2} \left\{ I_0\left(\frac{1}{2}a\left(1+x^2\right)\right) - I_1\left(\frac{1}{2}a\left(1+x^2\right)\right) \right\} \, dx.$$

2. **Failure of Fubini.** Evaluate these integrals:

(a)

$$\int_0^1 \int_0^1 \frac{x^2 - y^2}{(x^2 + y^2)^2} \, dx\, dy = -\frac{\pi}{4}$$

and

$$\int_0^1 \int_0^1 \frac{x^2 - y^2}{(x^2 + y^2)^2} \, dy\, dx = \frac{\pi}{4}$$

(b)

$$\int_0^1 \int_1^\infty \left(e^{-xy} - 2e^{-2xy}\right) dy\, dx - \int_1^\infty \int_0^1 \left(e^{-xy} - 2e^{-2xy}\right) dx\, dy = \ln 2$$

(c)

$$\int_0^\infty \int_0^\infty \frac{4xy - x^2 - y^2}{(x+y)^4} \, dy\, dx = \int_0^\infty \int_0^\infty \frac{4xy - x^2 - y^2}{(x+y)^4} \, dx\, dy = 0$$

but, for all $m, c > 0$

$$\int_0^{mc} \int_0^c \frac{4xy - x^2 - y^2}{(x+y)^4} \, dx\, dy = \frac{m}{(1+m)^2} \nrightarrow 0,$$

as $c \to \infty$.

In each case, explain why they differ without violating any known theorem.

3. **Failure of l'Hôpital's rule.** Evaluate these limits:

Let $f(x) = x + \cos(x)\sin(x)$ and $g(x) = e^{\sin(x)}(x + \cos(x)\sin(x))$.

Then $\lim_{x \to \infty} f(x)/g(x)$ does not exist although $\lim_{x \to \infty} f'(x)/g'(x) = 0$. This is a caution against carelessly dividing by zero!

4. **Various Fourier series evaluations.**

   (a) Compute the Fourier series of $t/2$, $|t|$, t^2, and $(t^3 - \pi^2 t)/3$ on $[-\pi, \pi]$.

   (b) Plot the 6th and 12th Fourier polynomials against the function in each case.

   (c) Compute enough Fourier coefficients of $\sin(x^3)$ on $[-\pi, \pi]$ to be convinced of Parseval's equation.

   (d) Compute the Fourier series of t^2 and $(t^3 - \pi^2 t)/3$ on $[0, 2\pi]$.

   (e) Use Parseval's equation with $(t^3 - \pi^2 t)/3$ to evaluate $\zeta(6)$. Then apply Parseval to $t^4/4$.

   (f) Show that

   $$\int_0^{\pi/2} \log(2\sin(t/2))\, dt = -G,$$

   where G is Catalan's constant.

   (g) Show that for $a > 0$,

   $$\cos(ax) = \frac{\sin(\pi a)}{\pi a} - 2\frac{\sin(\pi a)\, a \cos(x)}{(a^2 - 1)\pi} + 2\frac{\sin(\pi a)\, a \cos(2x)}{(a^2 - 4)\pi}$$
   $$- 2\frac{\sin(\pi a)\, a \cos(3x)}{(a^2 - 9)\pi} + \cdots. \tag{2.27}$$

   Similarly evaluate the Fourier series for $\exp(ax)$.

   (h) Substitute $x = \pi$ in (2.27) to obtain the partial fraction expansion for cot (compare the first example in Section 2.3.1) and integrate to recover the product formula for sin (justifying all steps).

   (i) Evaluate $\sum_{n \geq 0} 1/(4n^2 - 1)$.

5. **Two applications of Parseval's equation.** Use Parseval's equation in $L^2(\mathbb{R})$ to evaluate

$$\int_{-\infty}^{\infty} \frac{\sin^2(t)}{t^2}\, dt \;=\; \pi \qquad \text{and} \qquad \int_{-\infty}^{\infty} \frac{\sin^4(t)}{t^4}\, dt \;=\; \frac{2\pi}{3}.$$

See also Exercise 28.

6. **Lebesgue's function.**   An example of a continuous function with divergent Fourier series.

Construction. We follow Stromberg, page 557, and let $a_k = 2^{\sum_{j=1}^{k} j!}$ for $k \ge 0$ and define

$$f_n(x) \;=\; \sum_{k=1}^{n} \frac{\sin(a_k|x|)}{k}\, \chi_k(|x|)$$

on $[-\pi, \pi]$, where χ_k is the characteristic function of $[\pi/a_k, \pi/a_{k-1}]$, and extend f by 2π-periodicity onto $\mathbb{R}$. Then $f(x) = \lim_{n\to\infty} f_n(x)$ is continuous and the Fourier series is uniformly convergent on $[\delta, 2\pi - \delta]$ for $\delta > 0$, but $s_{a_k}(f, 0) \to \infty$ for $k \to \infty$.

Convergence on $[\delta, 2\pi - \delta]$. This is easy since $f = f_n$ on this interval for large n, so that the "Riemann localization principle" for Fourier

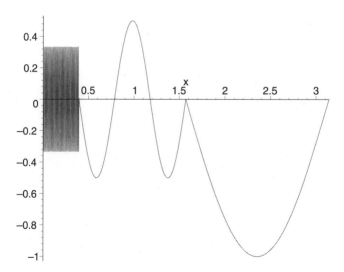

Figure 2.7. Approximation f_3 to Lebesgue's function.

series can be used: *If $f_1(t) = f_2(t)$ for every t in some nonvoid open interval I, then $|s_n(f_1,t) - s_n(f_2,t)| \to 0$ for every $t \in I$.*

Divergence at 0. The divergence estimate comes as follows:

(a) We start with Dirichlet's kernel: It can be proved that the n-th partial sum behaves like

$$s_n(f,0) = \frac{2}{\pi} \int_0^\pi f(t) \frac{\sin(nt)}{t} dt + \varepsilon_n$$

where $\varepsilon_n \to 0$.

(b) We estimate the first part of the integral as

$$\left| \int_0^{\pi/a_k} f(t) \frac{\sin(a_k t)}{t} dt \right| \leq \frac{\pi}{k+1},$$

since $|(\sin x)/x| \leq 1$ and $|f(t)| \leq 1/(k+1)$.

(c) We estimate the second part of the integral via

$$2 \int_{\pi/a_k}^{\pi/a_{k-1}} f(t) \frac{\sin(a_k t)}{t} dt + \frac{1}{k} \int_{\pi/a_k}^{\pi/a_{k-1}} \frac{\cos(2\,a_k t)}{t} dt = k! \frac{\ln 2}{k},$$

and the second term on the left, say I_k, is no bigger than $\frac{1}{2\pi k}$ on using the Bonnet second mean value theorem to write

$$|I_k| = \left| \frac{a_k}{k\pi} \int_{\pi/a_k}^{\psi} \cos(2\,a_k t)\, dt \right| \leq \frac{1}{2k\pi},$$

for some $\frac{a_{k-1}}{\pi} \leq \psi \leq \frac{a_k}{\pi}$.

(d) We estimate the third part of the integral as

$$\left| \int_{\pi/a_{k-1}}^{\pi} f(t) \frac{\sin(a_k t)}{t} dt \right| \leq \frac{a_{k-1}}{\pi} \left| \int_{\pi/a_{k-1}}^{\psi} f(t) \sin(a_k t)\, dt \right|$$

$$\leq \frac{a_{k-1}}{\pi} \left(\left| f(t) \frac{\cos(a_k t)}{a_k} \right|_{\pi/a_{k-1}}^{\psi} + \left| \int_\psi^\pi f'(t) \frac{\cos(a_k t)}{a_k} dt \right| \right)$$

$$\leq \frac{a_{k-1}}{a_k} \left(\frac{1}{\pi} + a_{k-1} \right) \to 0,$$

on using the mean value theorem again, and then applying integration by parts with the estimates that $|f(t)| \leq 1$, $|f'(t)| \leq a_{k-1}$. Thus, the dominant term is the first integral in (2.28) and $s_{a_k}(f,0) \to \infty$.

7. **Nonuniqueness of Fourier series.** Can a trigonometric series converge a.e. on R to a function $\varphi \in L^1(T)$ and yet not be the Fourier series of φ? This question was first answered in the affirmative with $\varphi = 0$ in 1916 by the Russian analyst D. E. Menshow. His counterexample involves the Cantor set. For more details, see [203], from which the above text was cited.

8. **Nowhere differentiable continuous functions.** The first and most famous example of a continuous, nowhere differentiable function was constructed by K. Weierstrass in 1872. (Tradition has it that Bolzano and Riemann constructed such examples before Weierstrass, but their examples did not become widely known.) Weierstrass's example was given in the form of a trigonometric series. We state it here as a series on $[0,1]$, not on $[-\pi, \pi]$ or $[0, 2\pi]$ as before, because this will simplify matters later; we analogously write $f \in L^1(0,1)$, and formulas and theorems on Fourier series are easily converted to this case. Weierstrass's example was

$$C_{a,b}(t) \;=\; \sum_{n=0}^{\infty} a^n \cos(b^n\, 2\pi t),$$

with $|a| < 1$ and integral $b > 1$. Weierstrass proved that $C_{a,b}$ is nowhere differentiable when $b \in 2N+1$ and $ab > 1 + 3\pi/2$. This result settled, once and for all, the question of whether such functions could exist (at the beginning of the 19th century, Ampére "proved" that every continuous function must be differentiable at some point). Some questions were left open, however: It is clear that $C_{a,b}$ is differentiable when $|a|\,b < 1$, since the series is then termwise differentiable. But what happens for $|a|\,b$ between 1 and $1 + 3\pi/2$? This question gave several mathematicians a headache, until in 1916, G. H. Hardy proved that both $C_{a,b}$ and the corresponding sine series

$$S_{a,b}(t) \;=\; \sum_{n=0}^{\infty} a^n \sin(b^n\, 2\pi t)$$

are nowhere differentiable whenever b is a real greater than 1, and $ab > 1$. In his paper, Hardy first treated the case when $b \in N$, i.e., when the functions are given by their Fourier series, and only afterwards treated the general case of arbitrary real b. Hardy's methods were not easy, not even in the Fourier case (where he used the Poisson kernel, among other things). In the ensuing years, several other, simpler proofs have been published. In the middle of the 20th century,

G. Freud and J.-P. Kahane gave conditions for the differentiability of lacunary Fourier series (where nonzero Fourier coefficients are spaced far apart), from which the nondifferentiability of Weierstrass's function follows. Another approach to the Weierstrass functions uses functional equations.

Prove:

(a) The Weierstrass sine series $f = S_{a,2}$ with $b = 2$ satisfies the system of two functional equations

$$f\left(\frac{t}{2}\right) = a\,f(t)+\sin(t), \quad f\left(\frac{t+1}{2}\right) = a\,f(t)-\sin(t) \quad (2.29)$$

for every $t \in [0,1]$. The cosine series $S_{a,2}$ satisfies an analogous system.

(b) The Weierstrass function is the only bounded solution of the respective system on $[0,1]$.

Hint for (b): Use Banach's fixed point theorem.

9. **Replicative functions.** Let D be an interval containing $(0,1)$. A function $f : D \to$ C is called *replicative* (on D) if it satisfies the functional equation

$$\frac{1}{p}\sum_{k=0}^{p-1} f\left(\frac{t+k}{p}\right) = u(p)\,f(t) \quad \text{for all } t \in D \text{ and } p \in \text{N}, \quad (2.30)$$

with a $u : \text{N} \to$ C (which turns out to be unique if $f \not\equiv 0$). This notion was introduced (with more generality) by D. E. Knuth in [152]. Examples are the cotangent ($\cot(\pi t)$ is replicative on $(0,1)$ with $u(p) = 1$), the Bernoulli polynomials ($B_m(t)$ is replicative on R with $u(p) = 1/p^m$), and derivatives of the Psi function (the m-th derivative of $\Psi = \Gamma'/\Gamma$ is replicative on R_+ with $u(p) = p^m$). Functions that are replicative and 1-periodic have multiplicative Fourier coefficients.

Theorem 2.29.

(a) *Let* $f : D \to$ C, $f \not\equiv 0$, *be replicative on* D *with* $u(p)$. *Then* u *is necessarily multiplicative, i.e.,* $u(mn) = u(m) \cdot u(n)$ *for all* $m, n \in \text{N}$.

(b) *Let* $f \in L^1(0,1)$ *be replicative on* $(0,1)$ *with a sequence* $u(p)$. *Then* $\widehat{f}_{mn} = u(n)\widehat{f}_m$.

This implies: If $u \equiv 1$, then $f(t) = \widehat{f}_0$. If $u \not\equiv 1$, then

$$f(t) \sim \widehat{f}_{-1} \sum_{n=-\infty}^{-1} u(-n)\, e^{2\pi int} + \widehat{f}_1 \sum_{n=1}^{\infty} u(n)\, e^{2\pi int}.$$

(c) Let u be multiplicative and assume that $f(t) = {}^{L}\sum_{n=1}^{\infty} u(n)\, e^{2\pi int}$ is pointwise convergent on $[0,1]$, where L is a linear summation method. Then f is replicative on $[0,1]$.

Part (c) of this theorem makes it easy to construct many different examples of replicative functions on $[0,1]$. Verify the following Fourier series:

(a) $\sum n^{-2}\sin(2\pi nt) = -2\pi \int_0^t \ln(2\sin(\pi x))\, dx$ on $[0,1]$,
$\sum n^{-2}\cos(2\pi nt) = B_2(t)$ on $[0,1]$.

(b) $\sum n^{-1}\sin(2\pi nt) = B_1(t)$ on $(0,1)$ and $= 0$ on $0,1$,
$\sum n^{-1}\cos(2\pi nt) = -\ln(2\sin(\pi t))$ on $(0,1)$ and $= \infty$ on $0,1$.

(c) ${}^{C_1}\sum \sin(2\pi nt) = \frac{1}{2}\cot(\pi t)$ on $(0,1)$ and $= 0$ on $0,1$,
${}^{C_1}\sum \cos(2\pi nt) = -\frac{1}{2}$ on $(0,1)$ and $= \infty$ on $0,1$,
where C_1 stands for Cesàro summation.

(d) ${}^{A}\sum n\sin(2\pi nt) = 0$ on $[0,1]$,
${}^{A}\sum n\cos(2\pi nt) = -1/(4\sin^2(\pi t))$ on $(0,1)$ and $= \infty$ on $0,1$,
where A stands for Abel summation.

(e) $\sum_{n=0}^{\infty} a^n \sin(p^n\, 2\pi t) = S_{a,p}(t)$ on $[0,1]$,
$\sum_{n=0}^{\infty} a^n \cos(p^n\, 2\pi t) = C_{a,p}(t)$ on $[0,1]$,
for p a prime, i.e., the nowhere differentiable Weierstrass functions can be replicative.

10. Conditions for integrability.

(a) As in Theorem 2.18, it is often useful to decide whether $\widehat{f} \in L^1(\mathbf{R})$ for a given $f \in L^1(\mathbf{R})$, without having to explicitly compute $\widehat{f}$. The usual conditions assume differentiability properties of f, since smoothness of f translates into shrinkage of $\widehat{f}$. Thus, $f \in C^2(\mathbf{R})$ is sufficient for $\widehat{f} \in L^1(\mathbf{R})$. However, this condition does not cover the Fejér kernel via $g(x) = \max\{1 - |x|, 0\}$, for example. A stronger condition, which is good for functions with bounded support, is given in the next theorem.

Theorem 2.30. *Let f be an absolutely continuous function on the real line with compact support and let f' be of bounded total* *variation on* R, *i.e.,* $V(f') < \infty$. *Then* $\widehat{f} \in L^1(R)$ *and*

$$\|\widehat{f}\|_1 \leq 4\sqrt{V(f')\,\|f\|_1}. \tag{2.31}$$

This theorem presents another experimental challenge: Is the constant "4" appearing there best possible? The answer is not known. Nonsystematic experimentation has found no value for the constant greater than π, which is achieved for the Fejér kernel. It is also not known if the "compact support" condition in the theorem is really needed.

Perform a systematic experiment on Theorem 2.30, in analogy to the experimentation for the uncertainty principle described in Section 5.2 of [43].

(b) A quite different condition is due to Chandrasekharan: *If* $f \in L^1(R)$, *continuous at* 0, *and satisfies* $\widehat{f} \geq 0$ *on* R, *then* $\widehat{f} \in L^1(R)$. The disadvantage of this condition is that it uses $\widehat{f}$ explicitly. It is applicable, however, to both the Fejér and the Poisson kernel.

11. **More kernels.** For each of the following kernels, decide whether (respectively, for which parameters) it is an approximate identity in $L^1(R)$. Note that sometimes a version of Theorem 2.18 with weakened assumptions (allowing more variety in the kernels) is needed.

(a) The *de la Vallée-Poussin kernel* V_m^n, depending on two integer parameters m, n with $n > m$, is defined by $V_m^n(t) = \sum_{k=-(n+m)}^{n+m} a_{m,k}^n e^{ikt}$, where

$$a_{m,k}^n = \begin{cases} 1, & \text{if } |k| \leq n - m, \\ \frac{n+m+1-|k|}{2m+1}, & \text{if } n - m \leq |k| \leq n + m, \\ 0, & \text{otherwise.} \end{cases}$$

Hint: Let m, n tend to infinity such that $n/m \to \lambda$.

(b) For $\alpha > 0$, the (C, α)-*kernel* $F_n^{(\alpha)}$ is defined as $F_n^{(\alpha)}(t) = \sum_{k=-n}^{n} a_{n,k}^{(\alpha)} e^{ikt}$ where

$$a_{n,k}^{(\alpha)} = \begin{cases} \frac{\Gamma(n-|k|+\alpha+1)\,\Gamma(n+1)}{\Gamma(n-|k|+1)\,\Gamma(n+\alpha+1)}, & \text{if } |k| \leq n + 1, \\ 0, & \text{otherwise.} \end{cases}$$

(c) For parameters $\alpha_0, \cdots, \alpha_p \in \mathbb{R}$ with $\alpha_0 + \cdots + \alpha_p = 1$, the *Blackman kernel* $H_n^{(\alpha_0, \cdots, \alpha_p)}$ is defined by $H_n^{(\alpha_0, \cdots, \alpha_p)}(t) = \sum_{k=-n}^{n} h_{n,k}^{(\alpha_0, \cdots, \alpha_p)} e^{ikt}$, where

$$h_{n,k}^{(\alpha_0, \cdots, \alpha_p)} = \sum_{j=0}^{p} \alpha_j \cos(jkt_n),$$

with $t_n = 2\pi/(2n+1)$.

(d) The *Fejér-Korovkin kernel* FK_n is defined as

$$FK_n(t) = \begin{cases} \dfrac{2\sin^2(\pi/(n+2))}{n+2} \left[\dfrac{\cos((n+2)t/2)}{\cos(\pi/(n+2)) - \cos t} \right]^2, \\[2em] (n+2)/2, \end{cases}$$

depending on whether $t \neq \pm\pi/(n+2) + 2j\pi$ or $t = \pm\pi/(n+2) + 2j\pi$, respectively. It can be written in the form $FK_n(t) = \sum_{k=-n}^{n} a_{n,k} e^{ikt}$ where

$$a_{n,k} = \frac{(n-|k|+3)\sin\frac{|k|+1}{n+2}\pi - (n-|k|+1)\sin\frac{|k|-1}{n+2}\pi}{2(n+2)\sin(\pi/(n+2))}.$$

(e) Finally, the *Jackson kernel* J_n is a rescaled version of the square of the Fejér kernel, namely

$$J_n(t) = \frac{3}{n(2n^2+1)} \left[\frac{\sin(nt/2)}{\sin(t/2)} \right]^4.$$

12. **The Haar basis.** As we mentioned in the text, the trigonometric functions e^{int} constitute an orthogonal basis for the space $L^2(\mathbb{T})$, so that L^2-statements follow from general Hilbert space theory. Bases other than the trigonometric are, of course, conceivable and in fact are used for the analysis of L^2-functions. Since about 15 years ago, certain bases of $L^2(\mathbb{R})$, called *wavelet bases*, have found widespread use in signal analysis. Such bases are constructed as follows. Take a $\psi \in L^2(\mathbb{R})$ and define $\psi_{j,n}(t) = 2^{n/2}\psi(2^n t - j)$. Then ψ is called an *orthogonal wavelet* if $\{\psi_{j,n} : j, n \in \mathbb{Z}\}$ is an orthonormal basis of $L^2(\mathbb{R})$.

Show: $\psi = \chi_{[0,1/2)} - \chi_{[1/2,1)}$ is an orthogonal wavelet. The associated basis $\{\psi_{j,n}\}$ is called the *Haar basis* of $L^2(\mathbb{R})$.

13. **The Schauder basis.** The foundation of the theory of bases in Banach spaces was laid by J. Schauder in the 1930s. A sequence (x_n) in a Banach space B is called a *basis of* B if for every $x \in B$, there exists a unique sequence of scalars (α_n) with

$$x = \sum_{n=1}^{\infty} \alpha_n x_n \quad \text{in } B.$$

The trigonometric functions are not a basis for $L^1(T)$ or for $C(T)$, although their span is dense in these spaces. The standard example of a basis for the space $C[0,1]$ is also due to Schauder (although G. Faber had used the same basis before Schauder, in a different analytical guise). This Faber-Schauder basis is the system of continuous functions $\{\sigma_{0,0}, \sigma_{1,0}\} \cup \{\sigma_{i,n} : n \in N, \ i = 0, \cdots, 2^{n-1} - 1\}$, where $\sigma_{0,0}(t) = 1 - t$, $\sigma_{1,0}(t) = t$, and the function $\sigma_{i,n}$ is the linear interpolation of the points

$$(0,0), \quad \left(\frac{i}{2^{n-1}}, 0\right), \quad \left(\frac{2i+1}{2^n}, 1\right), \quad \left(\frac{i+1}{2^{n-1}}, 0\right), \quad (1,0).$$

This system is a basis of the space $C[0,1]$, more precisely: *Every continuous function $f : [0,1] \to R$ has a unique, uniformly convergent expansion of the form*

$$f(x) = \gamma_{0,0}(f)\,\sigma_{0,0}(x) + \gamma_{1,0}(f)\,\sigma_{1,0}(x) + \sum_{n=1}^{\infty}\sum_{i=0}^{2^{n-1}-1} \gamma_{i,n}(f)\,\sigma_{i,n}(x),$$

where the coefficients $\gamma_{i,n}(f)$ are given by $\gamma_{0,0}(f) = f(0)$, $\gamma_{1,0}(f) = f(1)$, and

$$\gamma_{i,n}(f) = f\left(\frac{2i+1}{2^n}\right) - \frac{1}{2} f\left(\frac{i}{2^{n-1}}\right) - \frac{1}{2} f\left(\frac{i+1}{2^{n-1}}\right).$$

Knowing the Schauder basis expansion of a continuous function f can be useful in the analysis of f. For example, Faber proved in 1910 a criterion for differentiability of f in terms of its Schauder coefficients: If $f'(x_0) \in R$ *exists for some $x_0 \in [0,1]$, then*

$$\lim_{n \to \infty} 2^n \cdot \min\{|\gamma_{i,n}(f)| : i = 0, \cdots, 2^{n-1} - 1\} = 0. \qquad (2.32)$$

Interestingly, this condition can be used to prove nondifferentiability of the Weierstrass functions in an elementary way.

Prove:

(a) The Schauder coefficients of the Weierstrass sine series $f = S_{a,2}$ satisfy the recursion

$\gamma_{0,1}(f) = 0$,

$\gamma_{i,n+1}(f) = a\gamma_{i,n}(f) + \gamma_{i,n}(\sin)$ for $n \in \mathbb{N}$, $i = 0, \cdots, 2^{n-1} - 1$,

$\gamma_{i,n+1}(f) = a\gamma_{i-2^{n-1},n}(f) - \gamma_{i-2^{n-1},n}(\sin)$ for $n \in \mathbb{N}$, $i = 2^{n-1}$,

$\cdots, 2^n - 1$.

Hint: Use the functional equations (2.29).

(b) Faber's condition (2.32) is not satisfied for $f = S_{a,2}$. Thus this function is nowhere differentiable.

It is instructive to experiment with the recursion in (a): to plot the Schauder coefficients and Schauder approximations for the Weierstrass and for other functions that satisfy similar functional equations. Details of this method can be found in [121] and [122].

14. **Riemann-Lebesgue lemma.** Deduce the following from the Riemann-Lebesgue lemma for every Lebesgue integrable function f.

(a) For any real $\sigma(t)$

$$\lim_{t \to \infty} \int_{\mathbb{R}} f(x) \cos^2(tx + \sigma(t))\, dx = \frac{1}{2} \int_{\mathbb{R}} f(x)\, dx.$$

(b) The coefficients $\hat{f}(n) \to 0$ as $n \to \infty$.

Conclude that the trigonometric series $\sum_{n>1} \sin(nt)/\log(n)$ is not the Fourier series of any integrable function.

When a_n is convex, decreasing with limit zero and with $\sum_{n>0} a_n/n = \infty$, it is in fact the case that $\sum_{n>0} a_n \cos(nt)$ is a Fourier series of an integrable function, but $\sum_{n>0} a_n \sin(nt)$ is not [203, Chapter 8].

15. **A few Fourier transforms.** We have already seen many examples of Fourier transforms and their Laplace transform variants. The specialization to the Mellin transform is explored in the next chapter.

(a) Show that the L^2-Fourier transform of $(\sin t^2)/t$ is the expression $i\pi\left(S(x/\sqrt{2\pi}) - C(x/\sqrt{2\pi})\right)$, where the functions C and S are the Fresnel integrals,

$$C(x) = \int_0^x \cos(\frac{\pi}{2}t^2)\, dt, \qquad S(x) = \int_0^x \sin(\frac{\pi}{2}t^2)\, dt.$$

Determine the Fourier transform of $\cos(t^2)/t$. Find "sensible" Fourier transforms for $\sin(t^2)$ and $\cos(t^2)$, even though these functions are neither in $L^1(\mathbb{R})$ nor in $L^2(\mathbb{R})$.

(b) Show that for $a > 0$, the Fourier transform of $|t|\exp(-a|t|)$ is the function $2\left(a^2 - x^2\right)/\left(a^2 + x^2\right)^2$.

(c) Find the transform of $1/\left(a^2 + t^2\right)$ and of $1/t^\eta$ (for suitable η, and in a suitable sense).

(d) Find all square-integrable solutions to $\hat{f}/\sqrt{2\pi} = f$ (the fixed points of the normalized Fourier transform). Then experiment with the orbit of $f \mapsto \hat{f}/\sqrt{2\pi}$ for various choices f_0.

16. **The isoperimetric inequality.** The ancient Greek geometers knew already that a circle with given perimeter encloses a larger area than any polygon with the same perimeter. In 1841 Steiner extended this result to simple closed plane curves. Here we will sketch a Fourier series proof (due to Hurwitz), for simplicity restricted to (piecewise) C^1-curves. Thus, assume that we have a simple, closed C^1-curve $(x(t), y(t))$ in $\mathbb{R}^2$ of length $\int_{-\pi}^{\pi}(x'(s)^2 + y'(s)^2)^{1/2}\,ds = 2\pi$. Without loss of generality, we can assume that $x'(s)^2 + y'(s)^2 = 1$ for all s. We wish to minimize the area inside the curve, given by

$$A = \int_{-\pi}^{\pi} x(s)\,y'(s)\,ds.$$

Show: $A \geq \pi$ with equality if and only if $(x(t)-\widehat{x}_0)^2+(y(t)-\widehat{y}_0)^2 = 1$. This is the *isoperimetric inequality*.

Hint: Substitute Fourier series, transfer the formulas for derivatives from Section 2.4.2 to the $L^1(\mathbb{T})$-case, and use Parseval's equation.

17. **The maximum principle.** In like fashion, employ Poisson's kernel to heuristically deduce that the maximum principle discussed briefly in Section 6.5 applies.

18. **The heat equation.** The one-dimensional heat equation

$$\frac{\partial\phi}{\partial t}(x,t) = \frac{\pi}{4i}\frac{\partial^2\phi}{\partial x^2}(x,t),$$

is solved by the general theta function $\sum_{n\in\mathbb{Z}} x^n \exp(-\pi i\,tn^2)$. More usefully, when $G : \mathbb{R} \mapsto \mathbb{C}$ is continuous and bounded we may solve the equation

$$\frac{\partial\phi}{\partial t}(x,t) = K\frac{\partial^2\phi}{\partial x^2}(x,t),$$

with boundary condition $\phi(x,t) \to G(x)$ as $t \to 0^+$, by the infinitely differentiable function

$$G * E_{1/\sqrt{2Kt}}(x) = \frac{1}{2\sqrt{\pi Kt}} \int_R G(x-y) \exp(-y^2/2Kt)\, dy,$$

for x in R and $t > 0$.

19. **The easiest three-dimensional Watson integral.** We start with the easiest integral to evaluate. Let

$$W_3(w) = \int_0^\pi \int_0^\pi \int_0^\pi \frac{1}{1 - w\cos(x)\cos(y)\cos(z)}\, dx\, dy\, dz,$$

for suitable $w > 0$.

(a) Prove that

$$
\begin{aligned}
W_3(1) &= \int_0^\pi \int_0^\pi \int_0^\pi \frac{1}{1 - \cos(x)\cos(y)\cos(z)}\, dx\, dy\, dz \\
&= \frac{1}{4}\Gamma^4\left(\frac{1}{4}\right) = 4\pi\, \mathrm{K}\left(\frac{1}{\sqrt{2}}\right)
\end{aligned}
$$

via the binomial expansion and [44, Exercise 14, page 188].

(b) More generally,

$$W_3((2kk')^2) = \pi^3 \mathrm{F}\left(1/2, 1/2, 1/2; 1, 1; 4k^2\left(1-k^2\right)\right) = 4\pi\mathrm{K}^2(k).$$

20. **The harder three-dimensional Watson integrals.** We now describe results largely in Joyce and Zucker [146, 147], where more background can also be found. The following integral arises in Gaussian and spherical models of ferromagnetism and in the theory of random walks.

(a) One of the most impressive closed-form evaluations of a multiple integral is Watson's

$$
\begin{aligned}
W_1 &= \int_{-\pi}^\pi \int_{-\pi}^\pi \int_{-\pi}^\pi \frac{1}{3 - \cos(x) - \cos(y) - \cos(z)}\, dx\, dy\, dz \\
&= \frac{1}{96}(\sqrt{3}-1)\Gamma^2\left(\frac{1}{24}\right)\Gamma^2\left(\frac{11}{24}\right) \qquad (2.33) \\
&= 4\pi\left(18 + 12\sqrt{2} - 10\sqrt{3} - 7\sqrt{6}\right)\mathrm{K}^2(k_6),
\end{aligned}
$$

where $k_6 = (2-\sqrt{3})(\sqrt{3}-\sqrt{2})$ is the sixth singular value of Section 4.2. Note that $W_1 = \pi^3 \int_0^\infty \exp(-3t)I_0^3(t)\, dt$ allows for efficient computation [146] where the Bessel function $I_0(t)$ has been written as $(1/\pi) \int_0^\pi \exp(t \cos(\theta)) \, d\theta$. The evaluation (2.33), in its original form, is due to Watson and is really a tour de force. In the next exercise we describe a refined and simplified evaluation due to Joyce and Zucker [147].

(b) Similarly, the integral

$$W_2 = \int_0^\pi \int_0^\pi \int_0^\pi \frac{dx\,dy\,dz}{3 - \cos(x)\cos(y) - \cos(y)\cos(z) - \cos(z)\cos(x)}$$

$$= \sqrt{3}\,\pi\,\mathrm{K}^2\left(\sin\left(\frac{\pi}{12}\right)\right) = \frac{2^{1/3}}{8\,\pi}\,\beta^2\left(\frac{1}{3}, \frac{1}{3}\right), \qquad (2.34)$$

where $\sin(\pi/12) = k_3$ is the third singular value, again as in Section 4.2. Indeed, as we shall see in Exercise 21, (2.34) is easier and can be derived on the way to (2.33).

(c) The evaluation (2.34) then implies that

$$\frac{1}{\pi}\,W_2 = \int_0^\pi \int_0^\pi \frac{dy\,dz}{\sqrt{9 - 8\cos(y)\cos(z) - \cos^2(y) - \cos^2(z) + \cos^2(y)\cos^2(z)}}$$

on performing the innermost integration carefully.

(d) The expression inside the square root factors as $(\cos x \cos y + \cos x + \cos y - 3)(\cos x \cos y - \cos x - \cos y - 3)$. Upon substituting $s = x/2$, and $t = y/2$, one obtains

$$\int_0^{\frac{\pi}{2}} \int_0^{\frac{\pi}{2}} \frac{dy\,dx}{\sqrt{\left(1 - \sin^2(x)\sin^2(y)\right)\left(1 - \cos^2(x)\cos^2(y)\right)}} =$$

$$\frac{1}{4\pi} \sum_{m=0}^\infty \sum_{n=0}^\infty \beta^3\left(n + \frac{1}{2}, m + \frac{1}{2}\right)\binom{m+n}{n} = \sqrt{3}\,\mathrm{K}^2\left(\sin\left(\frac{\pi}{12}\right)\right).$$

21. **More about the Watson integrals.**

(a) For $a > 1, b > 1$, show that

$$\frac{1}{2} \int_0^\pi \frac{1}{\sqrt{(a + \cos(y))(b - \cos(y))}}\,dy = \frac{\mathrm{K}\left(\sqrt{\frac{2(b+a)}{(1+b)(1+a)}}\right)}{\sqrt{(1+b)(1+a)}}$$

$$= \int_0^{1/2\,\pi} \frac{1}{\sqrt{(1+b)(1+a)\cos^2(t) + (1-a)(1-b)\sin^2(t)}}\,dt.$$

(b) A beautiful, but harder to establish, identity is that

$$\int_0^{\frac{\pi}{2}} K\left(\sqrt{c^2 \cos^2(s) + \sin^2(s)}\right) ds = K\left(\sqrt{\frac{1-c}{2}}\right) K\left(\sqrt{\frac{1+c}{2}}\right),$$

(2.35)

or equivalently that

$$\int_0^{\frac{\pi}{2}} K\left(\sqrt{1 - (2kk')^2 \cos^2(\theta)}\right) d\theta = K(k) K(k')$$

with $k' = \sqrt{1 - k^2}$. Hence,

$$\int_0^{\frac{\pi}{2}} K\left(\sqrt{1 - (2k_N k'_N)^2 \cos^2(\theta)}\right) d\theta = \sqrt{N} K^2(k_N),$$

where k_N is the N-th singular value. This is especially pretty for $N = 1, 3, 7$ so that $2 k_N k'_N = 1, 1/2, 1/8$, respectively.

(c) Deduce that the face centered cubic (FCC) lattice for the Green's function evaluates as

$$\frac{1}{\pi} W_2 = \int_0^{\frac{\pi}{2}} K\left(\sqrt{\frac{3}{4} \cos^2(s) + \sin^2(s)}\right) ds = \sqrt{3} K^2(k_3).$$

(d) Correspondingly, Watson's evaluation for the simple cubic (SC) lattice relied on deriving

$$W_1 = \sqrt{2} \pi \int_0^\pi K\left(\frac{\cos(x) - 5}{2}\right) dx,$$

and the following extension of (2.35):

$$\int_0^{\frac{\pi}{2}} K\left(\sqrt{c^2 \cos^2(s) + d^2 \sin^2(s)}\right) ds =$$

$$K\left(\sqrt{\frac{1 - cd - \sqrt{(d^2 - 1)(c^2 - 1)}}{2}}\right)$$

$$\times K\left(\sqrt{\frac{1 + cd - \sqrt{(d^2 - 1)(c^2 - 1)}}{2}}\right).$$

(e) *The generalized Watson integrals.* Let

$$W_1(w_1) = \int_{-\pi}^{\pi}\int_{-\pi}^{\pi}\int_{-\pi}^{\pi} \frac{1}{3 - w_1\left(\cos(x) - \cos(y) - \cos(z)\right)}\, dx\, dy\, dz$$

$$W_2(w_2) =$$
$$\int_{0}^{\pi}\int_{0}^{\pi}\int_{0}^{\pi} \frac{dx\, dy\, dz}{3 - w_2\left(\cos(x)\cos(y) - \cos(y)\cos(z) - \cos(z)\cos(x)\right)}.$$

In a beautiful study, Joyce and Zucker [147], using the sort of elliptic and hypergeometric transformations we have explored, show fairly directly that

$$W_2(-w_1(3 + zw_1)/(1 - w_1)) = (1 - w_1)^{1/2}\, W_1(w_1).$$

Verify that, with $w_1 = -1$, this leads to a quite direct evaluation of (2.33) from (2.34).

(f) It is also true that

$$W_1 = \sqrt{2}\,\pi \int_{0}^{\frac{\pi}{2}} K\left(\frac{1}{2} + \frac{1}{2}\sin^2(t)\right) dt.$$

(g) *A more symmetric form.* Show that

$$\int_{0}^{\frac{\pi}{2}} K\left(\sqrt{1 - 4k^2(1 - k^2)\cos^2(x)}\right) dx$$
$$= \int_{0}^{\frac{\pi}{2}}\int_{0}^{\frac{\pi}{2}} \frac{dt\, dx}{\sqrt{\cos^2(t) + 4k^2(1 - k^2)\cos^2(x)\sin^2(t)}}$$

for $0 < k < 1$.

Hint: For (21d) consider $N = 3$ $(c^2 = 3/4)$ in (21c), and let a and b be defined as $a = (3 - \cos x)/(1 + \cos x)$ and $b = (3 + \cos x)/(1 - \cos x)$ in (21a).

22. **Watson integral and Burg entropy.** Consider the perturbed *Burg entropy* maximization problem

$$v(\alpha) = \sup_{p \geq 0}\{\log(p(x_1, x_2, x_3)) \mid \int_{0}^{1}\int_{0}^{1}\int_{0}^{1} p(x_1, x_2, x_3)dx_1 dx_2 dx_3 = 1,$$

and for $k = 1, 2, 3,$ $\int_{0}^{1}\int_{0}^{1}\int_{0}^{1} p(x_1, x_2, x_3)\cos(2\pi x_k)\,dx_1 dx_2 dx_3 = \alpha\},$

maximizing the log of a density p with given mean, and with the first three cosine moments fixed at a parameter value $0 \leq \alpha < 1$. It transpires that there is a parameter value $\overline{\alpha}$ such that below and at that value $v(\alpha)$ is attained, while above it is finite but unattained. This is interesting, because:

(a) The general method—maximizing $\int_T \log{(p(t))}\, dt$ subject to a finite number of trigonometric moments—is frequently used. In one or two dimensions, such spectral problems are always attained when feasible.

(b) There is no easy way to see that this problem qualitatively changes at $\overline{\alpha}$, but we can get an idea by considering

$$\overline{p}\,(x_1, x_2, x_3) = \frac{1/W_1}{3 - \sum_1^3 \cos{(2\pi x_i)}},$$

and checking that this is feasible for

$$\overline{\alpha} = 1 - 1/(3W_1) \approx 0.340537329550999142833$$

in terms of the first Watson integral, W_1.

(c) By using Fenchel duality [61] one can show that this $\overline{p}$ is optimal.

(d) Indeed, for all $\alpha \geq 0$ the only possible optimal solution is of the form

$$\overline{p}_\alpha\,(x_1, x_2, x_3) = \frac{1}{\lambda_\alpha^0 - \sum_1^3 \lambda_\alpha^i \cos{(2\pi x_i)}},$$

for some real numbers λ_α^i. Note that we have four coefficients to determine; using the four constraints we can solve for them. For $0 \leq \alpha \leq \overline{\alpha}$, the precise form is parameterized by the generalized Watson integral:

$$\overline{p}_\alpha\,(x_1, x_2, x_3) = \frac{1/W_1(w)}{3 - \sum_1^3 w \cos{(2\pi x_i)}},$$

and $\alpha = 1 - 1/(3W_1(w))$, as w ranges from zero to one. Note also that $W_1(w) = \pi^3 \int_0^\infty I_0^3(w\,t)\, e^{-3t}\, dt$ allows one to quickly obtain w from α numerically. For $\alpha > \overline{\alpha}$, no feasible reciprocal polynomial can stay positive. Full details are given in [60].

23. **A "momentary" recursion.** Choose $p \in \mathbb{N}$ and define polynomials $q_k(x)$ recursively by $q_0(x) = -1$ and $q_{n+1}(x) = q_n'(x) - x^{p-1}\, q_n(x)$. Give an explicit formula for $q_n(0)$ (and for $q_n(x)$).

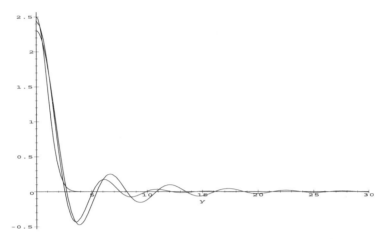

Figure 2.8. The oscillatory Fourier transforms $\widehat{f_2}, \widehat{f_8}, \widehat{f_{16}}$.

24. **The limit of certain Fourier transforms.** For $p \in 2\mathrm{N}$, let $f_p(t) = e^{-t^p/p}$. In Figure 2.8, the functions $\widehat{f_p}(x)$ are shown for $p = 2, 8, 16$. The figure suggests that there may be a limit function as $p \to \infty$. Identify this limit function!

25. **The Schilling equation.** The Schilling equation is the functional equation

$$4q\, f(qt) \;=\; f(t+1) + 2f(t) + f(t-1) \quad \text{for } t \in \mathrm{R}$$

with a parameter $q \in (0,1)$. It has its origin in physics, and although it has been studied intensively in recent years, there are still many open questions connected with it. The main question is to find values of q for which the Schilling equation has a nontrivial L^1-solution. Discuss this question!

Hint: If an L^1-function f satisfies (2.20), then a rescaled version of $f * f$ satisfies the Schilling equation.

26. **Another way to evaluate the sinc integral.** The evaluation of the integral $\int_0^\infty \sin y/y\, dy = \pi/2$ also follows on taking the limit, via Binet's mean value theorem [203, page 328], of the absolutely convergent integral

$$\int_0^\infty \frac{\sin y}{y^{1+\varepsilon}}\, dy = \frac{\pi}{2} \frac{\sec(\frac{\pi}{2}\varepsilon)}{\Gamma(1+\varepsilon)}.$$

Maple happily provides the second integral in a form which simplifies to that we have given. A proof based on the conventional Mellin transform follows.

(a) For $0 < \varepsilon < 1$, use the Γ-function to write

$$\int_0^\infty \frac{\sin y}{y^{1+\varepsilon}}\, dy = \frac{1}{\Gamma(\varepsilon+1)} \int_0^\infty dx \int_0^\infty \sin(x)\exp(-xt)t^\varepsilon\, dt.$$

(b) Interchange variables and evaluate the inner integral to $t^\varepsilon/(t^2+1)$.

(c) Then use the β-function to prove

$$\int_0^\infty \frac{t^\varepsilon}{t^2+1}\, dt = \beta\left(\frac{1}{2}-\frac{1}{\varepsilon}, \frac{1}{2}-\frac{1}{\varepsilon}\right) = \frac{\pi}{2}\sec\left(\frac{\pi}{2}\varepsilon\right).$$

Note:

$$\int_0^\infty \frac{\log^{2n}(s)}{s^2+1}\, ds = (-1)^n \left(\frac{\pi}{2}\right)^{2n+1} E_{2n}.$$

27. **An explicit formula for the sinc integrals.** Assume that $n \geq 1$ and $a_0, a_1, \cdots, a_n > 0$. For $\gamma = (\gamma_1, \cdots, \gamma_n) \in \{-1,1\}^n$ define

$$b_\gamma = a_0 + \sum_{k=1}^n \gamma_k a_k \quad \text{and} \quad \epsilon_\gamma = \prod_{k=1}^n \gamma_k.$$

Show:

(a) $$\sum_{\gamma \in \{-1,1\}^n} \epsilon_\gamma\, b_\gamma^r = \begin{cases} 0, & \text{for } r = 0, 1, \cdots, n-1, \\ 2^n\, n! \prod_{k=1}^n a_k, & \text{for } r = n, \end{cases}$$
where $b_\gamma^0 = 1$ even if $b_\gamma = 0$. *Hint:* Expand both sides of $e^{a_0 t}\prod_{k=1}^n\left(e^{a_k t}-e^{-a_k t}\right) = \sum_{\gamma \in \{-1,1\}^n} \epsilon_\gamma\, e^{b_\gamma t}$ into a power series in t and compare coefficients.

(b) $$\prod_{k=0}^n \sin(a_k x) = \frac{1}{2^n} \sum_{\gamma \in \{-1,1\}^n} \epsilon_\gamma \cos(b_\gamma x - \tfrac{\pi}{2}(n+1)).$$

(c) $$\int_0^\infty \prod_{k=0}^n \frac{\sin(a_k\, x)}{x}\, dx = \frac{\pi}{2}\frac{1}{2^n\, n!} \sum_{\gamma \in \{-1,1\}^n} \epsilon_\gamma\, b_\gamma^n\, \text{sign}(b_\gamma).$$

(d) $$\int_0^\infty \prod_{k=0}^n \text{sinc}(a_k\, x)\, dx = \frac{\pi}{2}\frac{1}{a_0}\left(1 - \frac{1}{2^{n-1}\, n!\, a_1 \cdots a_n} \sum_{b_\gamma < 0} \epsilon_\gamma b_\gamma^n\right).$$

(e) The first "bite." If $\sum_{k=1}^{n-1} a_k \leq a_0 < \sum_{k=1}^{n} a_k$, then

$$\int_0^\infty \prod_{k=0}^n \operatorname{sinc}(a_k\, x)\, dx \;=\; \frac{\pi}{2}\, \frac{1}{a_0}\left(1 - \frac{(a_1 + \cdots + a_n - a_0)^n}{2^{n-1}\, n!\, a_1 \cdots a_n}\right).$$

28. **A special sinc integral.** Evaluate

$$\int_0^\infty \operatorname{sinc}^n(x)\, dx \;=\; \frac{\pi}{2}\left(1 + \frac{1}{2^{n-2}} \sum_{1 \leq r \leq \frac{n}{2}} \frac{(-1)^r}{(r-1)!} \frac{(n-2r)^{n-1}}{(n-r)!}\right)$$

$$= \;\frac{\pi}{2^n\,(n-1)!} \sum_{0 \leq r \leq \frac{n}{2}} (-1)^r \binom{n}{r} (n-2r)^{n-1}.$$

In this way, confirm the results of Exercise 5.

29. **A strange cosine integral.** Let $C^*(x) = \cos(2x) \prod_{n=1}^{\infty} \cos(x/n)$. Show symbolically that $\int_0^\infty C^*(x)\, dx < \pi/8$, and show numerically that

$$0 \;<\; \frac{\pi}{8} - \int_0^\infty C^*(x)\, dx \;<\; 10^{-41}.$$

This is hard to distinguish numerically from $\pi/8$; compare Exercise 39.

30. **Multivariable sinc integrals.** For $x, y \in \mathrm{R}^m$ we write $x \cdot y$ to denote the dot product. Define the *sinc space* $\mathcal{S}^{m,n}$ to be the set of $m \times (m+n)$ matrices $S = (s_1\; s_2\; \cdots\; s_{m+n})$ of column vectors in R^m such that

$$\int_{\mathrm{R}^m} \left| \prod_{k=1}^{m+n} \operatorname{sinc}(s_k \cdot y) \right| dy \;<\; \infty,$$

and a function $\sigma : \mathcal{S}^{m,n} \to \mathrm{R}$ by

$$\sigma(S) \;=\; \int_{\mathrm{R}^m} \prod_{k=1}^{m+n} \operatorname{sinc}(s_k \cdot y)\, dy.$$

Correspondingly, define the *polyhedron space* $\mathcal{P}^{m,n}$ to be the complete set of $m \times (m+n)$ matrices $P = (p_1\; p_2\; \cdots p_{m+n})$ and a function $\nu : \mathcal{P}^{m,n} \to \mathrm{R}$ by

$$\nu(P) \;=\; \operatorname{Vol}\{x \in \mathrm{R}^n : |p_k \cdot x| \leq 1 \text{ for } k = 1, 2, \cdots, m+n\}.$$

(a) Note that by change of basis, for $S \in \mathcal{S}^{m,n}$ and $P \in \mathcal{P}^{m,n}$, we have

$$\sigma(S) = |\det(M)|\,\sigma(MS) \quad \text{and} \quad \nu(P) = |\det(N)|\,\nu(NP)$$

for nonsingular matrices M $(m \times m)$ and N $(n \times n)$.

(b) The following correspondence between multidimensional sinc integrals and volumes of polyhedra can be proved with some effort (see [34]): *If $n \geq m$, if A is a nonsingular $(m \times m)$-Matrix, and if B is any $(m \times n)$-matrix having m of its columns linearly independent, then*

$$\sigma(A|B) = \frac{\sigma(I^m|A^{-1}B)}{|\det(A)|} = \frac{\pi^m}{2^n}\frac{\nu(I^n|(A^{-1}B)^T)}{|\det(A)|}.$$

Similarly, if $n \geq m$, if C is a nonsingular $(n \times n)$-matrix, and if D is any $(n \times m)$-matrix such that $C^{-1}D$ has m linearly independent rows, then

$$\nu(C|D) = \frac{\nu(I^n|C^{-1}D)}{|\det(C)|} = \frac{2^n}{\pi^m}\frac{\sigma(I^m|(C^{-1}D)^T)}{|\det(C)|}.$$

(c) Use the theorem from (b) to determine (with the use of symbolic integration) the volume of $\{x \in \mathrm{R}^6 : |p_k \cdot x| \leq 1,\ k = 1, \cdots, 11\}$, where p_i is the i-th column of the matrix

$$P = \begin{pmatrix} 10 & 0 & 0 & 0 & 0 & 0 & 9 & 10 & -1 & -3 & 7 \\ 0 & 10 & 0 & 0 & 0 & 0 & -2 & -1 & -8 & 2 & -6 \\ 0 & 0 & 10 & 0 & 0 & 0 & -9 & 7 & -5 & 5 & 1 \\ 0 & 0 & 0 & 10 & 0 & 0 & 5 & -2 & -9 & -8 & -9 \\ 0 & 0 & 0 & 0 & 10 & 0 & -10 & -2 & -3 & 6 & -4 \\ 0 & 0 & 0 & 0 & 0 & 10 & -8 & 9 & 2 & 7 & -10 \end{pmatrix}.$$

Hint: $\nu(P) = (32/(5\pi^5)) \int_{\mathrm{R}^5} \prod_{k=1}^{11} \text{sinc}(s_i \cdot y)\,dy$, where

$$S = \begin{pmatrix} 10 & 0 & 0 & 0 & 0 & 9 & -2 & -9 & 5 & -10 & -8 \\ 0 & 10 & 0 & 0 & 0 & 10 & -1 & 7 & -2 & -2 & 9 \\ 0 & 0 & 10 & 0 & 0 & -1 & -8 & -5 & -9 & -3 & 2 \\ 0 & 0 & 0 & 10 & 0 & -3 & 2 & 5 & -8 & 6 & 7 \\ 0 & 0 & 0 & 0 & 10 & 7 & -6 & 1 & -9 & -4 & -10 \end{pmatrix}.$$

31. **Another iterated sinc integral.** For positive constants (a_i), evaluate

$$\int_{-\infty}^{\infty} \cdots \int_{-\infty}^{\infty} \frac{\sin(a_1 x_1)}{x_1} \cdots \frac{\sin(a_n x_n)}{x_n} \frac{\sin(a_1 x_1 + \cdots + a_n x_n)}{x_1 + \cdots + x_n} \, dx_1 \cdots dx_n.$$

Answer: $\pi^n \min(a_1, \ldots, a_n)$.

32. **Infinite series and Clausen's product.** For x and t appropriately restricted:

   (a) Use Clausen's product to obtain

$$\sum_{n=0}^{\infty} \frac{(t)_n (-t)_n}{(2n)!} (2x)^{2n} = \cos(2t \arcsin(x))$$

   and

$$-\frac{1}{2} \sum_{n=1}^{\infty} \frac{(t)_n (-t)_n}{(2n)!} \left(4 \sin^2 x\right)^n = \sin^2(t x).$$

   (b) Obtain the Taylor series

$$\arcsin^2(x) = \frac{1}{2} \sum_{n \geq 1} \frac{(2x)^{2n}}{n^2 \binom{2n}{n}}$$

   on taking an appropriate limit as $t \to 0$ (see also Exercise 16 of Chapter 1). Hence, show

$$\sum_{n \geq 1} \frac{1}{n^2 \binom{2n}{n}} = \frac{\pi^2}{18} \quad \text{and} \quad \sum_{n \geq 1} \frac{(-1)^n}{n^2 \binom{2n}{n}} = -2 \log^2\left(\frac{1 + \sqrt{5}}{2}\right).$$

   Evaluate $\sum_{n \geq 1} 3^n / \binom{2n}{n}$ and both of $\sum_{n \geq 1} (\pm 1)^n / \binom{2n}{n}$.

33. **Proof of the Korovkin theorems.** Prove Theorems 2.25 and 2.27.

34. **Korovkin by inequalities.** An interesting recent approach to the Korovkin theorems is given in [210]. Recall that a subset of a continuous function space is a *subalgebra* if it is closed under pointwise multiplication. Therein, the following elegant lemma is proven:

Lemma 2.31. *Suppose that $\mathcal{A}$ is a norm-closed subalgebra of $C[a, b]$ that contains 1. Let T be a positive linear operator on $\mathcal{A}$ such that $T(1) \leq 1$. Then*

(a) $\mathcal{E}(h) = T(h^2) - T(h)^2 \geq 0$,

(b) $|T(fg) - T(f)T(g)|^2 \leq \mathcal{E}(f)\,\mathcal{E}(g)$,

(c) $\|T(fg) - T(f)T(g)\|^2 \leq \|\mathcal{E}(f)\|\,\|\mathcal{E}(g)\|$,

(d) $\|T(fg) - T(f)T(g)\|^2 \leq \|\mathcal{E}(f)\|\,\|\mathcal{E}(g) + \mathcal{E}(k)\|$,

for all elements f, g, h and k in the algebra.

Proof. (a) is established by observing that $T\left((h + t\,1)^2\right) \geq 0$ for all real t. Then (b) follows with h replaced by $f + tg$, and (c) and (d) are easy consequences. $\qquad\square$

It is now a nice problem to show that the first and second Korovkin theorems follow—if one knows that the polynomials are dense in $C[a, b]$. Moreover, the same approach will yield:

Theorem 2.32. (Complex Korovkin theorem). *Let $D = \{z \in \mathbb{C} : |z| \leq 1\}$. Let T_n be positive linear operators on $C(D)$ such that $T_n(h) \rightrightarrows h$ for $h = 1, z$ and $|z|^2$. Then this holds for all h in $C(D)$.*

To prove this, it helps to observe that positive operators preserve conjugates: $T\left(\overline{h}\right) = \overline{T(h)}$ for all h in $C(D)$.

35. **Bézier curves.** The *Bézier curve* of degree n defined by $n + 1$ points $b_0, b_1, \ldots, b_n$ is exactly the Bernstein polynomial interpolating the values at k/n

$$\sum_{k=1}^{n} b_k \binom{n}{k} t^k (1 - t)^{n-k}. \tag{2.36}$$

Typically, parametric cubic Bézier curves in the plane such as

$$x(t) = -(1 - t)^3 - t(1 - t)^2 + \frac{3}{2} t^2 (1 - t) + t^3 \tag{2.37}$$

$$y(t) = \frac{1}{2}(1 - t)^3 + t(1 - t)^2 + \frac{3}{4} t^2 (1 - t) + \frac{1}{2} t^3$$

are fitted together for smoothing purposes. To compute the values, it is useful to observe Castlejau's algorithm that the basis functions $B_{n,k} = t \mapsto \binom{n}{k} t^k (1-t)^{n-k}$ satisfy the recursion $B_{n,-1} = B_{n-1,n} = 0$ and

$$B_{n,k}(t) = (1 - t)\,B_{n-1,k}(t) + t\,B_{n-1,k-1}(t),$$

for $0 \leq k \leq n$ and all real t.

36. **Bernstein polynomials.** Determine the appropriate Bernstein polynomials on $[-1, 1]$.

37. **Rate of approximation.** As we have seen, the rate of approximation is tied to the smoothness of the underlying function. In Lebesgue's proof of the Stone-Weierstrass Theorem, the main work is in showing that $|\cdot|$ can be uniformly approximated by polynomials on $[-1, 1]$. Plot the first few Bernstein polynomials and observe that the approximation is worst at zero, where $|t|$ is not differentiable.

38. **Korovkin kernels.** Apply the Korovkin theorems to the Poisson, Fejér-Korovkin, and Jackson kernels, respectively.

39. **Contriving coincidences.**

(a) A consequence of the theta transform, (2.15), in the form $s\,\theta_3^2\left(e^{-\pi s}\right) = \theta_3^2\left(e^{-\pi/s}\right)$, is that

$$\sum_{n \geq 1} e^{-(n/10)^2} \approx 5\,\Gamma\left(\frac{1}{2}\right) - \frac{1}{2}$$

and they agree through 427 digits, with similar more baroque estimates for higher powers of ten.

(b) The fact that $\alpha = \exp(\pi\sqrt{163}/3) \approx 640320$ lies deeper and relates to the fact that the only imaginary quadratic fields with unique factorization are $Q\left(\sqrt{-d}\right)$, with $d = 1, 2, 4, 7, 11, 19, 43, 67$, and 163.

(c) This leads to a spectacular "billion-digit" fraud

$$\sum_{n=1}^{\infty} \frac{[n\alpha]}{2^n} \approx 1280640.$$

As we saw this is explained by Theorem 1.3 and the fact that as a continued fraction,

$$\alpha = [640320, 1653264929, 30, 1, 321, 2, 1, 1, 1, 4, 3, 4, 2, \ldots].$$

(d) Determine the integers N_d such that

$$\sum_{n=1}^{\infty} \frac{[n\alpha_d]}{2^n} \approx N_d,$$

for $\alpha_d = \exp(\pi\sqrt{d}/3)$ with $d = 19, 43, 67, 163$, and determine the error in each case.

These examples signal the danger of inferring a symbolic identity from tools like PSLQ without knowing the context. That said, we know of nearly no cases where such spectacular deception has occurred without contrivance.

3 | Zeta Functions and Multizeta Values

I see some parallels between the shifts of fashion in mathematics and in music. In music, the popular new styles of jazz and rock became fashionable a little earlier than the new mathematical styles of chaos and complexity theory. Jazz and rock were long despised by classical musicians, but have emerged as art-forms more accessible than classical music to a wide section of the public. Jazz and rock are no longer to be despised as passing fads. Neither are chaos and complexity theory. But still, classical music and classical mathematics are not dead. Mozart lives, and so does Euler. When the wheel of fashion turns once more, quantum mechanics and hard analysis will once again be in style.

Freeman Dyson,
"Review of Nature's Numbers by Ian Stewart," 1996

The Riemann zeta function has already appeared in various contexts in earlier chapters. We start this chapter with gathering up its basic properties in one place. After some further discussion of special values of the function, we complete this chapter with a more detailed exploration of multiple zeta values (Euler sums), as introduced in Chapter 2 of [43].

The *zeta function* (of Riemann) is defined by the following series

$$\zeta(s) = \sum_{n=1}^{\infty} \frac{1}{n^s}, \tag{3.1}$$

for $\mathrm{Re}(s) > 1$. The estimate

$$\frac{1}{n+1} + \frac{1}{n+2} + \cdots + \frac{1}{2n} > \frac{n}{2n} = \frac{1}{2},$$

and the comparison test for series, show that ζ has a pole at $s = 1$. To go further, we introduce the *alternating zeta function*

$$\alpha(s) = \sum_{n=1}^{\infty} \frac{(-1)^{n+1}}{n^s}, \tag{3.2}$$

and note that $\alpha(1) = \log(2)$. This series clearly converges (and is analytic as a uniform limit) for $\text{Re}(s) > 0$. Moreover, regrouping the terms shows that

$$\alpha(s) = -\sum_{n=1}^{\infty} \frac{1}{(2n)^s} + \sum_{n=1}^{\infty} \frac{1}{(2n-1)^s} = 2^{-s}\zeta(s) - \sum_{n=1}^{\infty} \frac{1}{(2n-1)^s} \qquad (3.3)$$

$$\zeta(s) = \sum_{n=1}^{\infty} \frac{1}{(2n)^s} + \sum_{n=1}^{\infty} \frac{1}{(2n-1)^s} = 2^{-s}\zeta(s) + \sum_{n=1}^{\infty} \frac{1}{(2n-1)^s}.$$

Then (3.3) shows $\sum_{n=1}^{\infty} 1/(2n-1)^s = (1 - 2^{-s})\zeta(s)$ and that $\alpha(s) = (1 - 2^{1-s})\zeta(s)$. Thus,

$$\zeta(s) = \frac{\alpha(s)}{(1 - 2^{1-s})}, \qquad (3.4)$$

for $\text{Re}(s) > 0$ which provides an analytic continuation of ζ in the right half-plane, with $\zeta(1/2) = 0.6048986430\ldots$ We also note that $\alpha(2) = \zeta(2)/2 = \pi^2/12$.

3.1   Reflection and Continuation

There are various routes to extend ζ into the left halfplane. We choose to start with the function $\tau(t) = [\theta_3(e^{-\pi t}) - 1]/2$, where $\theta_3(q) = \sum_{n=-\infty}^{\infty} q^{n^2}$ as is discussed in Section 4.2. For $\text{Re}(s) > 1/2$, we choose to use the Mellin transform

$$M_s(\tau) = \int_0^{\infty} \tau(x) x^{s-1}\, dx$$

and write

$$M_s(\tau) = \sum_{n=1}^{\infty} n^{-2s} \pi^{-s} \int_0^{\infty} e^{-t} t^{s-1}\, dt = \frac{\Gamma(s)}{\pi^s} \zeta(2s). \qquad (3.5)$$

Hence

$$\Gamma\left(\frac{s}{2}\right) \zeta(s) \pi^{-s/2} = \int_0^{\infty} t^{s/2-1} \tau(t)\, dt \qquad (3.6)$$

$$= \int_1^{\infty} t^{s/2-1} \tau(t)\, dt + \int_0^1 t^{-1/2} \tau\left(\frac{1}{t}\right) t^{s/2-1}\, dt$$

$$+ \ \frac{1}{2} \int_0^1 (t^{-1/2} - 1) t^{s/2-1}\, dt. \qquad (3.7)$$

Here we have used the *theta transform* (2.15) to replace $\tau(t)$ by $\tau(1/t)$ on $[1, \infty)$. We deduce that

$$\Gamma\left(\frac{s}{2}\right)\zeta(s)\pi^{-s/2} = -\left(\frac{1}{s} + \frac{1}{1-s}\right) + \int_1^\infty \frac{t^{s/2} + t^{(1-s)/2}}{t}\tau(t)\, dt, \quad (3.8)$$

as we see on evaluating the final integral and sending $t \to 1/t$ in the second integral in (3.6). Because $\tau(t) = O(e^{-\pi t})$ as $t \to \infty$, the integral in (3.8) is an entire function of s, and as Γ has a simple pole at zero, we see that (3.8) extends ζ analytically with a single simple pole at $s = 1$.

Most beautifully, we note that (3.8) is left unchanged by the substitution $s \to 1 - s$ and so we obtain the famous functional equation or *reflection formula* for the Riemann zeta function:

$$\Gamma\left(\frac{s}{2}\right)\zeta(s)\pi^{-s/2} = \Gamma\left(\frac{1-s}{2}\right)\zeta(1-s)\pi^{(1-s)/2}. \quad (3.9)$$

Symmetry of ζ around the line $\mathrm{Re}(s) = 1/2$ is now apparent. We also note that (3.9) shows that $\zeta(-2n)$ is zero for even negative integers (because Γ has poles at those values). These are called "trivial" zeroes. As we shall see in the next section, $\zeta(-2n+1)$ is a Bernoulli number and thus rational.

The reflection formula must represent one of the most beautiful findings in mathematics. The British analyst G. N. Watson, discussing his response to equally beautiful formulae of the wonderful Indian mathematical genius Ramanujan (1887–1920), such as those in Section 3.2.3, describes

> a thrill which is indistinguishable from the thrill I feel when I enter the Sagrestia Nuovo of the Capella Medici and see before me the austere beauty of the four statues representing "Day," "Night," "Evening," and "Dawn" which Michelangelo has set over the tomb of Giuliano de'Medici and Lorenzo de'Medici. [213]

3.1.1   The Riemann Hypothesis

The mathematical centrality of the zeta function can hardly be overestimated. It plays a key role in Problem 8 (of 23) of Hilbert's famous 1900 lecture, and in Problem 5 (of 8) of the Millennium Problems posed by the Clay Foundation 100 years later. Central to the study of the zeta function is the *Riemann hypothesis*:

> The only nontrivial zeroes of $\zeta(s)$ for complex numbers $s = \sigma + i\gamma$ lie on $\sigma = 1/2$.

The importance and present status of the Riemann Hypothesis in prime number theory has already been discussed in Chapter 2 of [43], and is developed further in Exercise 2 at the end of this chapter. A related reason for the role of ζ in number theory comes from the following:

Lemma 3.1. (Euler product). *For* $\sigma = Re(s) > 1$,

$$\zeta(s) = \prod_p \left(1 - \frac{1}{p^s}\right)^{-1},\qquad(3.10)$$

where p *runs over the primes.*

Proof. This can be seen by expanding the finite product and using unique factorization:

$$\prod_{p \leq X} \left(1 - \frac{1}{p^s}\right) = \prod_{p \leq X} \left(1 + \frac{1}{p^s} + \frac{1}{p^{2s}} + \cdots\right) = \sum_{n \leq X} \frac{1}{n^s} + \mathcal{E}(s, X),\quad(3.11)$$

where $\mathcal{E}(s, X) \leq \sum_{n > X} 1/n^\sigma \to_X 0$. $\square$

From this we may derive again that for $Re(s) \geq 1$, the function $\zeta(s)$ has a pole only at $s = 1$.

3.2   Special Values of the Zeta Function

3.2.1   Zeta at Even Positive Integers

We have already evaluated $\zeta(2)$ in various ways. There are also several ways to find the closed form for $\zeta(2n)$. Let us start with the intuitive path followed by Euler. Euler intuited his product formula for π (1.11) from the analogy with the Fundamental Theorem of Algebra, and on writing down the Taylor series for $\sin(x)/x$, one is left to compare

$$\sum_{n=0}^{\infty} \frac{(-1)^n x^{2n}}{(2n + 1)!} = \prod_{n=1}^{\infty} \left(1 - \frac{x^2}{\pi^2 n^2}\right).\qquad(3.12)$$

Thus, one has

$$\zeta(2) = \sum_{n > 0} \frac{1}{n^2} = \frac{\pi^2}{6}.$$

Considering the next term, we have

$$\sum_{\substack{m,n>0\\m<n}} \frac{1}{n^2\,m^2} = \frac{\pi^4}{120}.$$

Now

$$\frac{\pi^4}{60} + \zeta(4) = 2 \sum_{\substack{m,n>0\\m\neq n}} \frac{1}{n^2\,m^2} + \sum_{m>0} \frac{1}{m^4} = \sum_{m,n>0} \frac{1}{n^2\,m^2} = \zeta(2)^2,$$

and

$$\zeta(4) = \frac{\pi^4}{36} - \frac{\pi^4}{60} = \frac{\pi^4}{90}.$$

One can continue in like fashion—by hand or in a computer algebra system—and obtain $\zeta(6) = \pi^6/945, \zeta(8) = \pi^8/9450, \zeta(10) = \pi^{10}/93555$, and we discover that $\zeta(2n) = q_n\,\pi^{2n}$, for some rational q_n, and it is clearly time to be better organized.

To do this we, introduce the even *Bernoulli numbers* via the generating function

$$\frac{z}{e^z-1} + \frac{z}{2} = \sum_{m=0}^{\infty} B_{2m} \frac{z^{2m}}{(2m)!}, \qquad |z| \leq 2\pi, \tag{3.13}$$

and define $B_1 = -1/2$, $B_{2n+1} = 0$ for $n > 0$. It is easy to discover, equivalently, that

$$\sum_{k=0}^{n} \binom{n+1}{k} B_k = 0, \tag{3.14}$$

and that

$$\pi z \cot(\pi z) = \sum_{m=0}^{\infty} (-1)^{m+1} B_{2m} \frac{(2\pi z)^{2m}}{(2m)!}. \tag{3.15}$$

Returning to (1.11) and differentiating logarithmically, we obtain

$$\pi \cot(\pi z) = \frac{1}{z} - \sum_{n=1}^{\infty} \frac{2z}{n^2 - z^2} = \sum_{n=1}^{\infty} \zeta(2n) z^{2n-1}, \tag{3.16}$$

where the second identity comes from repeated use of the geometric series. Comparing coefficients in (3.15) and (3.16) yields

$$\zeta(2m) = (-1)^m B_{2m} \frac{(2\pi)^{2m}}{2\,(2m)!}. \tag{3.17}$$

Using the reflection formula (3.9) we deduce that for nonnegative n

$$\zeta(-2n+1) = -\frac{B_{2n}}{2n}.$$

As the coefficient of B_{2n} in (3.14) is nonzero, namely $(n+1)$, equation (3.14) is a practical formula for generating Bernoulli numbers. The first few Bernoulli numbers are

$$\frac{1}{6}, -\frac{1}{30}, \frac{1}{42}, -\frac{1}{30}, \frac{5}{66}, -\frac{691}{2730}.$$

Note that if one did too little computation, one might come away with the impression that the numerator is always "1." Actually, the numerator is as hard to compute as the number, but a lovely theorem of Karl von Staudt and Thomas Clausen proves that the fractional parts of B_{2n} and $\sigma_{2n} = \sum_{p-1|2n} 1/p$ agree. Thus B_{2n} and σ_{2n} have the same denominators, and the later can be very quickly computed, even for large n. The first 15 even values are

$$1, 6, 30, 42, 30, 66, 2730, 6, 510, 798, 330, 138, 2730, 6, 870, 14322$$

and the mystery as to why terms re-occur is explained by von Staudt's result.

3.2.2   Zeta at Odd Positive Integers

It was only in 1976 that Roger Apéry proved that $\zeta(3)$ is irrational. As of the end of 2002, it is known that one of the next four odd zeta values is irrational and that infinitely many are, but we cannot prove that $\zeta(5)$ is. They certainly are not simple rational multiples of powers of π, a fact that can be determined by integer relation computations.

Thanks to Apéry, who used the series for $\zeta(3)$ below in his work, it is now well known that

$$\zeta(2) = 3\sum_{k=1}^{\infty} \frac{1}{k^2\binom{2k}{k}},$$

$$\zeta(3) = \frac{5}{2}\sum_{k=1}^{\infty} \frac{(-1)^{k-1}}{k^3\binom{2k}{k}},$$

$$\zeta(4) = \frac{36}{17}\sum_{k=1}^{\infty} \frac{1}{k^4\binom{2k}{k}}.$$

See [54] for further details. These results make it tempting to conjecture that

$$Z_5 = \zeta(5)/\sum_{k=1}^{\infty} \frac{(-1)^{k-1}}{k^5\binom{2k}{k}}$$

is a simple rational or algebraic number. Sadly (or happily), we may use PSLQ to determine that *if Z_5 satisfies a polynomial of degree ≤ 25, the Euclidean norm of coefficients exceeds 2×10^{37}*. This is based on a 1000 digit computation—even higher bounds can be obtained if desired. Hence, any relatively prime integers p and q satisfying

$$\zeta(5) \stackrel{?}{=} \frac{p}{q}\sum_{k=1}^{\infty} \frac{(-1)^{k+1}}{k^5\binom{2k}{k}}$$

have astronomically large q.

But a positive use of PSLQ yields in terms of *the first polylogarithms*:

$$\sum_{k=1}^{\infty} \frac{(-1)^{k+1}}{k^5\binom{2k}{k}} = 2\zeta(5) - \tfrac{4}{3}L^5 + \tfrac{8}{3}L^3\zeta(2) + 4L^2\zeta(3) \qquad (3.18)$$

$$+ 80\sum_{n>0}\left(\frac{1}{(2n)^5} - \frac{L}{(2n)^4}\right)\rho^{2n},$$

where $L = \log(\rho)$ and $\rho = (\sqrt{5}-1)/2$; with similar formulae for A_4 and A_6 (series with alternating signs), S_5, S_6 and S_7 (series with nonalternating signs) [54].

A less well known formula for $\zeta(5)$, due to Koecher, suggested generalizations for $\zeta(7), \zeta(9), \zeta(11), \cdots$. Again, the coefficients were found by integer relation algorithms. The technique of *bootstrapping* the earlier pattern kept size of the search space manageable. Note that the requisite sums converge relatively quickly, and so are easy to compute even to high precision, which is needed for large relation searches.

For example,

$$\zeta(7) = \frac{5}{2}\sum_{k=1}^{\infty} \frac{(-1)^{k+1}}{k^7\binom{2k}{k}} + \frac{25}{2}\sum_{k=1}^{\infty} \frac{(-1)^{k+1}}{k^3\binom{2k}{k}}\sum_{j=1}^{k-1}\frac{1}{j^4}. \qquad (3.19)$$

The authors of [36] were able, by finding integer relations for $n = 1, 2, \cdots, 10$, to encapsulate the formulae for $\zeta(4n+3)$ in a single conjectured generating function, entirely *ex machina*:

Theorem 3.2. *For any complex* $|z| < 1$, *we have, formally,*

$$\sum_{n=0}^{\infty} \zeta(4n+3)z^{4n} = \sum_{k=1}^{\infty} \frac{1}{k^3(1-z^4/k^4)} \tag{3.20}$$

$$= \frac{5}{2}\sum_{k=1}^{\infty} \frac{(-1)^{k-1}}{k^3\binom{2k}{k}(1-z^4/k^4)} \prod_{m=1}^{k-1} \frac{1+4z^4/m^4}{1-z^4/m^4}.$$

The first "=" is easy. The second is quite unexpected in its form! Thus, $z = 0$ yields Apéry's formula for $\zeta(3)$ and the coefficient of z^4 yields (3.19).

It is instructive to mention how Theorem 3.2 was discovered. The first ten cases show (3.20) has the form

$$\frac{5}{2}\sum_{k\geq 1} \frac{(-1)^{k-1}}{k^3\binom{2k}{k}} \frac{P_k(z)}{(1-z^4/k^4)},$$

for undetermined P_k; with abundant data to compute

$$\boxed{P_k(z) = \prod_{m=1}^{k-1} \frac{1+4z^4/m^4}{1-z^4/m^4}.}$$

Many reformulations of (3.20) were found, including a marvellous finite sum:

$$\boxed{\sum_{k=1}^{n} \frac{2n^2}{k^2} \frac{\prod_{i=1}^{n-1}(4k^4+i^4)}{\prod_{i=1,\,i\neq k}^{n}(k^4-i^4)} = \binom{2n}{n}.} \tag{3.21}$$

This was obtained via Gosper's *telescoping algorithm* of Wilf-Zeilberger type after a mistake in an electronic Petri dish—when a TEX "infty" was typed instead of "infinity" and *Maple* returned an answer that suggested it "knew" an algorithm for such finite sums.

This identity was subsequently proved by Almkvist and Granville [6], thus finishing the proof of (3.20) and giving a rapidly converging series for any $\zeta(4N+3)$, where N is positive integer. And perhaps shedding light on the irrationality of $\zeta(7)$?

Paul Erdős, when shown (3.21) shortly before his death, rushed off. Twenty minutes later he returned saying he did not know how to prove it, but if proven it would have implications for Apéry's result ("$\zeta(3)$ is irrational").

The failure to discover a similar function for $\zeta(4n+1)$ rests largely on the fact that *too many* relations were found by computer, and no candidate to behave like (3.2) was isolated to generalize the initial cases such as

$$\zeta(5) = 2\sum_{k=1}^{\infty} \frac{(-1)^{k+1}}{k^5\binom{2k}{k}} - \frac{5}{2}\sum_{k=1}^{\infty} \frac{(-1)^{k+1}}{k^3\binom{2k}{k}} \sum_{j=1}^{k-1} \frac{1}{j^2}. \qquad (3.22)$$

3.2.3   A Taste of Ramanujan

Ramanujan obtained almost analogous evaluations of $\zeta(2n+1)$. For $M \equiv 3 \bmod 4$,

$$\zeta(4N+3) = -2\sum_{k\geq 1} \frac{1}{k^{4N+3}\left(e^{2\pi k} - 1\right)}$$
$$+ \frac{2}{\pi}\left\{ \frac{4N+7}{4}\zeta(4N+4) - \sum_{k=1}^{N} \zeta(4k)\zeta(4N+4-4k) \right\},$$

where the interesting term is the rapidly convergent hyperbolic trigonometric series, while the term in braces is a rational multiple of π^{4N+4}. Correspondingly, for $M \equiv 1 \bmod 4$,

$$\zeta(4N+1) = -\frac{2}{N}\sum_{k\geq 1} \frac{(\pi k + N)e^{2\pi k} - N}{k^{4N+1}(e^{2\pi k} - 1)^2}$$
$$+ \frac{1}{2N\pi}\left\{ (2N+1)\zeta(4N+2) + \sum_{k=1}^{2N}(-1)^k 2k\zeta(2k)\zeta(4N+2-2k) \right\}.$$

In each case, only a finite set of $\zeta(2N)$ values is required, and the full precision value e^π is re-used throughout. The number e^π is the easiest transcendental number to rapidly compute (see Exercise 7 of Chapter 3 in [43]).

For $\zeta(4N+1)$, a "nicer" series has recently been decoded and then proved from a few PSLQ experiments of Plouffe. It is equivalent to:

$$\left\{ 2 - (-4)^{-N} \right\}\sum_{k=1}^{\infty} \frac{\coth(k\pi)}{k^{4N+1}} - (4)^{-2N}\sum_{k=1}^{\infty} \frac{\tanh(k\pi)}{k^{4N+1}} = Q_N \times \pi^{4N+1}. \qquad (3.23)$$

The quantity Q_N in (3.23) is an explicit rational:

$$Q_N = -\sum_{k=0}^{2N+1} \frac{B_{4N+2-2k}B_{2k}}{(4N+2-2k)!(2k)!} \times \left\{ (-1)^{\binom{k+1}{2}}(-4)^N 2^k + (-4)^k \right\}. \qquad (3.24)$$

This was also discovered using integer relation methods. For instance,

$$\frac{9}{4}\sum_{k=1}^{\infty}\frac{\coth(\pi k)}{k^5} - \frac{1}{16}\sum_{k=1}^{\infty}\frac{\tanh(\pi k)}{k^5} = \frac{5}{672}\pi^5.$$

On substituting

$$\tanh(x) = 1 - \frac{2}{\exp(2x)+1}, \qquad \coth(x) = 1 + \frac{2}{\exp(2x)-1}$$

in (3.23), one may solve for $\zeta(4N+1)$. For example:

$$\zeta(5) = \frac{1}{294}\pi^5 + \frac{2}{35}\sum_{k=1}^{\infty}\frac{1}{(1+e^{2k\pi})k^5} + \frac{72}{35}\sum_{k=1}^{\infty}\frac{1}{(1-e^{2k\pi})k^5},$$

and $\zeta(5) - \pi^5/294 = -0.0039555\ldots$

3.3   Other L-Series

The function

$$\beta(s) = \mathrm{L}_{-4}(s) = \sum_{n=0}^{\infty}\frac{(-1)^n}{(2n+1)^s}, \tag{3.25}$$

is sometimes known as the *Catalan zeta function*, since $\beta(2) = G$ is Catalan's constant, perhaps the simplest number whose irrationality is unproven. It is independent of ζ being based on the multiplicative character modulo 4 which takes values $0, 1, 0, -1$. It is the simplest example of a primitive Dirichlet L-series as arrives when one studies primes in arithmetic progression.

In this case, the generating function

$$\sec(z) = \sum_{m=0}^{\infty}\frac{|E_{2m}|z^{2m}}{(2m)!}, \qquad |z| \le \frac{\pi}{2}, \tag{3.26}$$

with $E_{2n+1} = 0$ for $n \ge 0$, defines the *Euler numbers*—some authors label our E_{2n} as E_n, which we met in Chapter 2 of [43]. Correspondingly,

$$\sum_{k=0}^{n}\binom{2n}{2k}E_{2k} = 0 \tag{3.27}$$

and

$$\pi\sec(\pi z) = \sum_{m=0}^{\infty}4^{m+1}\beta(2m+1)z^{2m}. \tag{3.28}$$

Comparing coefficients in (3.28) and (3.26) yields

$$\beta(2m+1) = |E_{2m}| \frac{\left(\frac{\pi}{2}\right)^{2m+1}}{2\,(2m)!}, \tag{3.29}$$

so that in this case, it is the odd values that are tractable:

$$\beta(1) = \frac{\pi}{4}, \quad \beta(3) = \frac{\pi^3}{32}, \quad \beta(5) = \frac{5\pi^5}{1536}, \quad \beta(7) = \frac{61\pi^7}{184320},$$

while the first six even Euler numbers are

$$1, -1, 5, -61, 1385, -50521, 2702765.$$

For integer k, an important class of Dirichlet series is given by

$$L_{\pm k}(s) = \sum_{n=1}^{\infty} \frac{\left(\frac{\pm k}{n}\right)}{n^s}, \tag{3.30}$$

where $\left(\frac{\pm k}{n}\right)$ is the Legendre-Jacobi symbol. When n is prime, the Legendre symbol is defined as $\left(\frac{m}{n}\right) = 1$ if m is a quadratic residue modulo n (i.e., m is a perfect square modulo n), and $= -1$ otherwise. When $n = p_1 p_2, \cdots p_r$ for odd primes p_i not necessarily distinct, then $\left(\frac{m}{n}\right)$ is defined as $\left(\frac{m}{p_1}\right) \left(\frac{m}{p_2}\right) \cdots \left(\frac{m}{p_r}\right)$. Using this notation,

$$L_{-8}(s) = 1 + \frac{1}{3^s} - \frac{1}{5^s} - \frac{1}{7^s} + \cdots$$

and

$$L_{+8}(s) = 1 - \frac{1}{3^s} - \frac{1}{5^s} + \frac{1}{7^s} + \cdots.$$

More generally, for any *multiplicative character* χ, one can define

$$L_\chi(s) = \sum_{n=1}^{\infty} \frac{\chi(n)}{n^s}. \tag{3.31}$$

The previous case corresponds to characters modulo k. Then $L_\chi(s)$ has a corresponding Euler product

$$L_\chi(s) = \prod_p \left(1 - \frac{\chi(p)}{p^s}\right)^{-1}, \tag{3.32}$$

where p runs over the primes.

For *primitive* characters modulo $d > 0$ (see [148, 44]), there is a functional equation analogous to that for zeta given in (3.9):

$$L_{-d}(s) = C(s) \cos\left(\frac{s\pi}{2}\right) L_{-d}(1-s) \tag{3.33}$$

$$L_{+d}(s) = C(s) \sin\left(\frac{s\pi}{2}\right) L_{+d}(1-s), \tag{3.34}$$

where $C(s) = 2^s \pi^{s-1} k^{-s+1/2} \Gamma(1-s)$. Moreover, the *Dirichlet class number formulae* are

$$L_{-d}(1) = \frac{2\pi}{\sqrt{d}} \frac{h(-d)}{w(d)} \tag{3.35}$$

$$L_d(1) = \frac{2h(d)}{\sqrt{d}} \log(\varepsilon(d)), \tag{3.36}$$

where $h(-d)$ is the class number of $Q(\sqrt{D})$, where $d = D$ when d is congruent to 1 modulo 4 and $d = -D$ otherwise, $\varepsilon(d)$ is a fundamental unit in $Q(\sqrt{D})$, and $w(d) = 2$ except that $w(1) = 4$ and $w(3) = 6$.

For our present purposes it suffices that we know that $L_{-d}(1)/\pi$ satisfies a quadratic equation and $L_d(1)/\pi \sqrt{d}$ is a rational multiple of the logarithm of a quadratic surd, as we can explore what the values are with integer relation methods—and hunt for evaluations at other odd integers s.

We illustrate with a few cases. Working with only about 20 digits precision, we find $L_{-3}(1) = \sqrt{3}\pi/9$, $L_{-3}(3) = 4\sqrt{3}\pi^3/343$, $L_{-8}(1) = \sqrt{2}\pi/4$, $L_{-8}(3) = 3\sqrt{2}\pi^3/128$, $L_{-12}(1) = \sqrt{3}\pi^3/6$, $L_{-12}(3) = \sqrt{3}\pi^3/45$ and in confirmation $L_{-4}(3) = \pi^3/32$. Also, $L_{-67}(1) = \sqrt{67}\pi/67$ and $L_{-163}(1) = \sqrt{163}\pi/163$ are particularly simple, since these are the largest cases where the corresponding imaginary quadratic field has unique factorization.

Similarly, $L_{+5}(1) = 2\sqrt{5}/5 \cdot \log((1+\sqrt{5})/2)$, $L_{+8}(1) = \sqrt{2}/2 \cdot \log(1+\sqrt{2})$, while $L_{+13}(1) = 2\sqrt{13}/13 \cdot \log((3 + \sqrt{13})/2)$ and $L_{+29}(1) = 2\sqrt{29}/29 \cdot \log((35 + \sqrt{29})/2)$.

3.4   Multizeta Values

Euler sums or *MZVs* ("multiple zeta values" or "multizeta values") are wonderful generalizations of the classical ζ function. For natural numbers $i_1, i_2, \cdots, i_k,$

$$\zeta(i_1, i_2, \cdots, i_k) = \sum_{n_1 > n_2 > \cdots > n_k > 0} \frac{1}{n_1^{i_1} n_2^{i_2} \cdots n_k^{i_k}}. \tag{3.37}$$

Thus $\zeta(a) = \sum_{n\geq 1} n^{-a}$ is as before, and

$$\zeta(a,b) = \sum_{n=1}^{\infty} \frac{1 + \frac{1}{2^b} + \cdots + \frac{1}{(n-1)^b}}{n^a}.$$

In general, the integer k is the sum's *depth*, and $i_1 + i_2 + \cdots + i_k$ is its *weight*.

This definition (3.37) clearly extends to alternating and character sums as we shall see below. MZVs have recently found interesting interpretations in fields as diverse as high energy physics, knot theory, and combinatorics. Such MZVs satisfy many striking identities, of which

$$\zeta(2,1) = \zeta(3) \qquad \text{and} \qquad 4\zeta(3,1) = \zeta(4) \qquad (3.38)$$

are the simplest. Thus these double zeta sums can be reduced to values of the classical zeta function! Does this happen just in a few special cases, or is this the tip of an iceberg, the beginning of a theory? We would like to answer questions such as: Which multiple zeta values can be reduced to simpler ones, i.e., to rational combinations of MZVs of lower depth? How many irreducible MZVs remain for given depth and weight? Can the relations between different MZVs be sorted, labeled and classified?

The needed computations quickly become very large in scale, requiring mixing fields, tools, and interfaces such as Reduce, C++, Fortran, Pari, Snap, etc. A high-precision *fast ζ-convolution* allows use of integer relation algorithms leading to important dimensional (reducibility) conjectures and amazing identities. An algorithm for computing these values is given in Section 7.5.1. See also the Euler-zeta computation tool at the URL http://www.cecm.sfu.ca/projects/ezface+. Euler himself found and partially proved theorems on reducibility of depth 2 to depth 1 zetas ($\zeta(6,2)$ is the lowest weight that is "irreducible").

3.4.1   Various Methods of Attack

One of the pleasures of work in this area is that so many methods are useful: combinatorial, analytic (complex and real), algebraic, number theoretic, and numerical. This leads to amazing identities and important dimensional (reducibility) conjectures. Almost certainly, the simplest of our dimensional conjectures are not provable by currently known mathematical techniques. As we emphasized above, we can't yet determine whether $\zeta(5), \zeta(7)$ or G are rational or irrational, much less transcendental.

A high point of this chapter is the proof of a conjecture of Zagier first published in [36, 52] to which we refer along with [51] for general information not given here. The proof we shall give in Section 3.7 is a refinement

due to Zagier of one found by Broadhurst, during the development of the
joint corpus in [63, 52]. The identity is

$$\zeta(\{3,1\}_n) = \frac{1}{2n+1}\zeta(\{2\}_{2n}) \qquad \left(= \frac{2\pi^{4n}}{(4n+2)!} \right). \qquad (3.39)$$

Here $\{s\}_n$ denotes the string s repeated n times. This is the *unique* non-commutative analogue of Euler's evaluation of $\zeta(2n)$.

We illustrate the diversity of the area with a deep conjecture that sits
as a very special case of various dimensional conjectures we make below
and provide evidence for:

Conjecture 3.3. (Drinfeld(1991)-Deligne). *The graded Lie algebra
of Grothendieck and Teichmuller has no more than one generator in odd
degrees, and no generators in even degrees.*

In the known nonreducible identities for Euler sums, all ζ-terms have
the same weight. This is of great importance for guided integer relation
searches, as it dramatically reduces the size of the search space. It is,
moreover, useful to consider more general sums:

$$\zeta(i_1, i_2, \cdots, i_k \, ; \, \sigma_1, \sigma_2, \cdots, \sigma_k) = \sum_{n_1 > n_2 > \cdots > n_k > 0} \frac{\sigma_1^{n_1} \sigma_2^{n_2} \cdots \sigma_k^{n_k}}{n_1^{i_1} n_2^{i_2} \cdots n_k^{i_k}}. \qquad (3.40)$$

For general complex σ_i, (3.40) defines *Eulerian polylogarithms*, while $\sigma_i \in \{1, -1\}$ produce *Euler sums*. We restrict the term *multi zeta value* (MZV)
to:

$$\zeta(i_1, i_2, \cdots, i_k) = \sum_{n_1 > n_2 > \cdots > n_k > 0} \frac{1}{n_1^{i_1} n_2^{i_2} \cdots n_k^{i_k}},$$

that is, when $\sigma_i \equiv 1$.

3.4.2   Reducibility and Dimensional Conjectures

As a first taste, we prove the following lemma due to Euler:

Lemma 3.4.
$$\zeta(a,b) + \zeta(b,a) = \zeta(a)\zeta(b) - \zeta(a+b),$$
for integer $a > 1, b \geq 1$. In particular,

$$\zeta(a,a) = \frac{1}{2}\zeta(a)^2 - \frac{1}{2}\zeta(2a)$$

reduces $\zeta(a,a)$.

Proof. Observe that

$$\zeta(a,b) + \zeta(b,a) + \zeta(a+b)$$

$$= \sum_{n=1}^{\infty} \sum_{m>n}^{\infty} \frac{1}{n^a \, m^b} + \sum_{n=1}^{\infty} \sum_{m<n}^{\infty} \frac{1}{n^a \, m^b} + \sum_{n=1}^{\infty} \sum_{n=m}^{\infty} \frac{1}{n^a \, m^b}$$

$$= \sum_{n=1}^{\infty} \sum_{n=1}^{\infty} \frac{1}{n^a \, m^b} = \zeta(a)\zeta(b). \tag{3.41}$$

$\square$

More complex versions of this sort of argument often involving partial fraction identities lead to sets of equations (and to the matrices in Section 4.5), and to algebraic proofs of many cognate MZV identities. For example,

$$\zeta(a,b,c) + \zeta(a,c,b) + \zeta(c,a,b) = \zeta(c)\zeta(a,b) - \zeta(a,b+c) - \zeta(a+c,b).$$

As a second taste, we show how a computer algebra system can "prove" Euler's first significant result in the area. Indeed, "generatingfunctionology" produces:

$$\frac{(-1)^m}{(m-1)!} \int_0^1 \frac{\log^{m-1}(t) \sum_{n>0} a_n \, t^n}{1-t} \, dt = \sum_{n>0} \frac{\sum_{k<n} a_k}{n^m},$$

and so represents $\zeta(m,1)$ as the integral

$$\begin{aligned}
\zeta(m,1) &= \frac{(-1)^m}{(m-1)!} \int_0^1 \frac{\log^{m-1}(t) \, \log(1-t)}{1-t} \, dt \\
&= \frac{(-1)^m}{2(m-1)!} \int_0^1 \frac{(m-1) \, \log^{m-2}(t) \, \log^2(1-t)}{t} \, dt. \quad (3.42)
\end{aligned}$$

Inspection of the definition of the beta function

$$\beta(x,y) = \int_0^1 t^{x-1}(1-t)^{y-1} \, dt \tag{3.43}$$

shows that the right side of (3.42) can be written as a β-function derivative

$$\zeta(m,1) = \frac{(-1)^m}{2(m-2)!} B_1^{(m-2)}(0), \tag{3.44}$$

where $B_1(y) = \frac{\partial^2}{\partial x^2} \beta(x,y)\Big|_{x=1}$. Since

$$\frac{\partial^2}{\partial x^2} \beta(x,y) = \beta(x,y) \left[(\Psi(x) - \Psi(x+y))^2 + (\Psi'(x) - \Psi'(x+y)) \right],$$

we have a *digamma* representation via

$$B_1(y) = \frac{1}{y}\left((-\gamma - \Psi(y+1))^2 + (\zeta(2) - \Psi'(y+1))\right).$$

Indeed, without going beyond (3.44), we may implement (3.44) in *Maple* or *Mathematica* very painlessly and *discover* its Riemann ζ-function reduction,

$$\begin{aligned}
\zeta(n,1) &= \sum_{k=1}^{\infty}\left(1 + \frac{1}{2} + \cdots + \frac{1}{k}\right)(k+1)^{-n} \\
&= \frac{n\zeta(n+1)}{2} - \frac{1}{2}\sum_{k=1}^{n-2}\zeta(n-k)\zeta(k+1),
\end{aligned}$$

from the first five or ten symbolic values. Moreover, each case computed is a proof. Note that one wants to ensure that *Maple* or *Mathematica* does not evaluate $\zeta(2n)$, otherwise when $n+1$ is even a convolution will not be exposed:

$$6\zeta^2(4) + 6\zeta(2)\zeta(6) = \frac{17}{9450}\pi^8 = 17\zeta(8).$$

To make the notion of reduction more explicit, a key problem is to find the *dimension* of a minimal generating set for a $(Q, +, \cdot)$-algebra that contains

1. all Euler sums of weight n and depth k, generated by Euler sums, $E_{n,k}$;

2. all MZVs of weight n and depth k, generated by Euler sums, $E_{n,k}$; or

3. all MZVs of weight n and depth k, generated by MZVs, $D_{n,k}$.

Conjectured generating functions, due to Broadhurst-Kreimer, Zagier, and others, are:

$$\prod_{n\geq 3}\prod_{k\geq 1}\left(1 - x^n y^k\right)^{E_{n,k}} \overset{?}{=} 1 - \frac{x^3 y}{(1-x^2)(1-xy)}$$

$$\prod_{n\geq 3}\prod_{k\geq 1}\left(1 - x^n y^k\right)^{M_{n,k}} \overset{?}{=} 1 - \frac{x^3 y}{1 - x^2}$$

$$\prod_{n\geq 3}\prod_{k\geq 1}\left(1 - x^n y^k\right)^{D_{n,k}} \overset{?}{=} 1 - \frac{x^3 y}{1 - x^2} + \frac{x^{12}y^2(1-y^2)}{(1-x^4)(1-x^6)}.$$

For $k = 2$, n odd, and $k = 3$, n even, the result implicit for $D_{n,k}$ can be proven by "elementary methods." Note that $D_{n,k}$ has a disconcertingly complicated conjectured rational generating function.

In the next example and elsewhere, we sometimes write $-s_i$ or $\overline{s_i}$ to denote alternation in the i-th position.

Example 3.5. Two MZV reductions.

MZV over MZVs: An example of a sum that reduces is

$$\zeta(4,1,3) = -\zeta(5,3) + \frac{71}{36}\zeta(8) - \frac{5}{2}\zeta(5)\zeta(3) + \frac{1}{2}\zeta(3)^2\zeta(2).$$

MZV over Euler sums: $\zeta(4,2,4,2)$ is irreducible as an MZV, but as an Euler sum we have

$$\zeta(4,2,4,2) = -\frac{1024}{27}\zeta(-9,-3) - \frac{267991}{5528}\zeta(12) - \frac{1040}{27}\zeta(9,3) - \frac{76}{3}\zeta(9)\zeta(3)$$

$$-\frac{160}{9}\zeta(7)\zeta(5) + 2\zeta(6)\zeta(3)^2 + 14\zeta(5,3)\zeta(4) + 70\zeta(5)\zeta(4)\zeta(3) - \frac{1}{6}\zeta(3)^4.$$

However, $\zeta(5,3)$, $\zeta(-9,-3)$ are irreducible over the Euler sums. □

$E_{n,k}$ k	1	2	3	4	5	6
n						
3	1					
4		1				
5	1		1			
6		1		1		
7	1		2		1	
8		2		2		1
9	1		3		3	
10		2		5		3
11	1		5		7	
12		3		8		9
13	1		7		14	
14		3		14		20
15	1		9		25	
16		4		20		42
17	1		12		42	
18		4		30		75
19	1		15		66	
20		5		40		132

Table 3.1. Values of $E_{n,k}$ for various n and k.

Tables 3.1, 3.2, and 3.3 give values of these three dimensions. The tools used included partial fractions, functional equations, and the "shuffles

$M_{n,k}$ k / n	1	2	3	4	5	6
3	1					
4						
5	1					
6						
7	1					
8		1				
9	1					
10		1				
11	1	1				
12		2				
13	1	2				
14		2	1			
15	1	3				
16		3	2			
17	1	5	1			
18		3	5			
19	1	7	3			
20		4	8	1		

Table 3.2. Values of $M_{n,k}$ for various n and k.

$D_{n,k}$ k / n	1	2	3	4	5	6
3	1					
4						
5	1					
6						
7	1					
8		1				
9	1					
10		1				
11	1	1				
12		1	1			
13	1	2				
14		2	1			
15	1	2	1			
16		2	3			
17	1	4	2			
18		2	5	1		
19	1	5	5			
20		3	7	3		

Table 3.3. Values of $D_{n,k}$ for various n and k.

algebra" [36, 52]. The generating functions have been confirmed *numerically*, and a sizable subset is proven symbolically, in the following ranges:

1. $E_{n,k}$ (with REDUCE and PSLQ): $k = 2$ and $n \leq 44$ and $k = 7$ and $n \leq 8$.

2. $M_{n,k}$ (with REDUCE and PSLQ): $k = 2$ and $n \leq 17$ and $k = 7$ and $n \leq 20$.

3. $D_{n,k}$, modulo a big prime (with REDUCE and Fortran): $k = 3$ and $n \leq 141$ and $k = 7$ and $n \leq 21$; (with Fortran): $k = 3$ and $n \leq 161$ and $k = 7$ and $n \leq 23$.

Hybrid code, based on symbolic evaluation of identities and on PSLQ, allowing exact reduction, has been run for (1) all alternating (Euler) sums to weight 9; and (2) all MZV's to weight 14. Thus all these evaluations are fully established, as are many more scattered in the literature.

3.5 Double Euler Sums

A natural first generalization of the ζ function, initially studied by Euler, is to let $\zeta(t, s) = \sum_{n=1}^{\infty} \sum_{k=1}^{n-1} k^{-s} n^{-t}$. Such sums, which we first encountered in Lemma 3.4, are called *double Euler sums*.

As a more detailed foray into Euler sums, it is our intention to prove the following evaluation, which is due to Euler, in a self-contained fashion:

$$\zeta(t, s) = \frac{1}{2} \left[\binom{s+t}{s} - 1 \right] \zeta(s+t) + \zeta(s)\zeta(t) \tag{3.45}$$
$$- \sum_{j=1}^{n} \left[\binom{2j-2}{s-1} + \binom{2j-2}{t-1} \right] \zeta(2j-1)\zeta(s+t-2j+1),$$

if s is odd, and

$$\zeta(t, s) = -\frac{1}{2} \left[\binom{s+t}{s} + 1 \right] \zeta(s+t) \tag{3.46}$$
$$+ \sum_{j=1}^{n} \left[\binom{2j-2}{s-1} + \binom{2j-2}{t-1} \right] \zeta(2j-1)\zeta(s+t-2j+1),$$

if s is even and $s+t$ odd (see [30]). The terms involving $\zeta(1)$ which he used here can be cancelled formally if $t > 1$. Euler obtained his evaluation by computing many examples ($s+t \leq 13$), and then extrapolating the general formula, without actual proof. The proof we give is not the simplest, but it allows us to play interestingly with combinatorial matrices, a subject we revisit in the next chapter.

We need the following lemma which was known already to Euler and can be proved by induction on $s + t$.

Lemma 3.6. *Define*

$$A_j^{(s,t)} = \binom{s+t-j-1}{s-j} \ and \ B_j^{(s,t)} = \binom{s+t-j-1}{t-j} .$$

Then we have the partial fraction decomposition

$$\frac{1}{x^s(1-x)^t} = \sum_{j=1}^{s} \frac{A_j^{(s,t)}}{x^j} + \sum_{j=1}^{t} \frac{B_j^{(s,t)}}{(1-x)^j},$$

for $s, t \geq 0$, $s + t \geq 1$.

We will now derive systems of linear equations for the values $\zeta(s,t)$, where $s + t = N$, a constant. First, as we have already seen (Lemma 3.4), there is a simple relation between $\zeta(s,t)$ and $\zeta(t,s)$:

$$\zeta(s,t) + \zeta(t,s) = \zeta(s)\zeta(t) - \zeta(s+t)$$

for $s, t \geq 2$. We will refer to this equation and its equivalents as "reflection formulas." It follows that $2\zeta(s,s) = \zeta^2(s) - \zeta(2s)$. Second, we have

$$\begin{aligned}
\zeta(s)\zeta(t) &= \left(\sum_{k=1}^{\infty} \frac{1}{k^s}\right) \cdot \left(\sum_{n=1}^{\infty} \frac{1}{n^t}\right) = \sum_{n=1}^{\infty} \sum_{k=1}^{n-1} \frac{1}{k^s(n-k)^t} \\
&= \sum_{n=1}^{\infty} \sum_{k=1}^{n-1} \left(\sum_{j=1}^{s} \frac{A_j^{(s,t)}}{n^{s+t-j}k^j} + \sum_{j=1}^{t} \frac{B_j^{(s,t)}}{n^{s+t-j}(n-k)^j}\right) \\
&= \sum_{j=1}^{s} A_j^{(s,t)} \zeta(s+t-j,j) + \sum_{j=1}^{t} B_j^{(s,t)} \zeta(s+t-j,j),
\end{aligned}$$

for $s, t \geq 2$, with $A_j^{(s,t)}$ and $B_j^{(s,t)}$ defined as in Lemma 3.6. We will refer to these equations as "decomposition formulas."

The following alternative version of the decomposition formulas can be used, together with the reflection formulas, to prove Euler's formulas (3.45

and 3.46) algebraically (which we do not do explicitly here):

$$\zeta(t,s) = (-1)^s \sum_{j=2}^{s} (-1)^j A_j^{(s,t)} \zeta(j)\zeta(s+t-j) + (-1)^s \sum_{j=2}^{t} B_j^{(s,t)} \zeta(s+t-j,j)$$
$$+ (-1)^s \binom{s+t-2}{s-1} (\zeta(s+t-1,1) + \zeta(s+t)).$$

We will now distinguish the two cases $s + t$ odd and $s + t$ even. First, we treat the case where $s + t = 2n + 1$. We have $2n - 2$ equations in the $2n - 2$ unknowns $\zeta(2, 2n - 1), \zeta(3, 2n - 2), \cdots, \zeta(2n - 3, 2)$. We can reduce the $\zeta(k, 2n+1-k)$ with $k > n$ to $\zeta(2n+1-k, k)$ by the reflection formulas. This leaves us with the $n - 1$ unknowns $\zeta(2, 2n - 1), \cdots, \zeta(n, n + 1)$. The matrix that corresponds to these equations has the entries

$$(A_j^{(k,2n+1-k)} + B_j^{(k,2n+1-k)} - B_{2n+1-k}^{(k,2n+1-k)})_{j,k=2,\cdots,n}.$$

However, it will simplify matters considerably if we augment this matrix by allowing j, k to run from 1 to n, and then multiply alternate rows by -1.

Define, therefore, the $n \times n$ matrices A, B, C, M by

$$A_{kj} = (-1)^{k+1} \binom{2n-j}{2n-k},$$
$$B_{kj} = (-1)^{k+1} \binom{2n-j}{k-1},$$
$$C_{kj} = (-1)^{k+1} \binom{j-1}{k-1}$$

$(k, j = 1, \cdots, n)$ and $M = A + B - C$. Define the n vector r by $r_1 = 0$ and

$$r_k =$$
$$(-1)^{k+1} \left[\zeta(k)\zeta(2n+1-k) + \sum_{i=k}^{n} \binom{i-1}{k-1}(\zeta(2n+1) - \zeta(2n+1-i)\zeta(i)) \right]$$

for $k = 2, \cdots, n$. We then have to solve the system $Mx = r$, where the vector x has $x_1 = 0$, and $x_k = \zeta(k, 2n+1-k)$ for $k > 1$.

We shall need the following lemma which is known and can be proved by induction on m.

Lemma 3.7. (i) *For $0 \le \mu \le m$,* $\displaystyle\sum_{i=0}^{\mu} \binom{m-\mu+i}{i} = \binom{m+1}{\mu}.$

(ii) *For $0 \leq \nu, \mu \leq m - 1$,*

$$\sum_{i=1}^{m}(-1)^{i+1}\binom{m-i}{\nu}\binom{m-\mu-1}{i-1} = \binom{\mu}{m-\nu-1}.$$

Setting $\mu = 0$ in (ii) and changing the order of summation yields

$$\sum_{i=1}^{m}(-1)^{i+1}\binom{i-1}{\nu-1}\binom{m-1}{i-1} = (-1)^{m+1}\delta_{m\nu}.$$

Lemma 3.7(ii) can be used to prove the following matrix identities.

$$A^2 = C^2 = I, \tag{3.47}$$

$$A = BC, \quad B = AC, \quad C = AB, \tag{3.48}$$

$$B^2 = CA, \quad CB = BA. \tag{3.49}$$

It follows from these identities that $B^3 = BCA = AA = I$, and that the matrix group generated by A, B, and C is the permutation group on 3 symbols. We shall not use this, but we will discuss this from a different perspective in Section 4.5, where we also prove (3.47), (3.48), and (3.49).

These matrix identities now allow us to show that M is invertible; in fact, we have

$$\begin{aligned}
M^2 &= AA + AB - AC + BA + BB - BC - CA - CB + CC \\
&= I + C - B - A + I = 2I - M,
\end{aligned}$$

so that

$$M^{-1} = \frac{1}{2}(M + I).$$

Thus, to prove Euler's formula, it remains to determine $M^{-1}r$. For this purpose, define $p_1 = p_{2n} = 0$, $p_k = \zeta(k)\zeta(2n+1-k)$ for $k = 2, \cdots, 2n-1$, $p = (p_k)_{k=1,\cdots,n}$, $\overline{p} = ((-1)^{k+1}p_k)_{k=1,\cdots,n}$, and $e = (1, \cdots, 1)$. Then $r_k = \overline{p}_k - (Cp)_k + \zeta(2n+1)(Ce)_k$. Now let $k \geq 2$. Then it follows that

$$\begin{aligned}
\left(M^{-1}r\right)_k &= \left(\frac{1}{2}(M+I)r\right)_k \\
&= \frac{1}{2}[((A+B-C+I)\overline{p})_k - ((A+B-C+I)Cp)_k \\
&\quad + \zeta(2n+1)((A+B-C+I)Ce)_k] \\
&= \frac{1}{2}[((A+B-C+I)\overline{p})_k - ((A+B+C-I)p)_k \\
&\quad + \zeta(2n+1)((A+B+C-I)e)_k]. \tag{3.50}
\end{aligned}$$

3.6 Duality Evaluations and Computations

For non-negative integers $s_1, \cdots, s_k$, we consider

$$\zeta_a(s_1, \cdots, s_k) = \sum_{n_j > n_{j+1} > 0} a^{-n_1} \prod_{j=1}^{k} n_j^{-s_j}, \qquad (3.51)$$

a special case of the *multidimensional polylogarithm*. Note that

$$\zeta_a(s) = \sum_{n>0} \frac{1}{a^n n^s} = \mathrm{Li}_s(a^{-1})$$

is the usual polylogarithm for $s \in \mathbf{N}$ and $|a| > 1$. We write $\zeta = \zeta_1$ and $\kappa = \zeta_2$.

We also define a *unit Euler sum* by

$$\rho(\sigma_1, \cdots, \sigma_k) = \sum_{n_j > n_{j+1} > 0} \prod_{j=1}^{k} \frac{\sigma_j^{n_j}}{n_j}.$$

Put

$$\omega_a = \frac{dx}{x - a}.$$

Integration over $0 \leq x_1 \leq x_2 \leq \cdots \leq x_s \leq 1$ allows us to write

$$\zeta_a(s_1, \cdots, s_k) = (-1)^k \int_0^1 \prod_{j=1}^{k} \omega_0^{s_j - 1} \omega_a,$$

and dually

$$\zeta_a(s_1, \cdots, s_k) = (-1)^{s+k} \int_0^1 \prod_{j=k}^{1} \omega_{1-a} \omega_1^{s_j - 1}$$

follows on changing $x \mapsto 1 - x$ at each level. So,

$$(-1)^k \zeta_a(s_1 + 2, \{1\}_{r_1}, \cdots, s_k + 2, \{1\}_{r_k}) = (-1)^r \int_0^1 \prod_{j=1}^{k} \omega_0^{s_j+1} w_a^{r_j+1}, \qquad (3.52)$$

where $s = s_1 + s_2 + \cdots + s_k$, and dually,

$$(-1)^k \zeta_a(s_1 + 2, \{1\}_{r_1}, \cdots, s_k + 2, \{1\}_{r_k}) = (-1)^s \int_0^1 \prod_{j=k}^{1} \omega_{1-a}^{r_j+1} w_1^{s_j+1} \qquad (3.53)$$

.

Theorem 3.8.

(a) *Setting* $a = 1$ *gives the "duality for MZVs":*

$$\zeta(s_1 + 2, \{1\}_{r_1}, \cdots, s_k + 2, \{1\}_{r_k}) = \zeta(r_k + 2, \{1\}_{s_k}, \cdots, r_1 + 2, \{1\}_{s_1}).$$
(3.54)

(b) *Setting* $a = 2$ *gives a corresponding "kappa-to-unit-Euler" duality:*

$$\kappa(s_1 + 2, \{1\}_{r_1}, \cdots, s_k + 2, \{1\}_{r_k}) \qquad (3.55)$$
$$= \quad (-1)^{r+k} \zeta(\overline{1}, \{1\}_{r_k}, \overline{1}, \{1\}_{s_k}, \cdots, \overline{1}, \{1\}_{r_1} \overline{1}, \{1\}_{s_1}).$$

(c) *A more general, less convenient, "kappa-to-unit-Euler" duality similarly derivable is*

$$\kappa(s_1, \cdots, s_k) = (-1)^k \rho(\tau_1, \tau_2/\tau_1, \tau_3/\tau_2, \cdots, \tau_s/\tau_{s-1}),$$

where $[\tau_1, \cdots, \tau_s] = [-1, \{1\}_{s_k-1}, \cdots, -1, \{1\}_{s_1-1}].$

For example, we immediately see from part 1(a) of Theorem 3.8 that $\zeta(2, 1, 1, 1) = \zeta(5)$, and that $\zeta(\{2, 1\}_n) = \zeta(\{3\}_n)$ for all n, while $\zeta(3, 1)$ is self dual. There is a profusion of nice specializations, some of which we now list.

Example 3.9. Some $\kappa \leftrightarrow \rho$ duality examples.

$$\kappa(1) \quad = \quad \sum_{n \geq 1} \frac{1}{n2^n} \quad = \quad -\log(1/2) \quad = \quad \sum_{n \geq 1} \frac{(-1)^{n+1}}{n} \quad = \quad -\zeta(\overline{1}),$$

$$\kappa(2) \quad = \quad \sum_{n \geq 1} \frac{1}{n^2 2^n} \quad = \quad \mathrm{Li}_2(1/2)$$

$$= \quad \sum_{n \geq 1} \frac{(-1)^{n+1}}{n} \sum_{k=1}^{n-1} \frac{(-1)^k}{k} \quad = \quad -\zeta(\overline{1}, \overline{1}),$$

$$\kappa(r + 2) \quad = \quad \sum_{n \geq 1} \frac{1}{n^{r+2} 2^n} \quad = \quad \mathrm{Li}_{r+2}(1/2) \quad = \quad -\zeta(\overline{1}, \overline{1}, \{1\}_r), \quad (r \geq 0)$$

$$\kappa(\{1\}_n) \quad = \quad (-1)^n \zeta(-1, \{1\}_{n-1}) \quad = \quad \frac{(\log 2)^n}{n!},$$

$$\kappa(2, \{1\}_n) \quad = \quad (-1)^{n+1} \zeta(-1, \{1\}_n, -1),$$

$$\kappa(\{1\}_{m+1}, 2, \{1\}_n) \quad = \quad (-1)^{m+n} \zeta(-1, \{1\}_n, \{-1\}_2, \{1\}_m),$$

$$\kappa(1, n+2) = \rho(-1, -1, \{1\}_n, -1),$$

$$\kappa(1, n) = \int_0^{\frac{1}{2}} \frac{\mathrm{Li}_n(z)}{1-z}\, dz.$$

In particular,

$$\kappa(1, 2) = \frac{5}{7}\mathrm{Li}_2\left(\frac{1}{2}\right)\mathrm{Li}_1\left(\frac{1}{2}\right) - \frac{2}{7}\mathrm{Li}_3\left(\frac{1}{2}\right) + \frac{5}{21}\mathrm{Li}_1\left(\frac{1}{2}\right)^3$$

$$\kappa(1, 3) = \mathrm{Li}_3\left(\frac{1}{2}\right)\mathrm{Li}_1\left(\frac{1}{2}\right) - \frac{1}{2}\mathrm{Li}_2\left(\frac{1}{2}\right)^2.$$

We note that differentiation proves $\kappa(0, \{1\}_n) = \kappa(\{1\}_n)$. Applied to $\zeta(n+2)$, this provides a lovely closed form for $\kappa(2, \{1\}_n)$. $\square$

Example 3.10. Two κ-reductions.

Every MZV of depth N is a sum of 2^N κ's of depth N, hence easily computed, using integral ideas similar to below:

A better method is to set $\omega_0 = dx/x$, $\omega_1 = -dx/(1-x)$. Then

$$\zeta(s_1, \cdots, s_k) = \sum_{n_j > n_{j+1} > 0} \prod_{j=1}^{k} n_j^{-s_j}$$

again has representation

$$\zeta(s_1, \cdots, s_k) = (-1)^k \int_0^1 \omega_0^{s_1-1}\omega_1 \cdots \omega_0^{s_k-1}\omega_1.$$

The domain, $1 > x_j > x_{j+1} > 0$, in $n = \sum_j s_j$ variables, splits into $n+1$ parts: each being a product of regions $1 > x_j > x_{j+1} > \lambda$, for first r variables, and $\lambda > x_j > x_{j+1} > 0$, for rest. The substitution $x_j \mapsto 1 - x_j$ replaces an integral of the former by the latter type, with λ replaced by $\bar{\lambda} = 1 - \lambda$.

Hence, let $S(\omega_0, \omega_1)$ be the n-string $\omega_0^{s_1-1}\omega_1 \cdots \omega_0^{s_k-1}\omega_1$ specifying a MZV. Let T_r denote the substring of the first r letters and U_{n-r} the complementary substring, on the last $n - r$ letters, so that $S = T_r U_{n-r}$ for $n \geq r \geq 0$. Then

$$\zeta(s_1, \cdots, s_k) = \int_0^1 S = \sum_{r=0}^{n} \pm \int_0^{\bar{\lambda}} \widetilde{T}_r \int_0^{\lambda} U_{n-r} \qquad (3.56)$$

where $\tilde{}$ indicates reversal of letter order.

The alternate polylogarithmic integral

$$\zeta_z(s_1, \cdots, s_k) = \sum_{n_j > n_{j+1} > 0} z^{-n_1} \prod_{j=1}^{k} n_j^{-s_j} = \int_0^z \omega_0^{s_1 - 1} \omega_1 \cdots \omega_0^{s_k - 1} \omega_1 \quad (3.57)$$

applied to the right side of (3.56) produces the MZV as the scalar product of two vectors, composed of ζ_z-values with $z = p$ and $z = q$, for any desired $p, q > 1$ with $1/p + 1/q = 1$. We usually set $p = q = 2$, i.e., $\zeta_2 = \kappa$.   □

Example 3.11. Hölder convolution example.

For any $1/p + 1/q = 1$,

$$\underline{\zeta(2, 1, 2, 1, 1, 1)} = \zeta_p(2, 1, 2, 1, 1, 1)$$
$$+\zeta_p(1, 1, 2, 1, 1, 1)\zeta_q(1) + \zeta_p(1, 2, 1, 1, 1)\zeta_q(2)$$
$$+\zeta_p(2, 1, 1, 1)\zeta_q(3) + \zeta_p(1, 1, 1, 1)\zeta_q(1, 3)$$
$$+\zeta_p(1, 1, 1)\zeta_q(2, 3) + \zeta_p(1, 1)\zeta_q(3, 3)$$
$$+\zeta_p(1)\zeta_q(4, 3) + \zeta_q(5, 3) = \underline{\zeta(5, 3)}$$

This uses a homogenous combination of $2s$ polylogs of no higher depth, where s is the weight of the MZV. It also provides another duality result on letting $q \to \infty$.   □

The method, which is called Hölder convolution, because it relies on complementary p and q as in Hölder's inequality of Chapter 5 of [43], is easily programmed (see Section 7.5.1), as the following code partially illustrates:

```
Seq := proc(s, t) local k, n;
    if 1 < s[1] then
[s[1] - 1, seq(s[k], k = 2 .. nops(s))], [1, op(t)]
    else [seq(s[k], k = 2 .. nops(s))],
        [t[1] + 1, seq(t[k], k = 2 .. nops(t))]
    fi end

SEQ := proc(a) local w, k, s;
    w := convert(a, '+'); s := a, [];
    for k to w do s := Seq(s); print(s, k) od;
    s[2] end
```

```
>SEQ([5,3]);
```

$$
\begin{array}{l}
[4,\ 3],\ [1],\ 1\\
[3,\ 3],\ [1,\ 1],\ 2\\
[2,\ 3],\ [1,\ 1,\ 1],\ 3\\
[1,\ 3],\ [1,\ 1,\ 1,\ 1],\ 4\\
[3],\ [2,\ 1,\ 1,\ 1],\ 5\\
[2],\ [1,\ 2,\ 1,\ 1,\ 1],\ 6\\
[1],\ [1,\ 1,\ 2,\ 1,\ 1,\ 1],\ 7\\
[],\ [2,\ 1,\ 2,\ 1,\ 1,\ 1],\ 8
\end{array}
$$

$$
[2,\ 1,\ 2,\ 1,\ 1,\ 1]
$$

The time to compute D digits for MZV $\zeta(s_1,\cdots,s_k)$, of weight n, is roughly $c(n)D$ precision D multiplications (with $c(n) \propto n$, for large n, whatever the depth, k). This idea extends reasonably to all Euler sums. Thus computing 100 digits of $\zeta(5,3)$ takes only a fraction of a second on a 2003-era system, and $1,000$ digits takes only a few seconds. The MP-FUN multiprecision software (a previous version of the ARPREC arbitrary precision computation package described in Chapter 6 of [43]) required 47 minutes on a 2000-era system to compute $20,000$ digits.

3.7 Proof of the Zagier Conjecture

For $r \geq 1$ and $n_1, \cdots, n_r \geq 1$, again specialize the *polylogarithm* to one variable (we write $L(n_1, \cdots, n_r; x) = \zeta_{x^{-1}}(n_1, \ldots, n_r)$ to highlight dependence on x; note that this is a different usage than introduced in Section 1.7)

$$
L(n_1, \ldots, n_r; x) = \sum_{0 < m_r < \ldots < m_1} \frac{x^{m_1}}{m_1^{n_1} \cdots m_r^{n_r}}.
$$

Hence,

$$
L(n; x) = \frac{x}{1^n} + \frac{x^2}{2^n} + \frac{x^3}{3^n} + \cdots
$$

is the classical *polylogarithm* (see Section 1.7), while

$$
L(n, m; x) = \frac{1}{1^m}\frac{x^2}{2^n} + \left(\frac{1}{1^m} + \frac{1}{2^m}\right)\frac{x^3}{3^n} + \left(\frac{1}{1^m} + \frac{1}{2^m} + \frac{1}{3^m}\right)\frac{x^4}{4^n} + \cdots,
$$

and

$$
L(n, m, l; x) = \frac{1}{1^l}\frac{1}{2^m}\frac{x^3}{3^n} + \left(\frac{1}{1^l}\frac{1}{2^m} + \frac{1}{1^l}\frac{1}{3^m} + \frac{1}{2^l}\frac{1}{3^m}\right)\frac{x^4}{4^n} + \cdots.
$$

These series converge absolutely for $|x| < 1$ (conditionally on $|x| = 1$ unless $n_1 = 1$ and $x = 1$). These polylogarithms are *determined uniquely* by the differential equations

$$\frac{d}{dx} L(n_1, \cdots, n_r; x) = \frac{1}{x} L(n_1 - 1, n_2, \cdots, n_r; x),$$

if $n_r \geq 2$; while for $n_r = 1$,

$$\frac{d}{dx} L(n_1, \cdots, n_r; x) = \frac{1}{1-x} L(n_2, \cdots, n_r; x),$$

with the initial conditions $L(n_1, \cdots, n_r; 0) = 0$ for $r \geq 1$ and $L(\emptyset; x) \equiv 1$.

It transpires that if $\overline{s} = (s_1, s_2, \cdots, s_r)$ and $w = \sum s_i$, every *periodic* polylogarithm leads to a function

$$L_{\overline{s}}(x, t) = \sum_{n=0}^{\infty} L(\{\overline{s}\}_n; x) t^{wn},$$

which solves an algebraic ordinary differential equation in x, and leads to nice recurrence relations.

In the simplest case, with $r = 1$, the ODE is $D_s F = t^s F$ where

$$D_s = \left((1-x) \frac{d}{dx} \right)^1 \left(x \frac{d}{dx} \right)^{s-1}$$

and the solution (by series) is a generalized hypergeometric function,

$$L_{\overline{s}}(x, t) = 1 + \sum_{n \geq 1} x^n \frac{t^s}{n^s} \prod_{k=1}^{n-1} \left(1 + \frac{t^s}{k^s} \right),$$

as follows from considering $D_s(x^n)$. Similarly, if we define the polylogarithms L for negative integers by replacing $1/m_j^{n_j}$ in the sum for L by $(-1)^{m_j}/m_j^{|n_j|}$ if $n_j < 0$, and if we note that the differential equations above then slightly change, then we get, again for $r = 1$,

$$L_{-s}(x, t) = 1 + \sum_{n \geq 1} (-x)^n \frac{t^s}{n^s} \prod_{k=1}^{n-1} \left(1 + (-1)^k \frac{t^s}{k^s} \right),$$

and $L_{\overline{-1}}(2x - 1, t)$ solves a hypergeometric ODE.

Indeed,

$$L_{\overline{-1}}(1, t) = \frac{2}{\beta(1 + t/2, 1/2 - t/2)}.$$

We correspondingly obtain ODEs for eventually periodic Euler sums. Thus $L_{\overline{-2},1}(x,t)$ is a solution of...

$$t^6 F = x^2(x-1)^2(x+1)^2 D^6 F + x(x-1)(x+1)(15x^2 - 6x - 7) D^5 F$$
$$+ (x-1)(65x^3 + 14x^2 - 41x - 8)D^4 F + (x-1)(90x^2 - 11x - 27)D^3 F$$
$$+ (x-1)(31x - 10) D^2 F + (x-1) DF.$$

This leads to a four-term recursion for $F = \sum c_n(t)x^n$ with initial values $c_0 = 1, c_1 = 0, c_2 = t^3/4, c_3 = -t^3/6$, and the ODE can be simplified.

We are now ready to prove Zagier's conjecture. Again, let $F(a, b; c; x)$ denote the *hypergeometric function*. Then:

Theorem 3.12. *For* $|x|, |t| < 1$,

$$\sum_{n=0}^{\infty} L(\underbrace{3,1,3,1,\cdots,3,1}_{n-fold}; x) \, t^{4n} = \quad\quad\quad\quad (3.58)$$

$$F\left(\frac{t(1+i)}{2}, \frac{-t(1+i)}{2}; 1; x\right) \quad \times \quad F\left(\frac{t(1-i)}{2}, \frac{-t(1-i)}{2}; 1; x\right).$$

Proof. Both sides of the putative identity start

$$1 + \frac{t^4}{8} x^2 + \frac{t^4}{18} x^3 + \frac{t^8 + 44t^4}{1536} x^4 + \cdots$$

and are *annihilated* by the differential operator

$$D_{31} = \left((1-x)\frac{d}{dx}\right)^2 \left(x\frac{d}{dx}\right)^2 - t^4.$$

Once discovered, and it was discovered after much computational evidence, this can be checked variously in *Mathematica* or *Maple* (e.g., in the package *gfun*). $\quad\square$

Maple code:

```
deq:=proc(F) D(D(F))+A*D(F)+B*F; end;
eqns:= {(deq(H))(x),D((deq(H)))(x),D(D((deq(H))))(x)};
for p from 0 to 4 do eqns:=subs(('@@'(D,p))(H)(x)=y[p],eqns); od:
yi_sol:=solve(eqns,{seq(y[i], i=0..4)});
id:=x->x;    T:=x->t;
# The annihilator to be checked for any product
```

```
A31:=proc(F) (1-id)*D((1-id)*D(id*D(id*D(F)))) - T^4*F; end;
Z:=expand(A31(F1*F2)(x)); for p from 0 to 4 do
   Z:=subs(('@@'(D,p))(F1)(x)=y[1,p],
          ('@@'(D,p))(F2)(x)=y[2,p],Z); od:
```

The annihilator to be checked for the given product is

```
a[1]:= x -> 1/x; b[1]:= x -> I*t^2/2*1/(x*(1-x)); a[2]:= x ->1/x;
b[2]:= x -> -I*t^2/2*1/(x*(1-x));
for i from 1 to 2 do for o from 2 to 4 do
   y[i,o]:=subs(subs(A=a[i],B=b[i],
   y[0]=y[i,0],y[1]=y[i,1],yi_sol),y[o]): od: od:
```

```
normal(Z);
```

The code returns zero showing the hypergeometric product solves the differential equation.

Corollary 3.13. (Zagier conjecture).

$$\zeta(\underbrace{3,1,3,1,\cdots,3,1}_{n-fold}) = \frac{2\,\pi^{4n}}{(4n+2)!}. \tag{3.59}$$

Proof. We have

$$F(a,-a;1;1) = \frac{1}{\Gamma(1-a)\Gamma(1+a)} = \frac{\sin \pi a}{\pi a},$$

where the first equality comes from Gauss's evaluation of $F(a,b;c;1)$ (see (6.52)) and the second was proved in Section 5.4 of [43]. Hence, setting $x=1$ in (3.58) produces

$$F\left(\frac{t(1+i)}{2}, \frac{-t(1+i)}{2}; 1; 1\right) F\left(\frac{t(1-i)}{2}, \frac{-t(1-i)}{2}; 1; 1\right)$$

$$= \frac{2}{\pi^2 t^2} \sin\left(\frac{1+i}{2}\pi t\right) \sin\left(\frac{1-i}{2}\pi t\right)$$

$$= \frac{\cosh \pi t - \cos \pi t}{\pi^2 t^2} = \sum_{n=0}^{\infty} \frac{2\pi^{4n} t^{4n}}{(4n+2)!}$$

on using the Taylor series of cos and cosh. Comparing coefficients in (3.58) completes the proof.   □

If one suspects that Corollary 3.13 holds, once one can compute these sums well, it is very easy to verify many cases numerically and be entirely convinced.

3.8 Extensions and Discoveries

It is possible to arrive at the result without differential equations, just combinatorial manipulations of the iterated integral representations. This can be generalized to

$$\sum_{s \in \mathcal{I}} \zeta(s) = \frac{\pi^{4n+2}}{(4n+3)!} \tag{3.60}$$

where s runs over the set $\mathcal{I}$ of all $2n+1$ possible insertions of the number 2 in the string $\{3,1\}_n$. Actually, (3.60) is just the beginning of a large family of conjectured identities that were determined intensively using PSLQ, not all of which are proved. Other broad generalizations are discussed in Exercise 15.

Compare the much easier result, which follows from the product formula for sin

$$\sum_{n=0}^{\infty} L(\{2\}_n; x) \, t^{2n} = F(it, -it; 1; x)$$

and, more generally,

$$\sum_{n=0}^{\infty} L(\{p\}_n; x) \, t^{pn} = {}_pF_{p-1}(-\omega t, -\omega^3 t, \cdots, -\omega^{2p-1} t; 1, \cdots, 1; x)$$

where $\omega^p = -1$. In each case, expanding the right-hand side as a power series is easy.

The amazing factorizations in the result for $\zeta(\{3,1\}_n)$ and

$$\zeta(\{3,1\}_n) = 4^{-n} \zeta(\{4\}_n) = \frac{1}{2n+1} \zeta(\{2,2\}_n)$$

beg the question, "What other deep Clausen-like hypergeometric factorizations lurk within?"

Broadhurst and Lisoněk used one of the present authors' implementation of PSLQ to search for Zagier generalizations. They found that "cycles" such as

$$Z(m_1, m_2, \cdots, m_{2n+1}) = \zeta(\{2\}_{m_1}, 3, \{2\}_{m_2}, 1, \{2\}_{m_3}, 3, \cdots, 1, \{2\}_{m_{2n+1}})$$

participate in many such identities. Checking PSLQ input vectors from all Z values of fixed weight ($2K$, say), along with the value $\zeta(\{2\}_K)$, detected many identities, from which general patterns were "obvious." This led to a *conjecture* (among many):

$$\sum_{i=0}^{2n} Z(C^i S) \stackrel{?}{=} \zeta(\{2\}_{M+2n}) \tag{3.61}$$

for S a string of $2n+1$ numbers summing to M, and $C^i S$ the cyclic shift of S by i places. Zagier's identity is the case of (3.61) with entries of S zero.

The symmetry in (3.61) highlighted that Zagier-type identities have serious combinatorial content. For $M = 0, 1$ we could reduce (3.61) to evaluation of combinatorial sums; and thence to truly combinatorial proofs. For $M \geq 2$, we have no proofs, but very strong evidence.

Perhaps the most striking conjecture (open indeed for $n > 2$) is tantalizingly easy to state, and to numerically verify, and has eluded proof for five years since its numerical discovery:

$$8^n \, \zeta(\{-2, 1\}_n) \overset{?}{=} \zeta(\{2, 1\}_n), \qquad\qquad (3.62)$$

or equivalently that the functions

$$L_{\overline{-2,1}}(1, 2t) = L_{\overline{2,1}}(1, t) \qquad (= L_{\overline{3}}(1, t))$$

agree for small t. It appears to be the unique identification of an Euler sum with a distinct MZV of its type. Can just the case $n = 2$ be proven symbolically, as is the case for $n = 1$ (see Exercise 16)?

To sum up, our simplest conjectures (on the number of irreducibles) are still beyond present proof techniques. Is $\zeta(5)$ or G rational? Such questions may or may not be close to proof! Thus, the field appears wide open for numerical exploration.

Dimensional conjectures sometimes involve finding integer relations between hundreds of quantities and so demanding precision of thousands of digits—often of hard-to-compute objects. In that vein, one of the present authors and Broadhurst recently found a *polylogarithmic ladder* of length 17 (a record) with such "ultra-PSLQ" computation [19].

3.9   Multi-Clausen Values

We finish this chapter by returning briefly to binomial sums first detailed in Section 1.7. The study of so-called *Deligne words* for multiple integrals generating *Multiple Clausen (or Multi-Clausen) Values* at $\pi/3$, such as

$$\mu(a) = \sum_{n>0} \frac{\cos(n\frac{\pi}{3})}{n^a}, \qquad \mu(a, b) = \sum_{n>m>0} \frac{\sin(n\frac{\pi}{3})}{n^a m^b},$$

seems quite fundamental. It leads to results like

$$S_3 = \sum_{k=1}^{\infty} \frac{1}{k^3 \binom{2k}{k}} = \frac{2\pi}{3}\mu(2) - \frac{4}{3}\zeta(3),$$

$$S_5 = \sum_{k=1}^{\infty} \frac{1}{k^5 \binom{2k}{k}} = 2\pi\mu(4) - \frac{19}{3}\zeta(5) + \frac{2}{3}\zeta(2)\zeta(3),$$

$$S_6 = \sum_{k=1}^{\infty} \frac{1}{k^6 \binom{2k}{k}} = -\frac{4\pi}{3}\mu(4,1) + \frac{3341}{1296}\zeta(6) - \frac{4}{3}\zeta(3)^2.$$

Much more is detailed in [54]. This includes a generalization of MZV duality, and finishes with an accounting of alternating sums:

$$A_N = \sum_{k=1}^{\infty} \frac{(-1)^{k+1}}{k^N \binom{2k}{k}}.$$

In this setting, it is more convenient to work with the set

$$\mathcal{K}_k = \{L_k(\rho^p) \mid p \in \mathcal{C}\},$$

where $\rho = (\sqrt{5} - 1)/2$, of *Kummer-type polylogarithms* of the form

$$L_k(x) = \frac{1}{(k-1)!} \int_0^x \frac{(-\log|y|)^{k-1} dy}{1-y} = \sum_{r=0}^{k-1} \frac{(-\log|x|)^r}{r!} \mathrm{Li}_{k-r}(x),$$

where as before $\mathrm{Li}_k(x) = \sum_{n>0} x^n/n^k$.

$$\widetilde{L}_k(x) = L_k(x) - L_k(-x) = 2L_k(x) - 2^{1-k}L_k(x^2).$$

Then A_3 is Apéry's sum, A_5 was expressed in (3.18), and

$$A_4 = 4\widetilde{L}_4(\rho) - \tfrac{1}{2}L^4 - 7\zeta(4).$$

In fact, there are five integer relations between $\mathcal{K}_4$, L^4, $\zeta(4)$, and A_4. Another simple example is

$$A_4 = \tfrac{16}{9}\widetilde{L}_4(\rho^3) - 2L^4 - \tfrac{23}{9}\zeta(4).$$

For $9 \geq N \geq 6$ one gets corresponding integer relations, but not enough to obtain closed forms for A_N.

3.10   Commentary and Additional Examples

1. **A binomial generalization of zeta.** Show that

$$\sum_{k=0}^{\infty} \frac{\binom{2k}{k}}{\binom{2k+4n}{k+2n}(2k+4n+1)(2k+2n+1)} = \binom{2n}{n}^2 \frac{\pi^2}{2^{8n+3}}, \quad (3.63)$$

for each integer $n \geq 0$.

Hint: This is a sum that *Maple* can evaluate in closed form.

Reason: The summand on the left of (3.63) equals

$$\frac{(2k)!(k+2n)!^2}{k!^2(2k+4n)!(2k+4n+1)(2k+2n+1)}$$

$$= \frac{[(k+1)(k+2)\cdots(k+2n)]^2}{(2k+2n+1)[(2k+1)(2k+2)\cdots[2k+4n+1]]}$$

$$= \frac{(k+1)(k+2)\cdots(k+2n)}{2^{2n}(2k+2n+1)[(2k+1)(2k+3)\cdots(2k+4n+1)]} = f_n(k),$$

where f_n is a rational function with numerator and denominator of degrees $2n$ and $2n+2$, respectively. The partial fraction expansion of f_n is thus

$$f_n(x) = \frac{a}{(2x+2n+1)^2} + \sum_{i=0}^{2n} \frac{b_i}{(2x+2i+1)}.$$

But $f_n(x) = -f_n(-2n-1-x)$ and it follows that $b_i = -b_{2n-i}$ for each i. In particular, $b_n = 0$. Also,

$$
\begin{aligned}
a &= \lim_{x \to -n-1/2} (2x+2n+1)^2 f_n(x) \\
&= \frac{(-n+1/2)(-n+3/2)\cdots(n-1/2)}{2^{2n}(-2n)(-2n+2)\cdots(-2)\cdot 2\cdot 4\cdots(2n)} \\
&= \frac{[1\cdot 3\cdots(2n-1)]^2}{2^{6n}n!^2} = \frac{(2n)!^2}{2^{8n}n!^4} = \frac{1}{2^{8n}}\binom{2n}{n}^2.
\end{aligned}
$$

For $-n \leq k < 0$, $f_n(k) = 0$ so that the sum in question equals

$$
\sum_{k=-n}^{\infty} f_n(k) = \frac{1}{2^{8n}}\binom{2n}{n}^2 \sum_{k=-n}^{\infty} \frac{1}{(2n+2k+1)^2}
$$

$$
+ \sum_{i=0}^{n-1} b_i \sum_{k=-n}^{\infty} \left(\frac{1}{2k+2i+1} - \frac{1}{2k+4n-2i+1} \right).
$$

The sum

$$\sum_{k=-n}^{\infty} \left(\frac{1}{2k + 2i + 1} - \frac{1}{2k + 4n - 2i + 1} \right)$$

telescopes and equals

$$\sum_{k=-n}^{n-2i-1} \frac{1}{2k + 2i + 1} = 0,$$

since in this sum the k-th term cancels with the $-2i - k - 1$-st term.

As

$$\sum_{k=-n}^{\infty} \frac{1}{(2n + 2k + 1)^2} = \sum_{j=0}^{\infty} \frac{1}{(2j + 1)^2} = \frac{\pi^2}{8},$$

the identity (3.63) follows.

2. **The Riemann-Siegel formula.** In this exercise, we derive the principle term of an asymptotic formula that Riemann used to compute the roots of the zeta function, $\zeta(s) = \sum_{n=1}^{\infty} n^{-s}$ ($s \in \mathbb{C}$) whenever this sum converges. We then use the asymptotic formula to find roots of $\zeta(s)$.

(i) We first derive an integral representation of $\zeta(s)$ which remains valid for all $s \in \mathbb{C}$, as Riemann states without proof in his famous 1859 paper [187]. Use the Mellin transform

$$\int_0^{\infty} e^{-nx} x^{s-1} \, dx = \frac{\Gamma(s-1)}{n^s},$$

together with the Gamma function identities from Chapter 5 of [43]

$$\Gamma(s) = s\Gamma(s-1), \qquad \frac{\pi s}{\Gamma(s)\Gamma(-s)} = \sin \pi s, \qquad (3.64)$$

and the standard identity $e^{-Nx}(e^x - 1)^{-1} = \sum_{n=N+1}^{\infty} e^{-nx}$, to show that

$$\zeta(s) = \sum_{n=1}^{N} n^{-s} + \Gamma(-s)\pi^{-1} \sin(\pi s) \int_0^{\infty} \frac{x^{(s-1)} e^{-Nx}}{e^x - 1} \, dx. \quad (3.65)$$

Now, using the residue theorem and the identities (3.64), show that

$$\zeta(s) = \sum_{n=1}^{N} n^{-s} + \Gamma(-s)(2\pi)^{s-1} 2\sin\frac{\pi s}{2} \sum_{n=1}^{N} n^{-(1-s)}$$

$$+ \frac{\Gamma(-s)}{2\pi i} \int_{C_N} \frac{(-x)^s e^{-Nx}}{x(e^x - 1)}\, dx, \qquad (3.66)$$

where C_N is the contour whose path descends the real axis (or, technically, just above the real axis) from $+\infty$, traces the circle of radius $2\pi(N+1)$ counterclockwise and returns to $+\infty$, and where $(-x)^s = \exp[s\,\log(-x)]$ is defined to be the branch that is real for positive real x.

(ii) Next consider the auxiliary function $Z(t)$ defined by

$$Z(t) = e^{i\theta(t)}\zeta(1/2+it), \quad \theta(t) = i\,\log\left(\Gamma(it/2 - 3/4)\right) - t/2\log\pi.$$

Define $\xi(s) = \Gamma(s/2)(s-1)\pi^{-s/2}\zeta(s)$. Again using the identity (3.64), show that, $\xi(1/2 + it) = r(t)Z(t)$, where

$$r(t) = -e^{\mathrm{Re}\,\log\Gamma(i\,t/2-3/4)} \frac{t^2 + 1/4}{2\pi^{1/4}}.$$

Substitute the integral expression (3.65) for $\zeta(s)$ into the definition of $\xi(s)$ above and use the identities $\theta(-t) = -\theta(t)$, $r(-t) = r(t)$, and $2i\sin(\pi s/2) = -e^{i\pi/4}e^{t\pi/2}(1 - ie^{-t\pi})$ to show that $Z(t) = Z_0(t) + R(t)$ where

$$Z_0(t) = \sum_{n=1}^{N} n^{-1/2} 2\cos(\theta(t) - t\,\log n), \qquad (3.67)$$

and

$$R(t) = \frac{e^{-i\theta(t)}e^{-t\pi/2}}{(2\pi)^{1/2}(2\pi)^{it}e^{-i\pi/4}(1 - ie^{-t\pi})} \int_{C_N} \frac{(-x)^{it-1/2}e^{-Nx}}{e^x - 1}\, dx. \qquad (3.68)$$

To evaluate the remainder term R, Riemann used the asymptotic expansion

$$R(t) \approx \widetilde{R}(t) = (-1)^{N-1}\left(\frac{t}{2\pi}\right)^{-1/4} \sum_{j}\left(\frac{t}{2\pi}\right)^{-j/2} C_j(p), \qquad (3.69)$$

where the integer N is chosen to be the integer part of $(t/2\pi)^{1/2}$, and p is the fractional part. The first three terms of the expansion are

$$C_0(p) = \Psi(p) = \frac{\cos(2\pi(p^2 - p - \frac{1}{16}))}{\cos(2\pi p)},$$

$$C_1(p) = -\frac{1}{2^5 3\pi^2} \Psi^{(3)}(p),$$

$$C_2(p) = \frac{1}{2^{11} 3^4 \pi^4} \Psi^{(6)}(p) + \frac{1}{2^6 \pi^2} \Psi^{(2)}(p).$$

The notation $\Psi^{(n)}$ indicates the n-th derivative. The expansion (3.69) is the Riemann-Siegel formula, so named for its originator and the mathematician Carl Siegel [196], who discovered the formula in Riemann's working papers some 70 years after the publication of Riemann's original paper. If the C_1 term is the first term omitted in the Riemann-Siegel formula, then Titchmarsh [206, pg. 331] showed that for $t > 250\pi$, the error $|R(t) - \widetilde{R}(t)|$ is bounded by $(3/2)(t/2\pi)^{(-3/4)}$. The formula refined is still in use today. See also [93] for an Exercise (1.59) that leads the reader through zeta calculations, including some explicit parts on Riemann-Siegel.

(iii) Using the Riemann-Siegel formula and Titchmarsh's estimate for the error, give a numerical proof for the existence of zeroes of $\zeta(1/2 + it)$ in the interval $t \in [999.784, 999.799]$. Note that only 12 terms in the main sum (3.67) are needed to calculate the estimate for $Z(t)$. To achieve comparable accuracy using the alternative Euler-Maclaurin formula (see [114, Chapter 6] or Chapter 7 in this volume) would require hundreds of terms.

3. **Some quadratic zeta functions**. Explore evaluations of sums of the form

$$\zeta(a, b, c) = \sum_{n, m \geq 0}' \frac{n^{2a} m^{2b}}{(n^2 + m^2)^c}$$

for a, b, c nonnegative integers. (Here, as before, the summation avoids poles of the summand.)

Solution:

(a) The following identity expresses $\zeta(a, 0, c)$ by using a Bessel function expansion of the normalized Mellin transform

$$M_c(f) = \frac{1}{\Gamma(c)} \int_0^\infty f(t) t^{c-1} \, dt$$

of the function

$$t \to \sum_{n,m} n^{2a} q^{n^2+m^2} = \sum_n n^{2a} q^{n^2} \theta_3(q),$$

with $q = \exp(-t)$, after using the theta transform (2.15)

$$\theta_3(\exp(\pi t)) = \sqrt{\frac{\pi}{t}} \theta_3 \left(\exp \left(\frac{\pi}{t} \right) \right).$$

This leads to the identity

$$\zeta(a,0,c) = 2\delta_{0a}\zeta(2c) + 2\beta \left(c - \tfrac{1}{2}, \tfrac{1}{2} \right) \zeta(2c - 2a - 1)$$
$$+4 \sum_{p \geq 1} \sigma_{[2a+1-2c]}(p) \mathcal{E}_c(p).$$

$$(3.70)$$

This is valid for real ac with $d = c - a > 1$ and presumably provides an analytic continuation of $\zeta(a,0,c)$, for the sum over positive integers. Here σ is a divisor function

$$\sigma_{[d]}(p) = \sum_{n|p} n^d,$$

and

$$\mathcal{E}_c(p) = \frac{\sqrt{\pi}}{\Gamma(c)} 2(\pi p)^{c-1/2} K_{(c-1/2)}(2\pi p)$$

is derived from the modified Bessel function of half integer order, $K_{(c-1/2)}$, of the second kind. When $c = N$ is integer, $\mathcal{E}_N(p)$ is of the form

$$\pi \exp(-2\pi p) P_N(\pi p),$$

where P_N is a rational polynomial of degree $N-1$ with positive coefficients: $P_1(x) = 1, P_2(x) = x + 1/2, P_3(x) = x^2/2 + 3/4x + 3/8, P_4(x) = 1/6x^3 + 1/2x^2 + 5/8x + 5/16$. In general,

$$P_N(x) = \sum_{k=0}^{N-1} \frac{\binom{N+k-1}{N-1} x^{N-1-k}}{(N-k-1)! 4^k}.$$

This allows one to very efficiently compute $\zeta(a,0,c)$ via (3.70), using roughly $D/4$ terms for D digits. Note also that for fixed c and variable a, only the powers in $\sigma_{[2a+1-2c]}$ vary, so most of the computation can be saved.

(b) Then the general integer case follows from

$$\zeta(a,b,c) = \sum_{k=0}^{e} (-1)^{e-k} \binom{e}{k} \zeta(a+b-k, 0, c-k),$$

where $e = \min(a,b)$.

(c) Similar developments are possible for the more general form

$$\zeta_N(a,b,c) = \sum_{n,m}' \frac{n^{2a}m^{2b}}{(Nn^2+m^2)^c},$$

with $N > 0$. We write $\zeta = \zeta_1$. For example,

$$\zeta_N(a,0,c) = 2\,\delta_{0a}\zeta(2c) + N^{(1/2-c)}\left[2\beta\left(c-\frac{1}{2},\frac{1}{2}\right)\right. \tag{3.71}$$

$$\left.\zeta(2c-2a-1) + 4\sum_{p\geq 1}\sigma_{[2a+1-2c]}(p)\mathcal{E}_c(\sqrt{N}p)\right].$$

Note that we now lose a symmetry (and apparently have many fewer closed forms), but have: $\zeta_{1/N}(a,b,c) = N^c\zeta_N(b,a,c)$. Also, for $N = 1, 2, 3$, and especially for those with *disjoint discriminants*, many special values may be computed via elliptic integrals in the corresponding singular values. This leads to closed forms such as

$$\zeta(0,0,c) = 4\zeta(c)L_{-4}(c)$$

and

$$\zeta_2(0,0,c) = 2\zeta(c)L_{-8}(c),$$

where $L_\sigma(c) = \sum_{n\geq 1}\left(\frac{\sigma}{n}\right)n^{-c}$ is the corresponding primitive L-series and $\left(\frac{\sigma}{n}\right)$ is the Legendre-Jacobi symbol. We obtain

$$\zeta(2,0,4) = \frac{1}{4}\pi^2 G + \frac{1}{480}\frac{\pi^6}{\Gamma(3/4)^8},$$

and

$$\zeta(2,2,6) = \frac{1}{64}\pi^2 G - \frac{1}{1920}\frac{\pi^6}{\Gamma(3/4)^8} + \frac{1}{92160}\frac{\pi^{10}}{\Gamma(3/4)^{16}},$$

where $G = L_{-4}(2)$ is Catalan's constant. There is a similar expression for $\zeta(a,b,a+b+2)$ for all integers $a,b \geq 0$. Then also we have

$$\zeta_2(1,1,4) = \frac{1}{8}\zeta(2)L_{-8}(2) - \frac{1}{18}\left(3-2\sqrt{2}\right)\left(\frac{1}{8}\beta\left(\frac{1}{8},\frac{1}{8}\right)\right)^4,$$

and there is a similar evaluation of $\zeta_P(1,1,4)$ and $\zeta_{2P}(1,1,4)$ when P is respectively of "type 1" $((1),5,13,21,33\cdots)$ or "type 2" $(1,3,5,11,15,\cdots)$, as described in Section 9.2 of [44]. Hence,

$$\zeta_6(0,0,c) = \zeta(c)\,L_{-24}(c) + L_{-3}(c)\,L_8(c)$$

and

$$\zeta_6(1,1,4) = \frac{1}{12}\zeta_6(0,0,2)$$
$$- \frac{1}{15}\,2^{1/3}\,(35 + 16\sqrt{3} - 20\sqrt{2} - 14\sqrt{6})\left(\frac{1}{24}\,\beta\left(\frac{1}{24},\frac{1}{24}\right)\right)^4,$$

while

$$\zeta_{10}(1,1,4) = \frac{1}{80}\,(\zeta(2)\,L_{-40}(2) + L_5(2)\,L_{-8}(2))$$
$$- \left(7725 + 3452\sqrt{5} - 5460\sqrt{2} - 2442\sqrt{10}\right)$$
$$\times \frac{(\frac{1}{40}\beta)^4(\frac{1}{40},\frac{1}{40})\beta^4(\frac{9}{40},\frac{9}{40})}{120\,\beta^4(\frac{3}{8},\frac{3}{8})}.$$

For comparison, we note that we may also write

$$\zeta_1(1,1,4) = \frac{1}{2}\,\zeta(2)\,L_{-4}(2) - \frac{1}{30}\left(\frac{1}{4}\,\beta\left(\frac{1}{4},\frac{1}{4}\right)\right)^4,$$

and may use elliptic transformation formulae to derive

$$\zeta_4(1,1,4) = \frac{1}{32}\,\zeta(2)\,L_{-4}(2) - \frac{11}{64}\,\zeta_1(1,1,4).$$

Hence,

$$\sideset{}{'}\sum_{n,m}(-1)^n\,\frac{n^2 m^2}{(n^2+m^2)^4} = 8\,\zeta_4(1,1,4) - \zeta_1(1,1,4)$$

$$= \frac{1}{16}\,\zeta(2)\,L_{-4}(2) + \frac{1}{80}\left(\frac{1}{4}\,\beta\left(\frac{1}{4},\frac{1}{4}\right)\right)^4,$$

and

$$\sideset{}{'}\sum_{n,m}(-1)^{n+m}\,\frac{n^2 m^2}{(n^2+m^2)^4} = 4\sideset{}{'}\sum_{n,m}(-1)^n\,\frac{n^2 m^2}{(n^2+m^2)^4}.$$

Note also that $L_5(2) = 4\sqrt{5}\pi^2/125$ and $L_8(2) = \sqrt{2}\pi^2/16$.

4. **A multizeta evaluation.** Consider

$$\sigma_{n,m} = \sum_{s \in \mathcal{S}(n,m)} \zeta(s_1, s_2, \cdots, s_m)$$

summed over all strings of length m consisting of nonnegative integers adding up to n, with $s_1 > 1$ to insure convergence. Determine $\sigma_{n,m}$ as a multiple of $\zeta(n)$.

Hint: Note that $\zeta(2,1) = \zeta(3)$, $\zeta(3,1) + \zeta(2,2) = \zeta(4)$, etc. After working out a few more examples, the pattern can easily be observed. See also Item 11.

5. **Harmonic numbers.** Prove that

$$H_n = \sum_{k=1}^{n} \frac{1}{k}$$

is never integer for $n > 1$. *Hint*: A slick proof uses Bertrand's postulate: the proven fact that *there is always a prime p in the interval* $(n/2, n]$. Now write

$$H_n = \frac{n!/1 + n!/2 + \cdots + n!/n}{n!}.$$

Then p divides the numerator and all but one term of the denominator $n!/p$ (as $p > n/2$). One can alternately establish the result using more elementary methods.

6. **Calabi's unifying integral.** It is possible to unify the L-series evaluations in terms of Bernoulli numbers (3.13) and Euler numbers (3.26), as follows. Consider, for integer $N \geq 1$

$$\sigma(N) = \sum_{k=-\infty}^{\infty} \frac{1}{(4k+1)^N}.$$

(a) Show that the ordinary generating function of $\sigma(N)$ is

$$\Sigma(z) = \frac{\pi z}{4} \left(\sec\left(\frac{\pi z}{2}\right) + \tan\left(\frac{\pi z}{2}\right) \right),$$

and obtain the explicit relations between the Euler numbers, the Bernoulli numbers, and $\sigma(N)$.

(b) Show that $\sigma(1) = \int_0^1 1/(1+x^2)\, dx = \arctan(1) = \pi/4$, and that

$$\sigma(2) = \sum_{k=-\infty}^{\infty} \frac{1}{(4k+1)^2} = \sum_{k=0}^{\infty} \frac{1}{(2k+1)^2} = \int_0^1 \int_0^1 \frac{dx\, dy}{1 - (xy)^2} = \frac{\pi^2}{8}.$$

(See also Exercise 11, parts (a) and (b), of Chapter 2.)

(c) More generally for $N > 1$, show that

$$\sigma(N) = \sum_{k=0}^{\infty} \frac{(-1)^{kN}}{(2k+1)^N} = \int_0^1 \cdots \int_0^1 \frac{dx_1 \cdots dx_N}{1 \pm (x_1 \cdots x_N)^2} = \left(\frac{\pi}{2}\right)^N \nu(N),$$

where $\nu(N)$ is the volume, necessarily rational, of the polytope

$$P_N = \{(v_1, v_2, \ldots, v_N) : v_k \geq 0, v_k + v_{k+1} \leq 1, (1 \leq k \leq N)\}.$$

One way to obtain the volume is to use, and justify, the (one-to-one) change of variables

$$x_k = \frac{\sin u_k}{\cos u_{k+1}} \qquad (1 \leq k \leq N),$$

where we set $u_{N+1} = u_1, v_{N+1} = v_1$.

This and more may be explored in Elkies' recent article [115]. It is a nice computer algebra challenge to symbolically invert the substitution for $N = 2, 3, 4$.

7. **A multi-dimensional polylogarithm extension.** A useful specialization of the general multidimensional polylogarithm, which is at the same time an extension of the polylogarithm, is the case in which each $b_j = b$. Under these circumstances, we write

$$\lambda_b(s_1, \cdots, s_k) \;=\; \sum_{\nu_1, \nu_2, \cdots, \nu_k \geq 1} \prod_{j=1}^{k} b^{-\nu_j} \left(\sum_{i=j}^{k} \nu_i\right)^{-s_j} . \quad (3.72)$$

When $b = \pm 1$, this is an Euler sum. Let $|p| \geq 1$. The double generating function equality

$$1 - \sum_{m=0}^{\infty} \sum_{n=0}^{\infty} x^{m+1} y^{n+1} \lambda_p(m+2, \{1\}_n) = F\left(y, -x; 1-x; \frac{1}{p}\right) \quad (3.73)$$

holds. Note that when $p = 1$, the symmetry of the hypergeometric function produces a case of MZV duality: $\zeta(m+2, \{1\}_n) = \zeta(n+2, \{1\}_m)$, for all m and n, because

$$F(y, -x; 1-x; 1) \;=\; \frac{\Gamma(1-x)\Gamma(1-y)}{\Gamma(1-x-y)} \quad (3.74)$$

$$\;=\; \exp\left\{\sum_{k=2}^{\infty} (x^k + y^k - (x+y)^k)\frac{\zeta(k)}{k}\right\}.$$

Expanding the rightmost function gives a closed form for $\zeta(m + 2, \{1\}_n)$.

Proof. (of (3.73)) By definition of λ_p,

$$\sum_{m=0}^{\infty} \sum_{n=0}^{\infty} x^{m+1} y^{n+1} \lambda_p(m+2, \{1\}_n) = y \sum_{m=0}^{\infty} x^{m+1} \sum_{k=1}^{\infty} \frac{1}{k^{m+2} p^k} \prod_{j=1}^{k-1} \left(1 + \frac{y}{j}\right)$$

$$= \sum_{m=0}^{\infty} x^{m+1} \sum_{k=1}^{\infty} \frac{(y)_k}{k^{m+1} k! p^k}$$

$$= \sum_{k=1}^{\infty} \frac{(y)_k}{k! p^k} \left(\frac{x}{k - x}\right)$$

$$= -\sum_{k=1}^{\infty} \frac{(y)_k (-x)_k}{k! p^k (1 - x)_k}$$

$$= 1 - \mathrm{F}\left(y, -x; 1 - x; \frac{1}{p}\right)$$

as claimed.   □

8. **A symbolic multidimensional zeta evaluation.** Use

$$1 - \exp \sum_{k=2}^{\infty} \zeta(k) \frac{\left(x^k + y^k - (x+y)^k\right)}{k}$$

in (3.74) to compute $\zeta(n, \{1\}_m)$ symbolically for $n + m < 9$.

9. **Three proofs of an identity.**

$$\sum_{n>0} \frac{H_n^2}{n^2} = \frac{17}{4} \zeta(4).$$

Here, as before $H_n = \sum_{k=1}^{n} 1/k$.

(a) *Fourier analysis proof.* As in Chapter 2 of this volume, obtain the Fourier series of the function whose square integral is given by

$$\frac{1}{2\pi} \int_0^{\pi} (\pi - t)^2 \log^2\left(2 \sin \frac{t}{2}\right) dt = \sum_{n=1}^{\infty} \frac{\left(\sum_{k=1}^n \frac{1}{k}\right)^2}{(n+1)^2}. \qquad (3.75)$$

(b) *Algebraic proof.* Write

$$\sum_{n>0} \frac{H_n^2}{n^2} = \sum_{n>0} \frac{\left(H_{n-1} + \frac{1}{n}\right)^2}{n^2} \tag{3.76}$$

$$= \sum_{n>0} \frac{H_{n-1}^2}{n^2} + 2\zeta(3,1) + \zeta(4)$$

$$= 2\zeta(2,1,1) + \zeta(4) + 2\zeta(3,1) + \zeta(4)$$

$$= 2\zeta(2,1,1) + \zeta(4) + 2\zeta(3,1) + \zeta(4) = \left(4 + \frac{1}{4}\right)\zeta(4),$$

since $\zeta(2,1,1) = \zeta(4)$ (by MZV duality or the $\zeta(m+2, \{1\}_n)$ special case), and $4\zeta(3,1) = \zeta(4)$ (by the first case of Zagier's evaluation or by the double Euler sum evaluation).

(c) *Residue theory proof.* Apply residue theory to

$$\phi_{p,q}(s) = \frac{\pi}{2} \frac{\cot(\pi s)}{s^q} \frac{\Psi^{(p-1)}(-s)}{(p-1)!}$$

to obtain $\zeta(p,q)$ for $p + q < 8$, by integrating over circles of radius $R \to \infty$, centered at the origin [117].

10. **Reduction to zeta values.** Reduce

$$\sum_{n>0} \frac{H_n^3}{n^3}$$

and

$$\sum_{n>0} \frac{H_n^2}{n^4}$$

to Riemann zeta values.

11. **The Ohno duality theorem.** Ohno [174] provides the following beautiful generalization of MZV duality (see Theorem 3.8). Define, for integers $k_1 \geq 2, k_2, \cdots, k_n \geq 1, l \geq 0$,

$$Z(k_1, k_2, \cdots, k_n; l) = \sum_{\substack{c_1 + c_2 + \cdots c_n = l \\ c_i \geq 0}} \zeta(k_1 + c_1, k_2 + c_2, \cdots, k_n + c_n).$$

Also define, for integer $a_1, b_1, a_2, b_2, \cdots, a_s, b_s \geq 1$, the dual sequences

$$\mathbf{k} = (b_1 + 1, \{1\}_{a_1-1}, b_2 + 1, \{1\}_{a_2-1}, \cdots, b_s + 1, \{1\}_{a_s-1})$$

and

$$\mathbf{k}' = (a_s + 1, \{1\}_{b_s - 1}, a_{s-1} + 1, \{1\}_{b_{s-1} - 1}, \cdots, a_1 + 1, \{1\}_{b_1 - 1}).$$

Then for all such $\mathbf{k}, \mathbf{k}'$ and l,

$$Z(\mathbf{k}, l) = Z(\mathbf{k}'; l). \tag{3.77}$$

(a) Recover the MZV duality result from (3.77) with $l = 0$.

(b) Apply (3.77) with $k = n + 1$ and $\mathbf{k}' = (2, \{1\}_{n-1})$ to obtain an evaluation of the sum of all legal ζ-values of length n with weights summing to $n + l + 1$.

(c) Deduce

$$\frac{1}{\Gamma(s)} \int_0^\infty \frac{t^s - 1}{e^t - 1} \operatorname{Li}_k \left(1 - e^{-t}\right) dt = \zeta(k + 1, \{1\}_{s-1}).$$

12. **The Ohno-Zagier generating function.** Ohno and Zagier provide the following impressive generating function. For multi-indices of the form $\mathbf{k} = (k_1, k_2, \cdots, k_n)$, with $k_i > 0$, let $I_0(k, n, s)$ denote those *admissible multi-indices* of weight k, depth n, and height $s = \#\{i : k_i > 1\}$. Let

$$G_0(k, n, s) = \sum_{\mathbf{k} \in I_0(k, n, s)} \zeta(\mathbf{k}).$$

Note that $I_0(k, n, s)$ is nonempty exactly if $s > 0, n \geq s$, and $k \geq n+s$. Denote the generating function

$$\Phi_0(x, y, z) = \sum_{k, n, s} G_0(k, n, s) \, x^{k-n-s} y^{n-s} z^{s-1}.$$

Then

$$(xy - z) \, \Phi_0(x, y, z) = \left(1 - \exp\left(\sum_{n>1} (x^n + y^n - \alpha^n - \beta^n) \frac{\zeta(n)}{n}\right)\right), \tag{3.78}$$

where α, β are the roots of $t^2 - t(x + y) = z$.

(a) Deduce that all the coefficients $G_0(k, n, s)$ of Φ_0 are polynomials in $\zeta(2), \zeta(3), \cdots$ with rational coefficients.

(b) Show (3.78) is equivalent to

$$1 - (xy - z) \, \Phi_0(x, y, z) = \prod_{m \geq 1} \left(1 - \frac{xy - z}{(m - x)(m - y)}\right).$$

13. **Bertrand's postulate.** Show that $\binom{2n}{n}$ is even for $n > 0$ and use this to prove Bertrand's postulate that there is always a prime between n and $2n$.

14. **MZV stuffles.** Define a binary operation mapping pairs of ordered lists $u = (u_1, \cdots, u_m)$ and $v = (v_1, \cdots, v_n)$ (for non-negative integers m and n) into multisets of ordered lists by the recursion

$$\begin{cases} () * u & = u * () = \{u\}, \\ au * bv & = a(u * bv) \cup b(au * v) \cup (a+b)(u * v), \end{cases}$$

where, for example, $au = (a, u_1, \cdots, u_m)$ and more generally, if M is a multiset of ordered lists, then aM denotes the multiset obtained by placing a at the front of each list in M.

(a) Show that

$$\zeta(u)\zeta(v) = \sum_{w \in u * v} \zeta(w).$$

(b) Let $f(|u|, |v|)$ denote the number of lists (counting multiplicity) in $u * v$. Show that the formal power series identity

$$\sum_{m=0}^{\infty} \sum_{n=0}^{\infty} f(m, n) x^m y^n = \frac{1}{1 - x - y - xy}$$

holds.

(c) Hence, show [70] that

$$\begin{aligned} f(m, n) & = \sum_{k=0}^{m} \binom{m}{k}\binom{n+k}{m} = \sum_{k=0}^{\min(m,n)} \binom{n}{k}\binom{m}{k} 2^k \\ & = \left| \left\{ (b_1, \cdots, b_m) \in \mathbf{Z}^m : \sum_{j=1}^{m} |b_j| \leq n \right\} \right| \\ & = \left| \left\{ (b_1, \cdots, b_n) \in \mathbf{Z}^n : \sum_{j=1}^{n} |b_j| \leq m \right\} \right|. \end{aligned}$$

(The proof that the two sets of lattice points are the same size is an exercise in Pólya-Szegö's *Problems and Theorems in Analysis*.) There is an analogous, but simpler, result to (a) for integral "shuffles."

15. **Extensions of Zagier's identity.** We sketch some quite broad extensions of the method of Theorem 3.12. Let f and g be differentiable univariate functions, and define differential operators $D_f = f(x)d/dx$, $D_g = g(x)d/dx$. Fix a constant t and let U and V be sets of solutions to the respective differential equations

$$(D_f D_g - t)u = 0, \qquad (D_f D_g + t)v = 0.$$

(a) Prove [72, 70] that $UV = \{uv : u \in U, v \in V\}$ is a set of solutions to the differential equation

$$(D_f^2 D_g^2 + 4t^2)w = 0,$$

and moreover [71], UV is a basis iff U and V are bases for the solution spaces of their respective equations.

Hint: With obvious notation, prove the modified Wronskian determinant identity [71]

$$\begin{vmatrix} u_1 v_1 & u_2 v_1 & u_1 v_2 & u_2 v_2 \\ D_g u_1 v_1 & D_g u_2 v_1 & D_g u_1 v_2 & D_g u_2 v_2 \\ D_g^2 u_1 v_1 & D_g^2 u_2 v_1 & D_g^2 u_1 v_2 & D_g^2 u_2 v_2 \\ D_f D_g^2 u_1 v_1 & D_f D_g^2 u_2 v_1 & D_f D_g^2 u_1 v_2 & D_f D_g^2 u_2 v_2 \end{vmatrix}$$

$$= 8t \begin{vmatrix} u_1 & u_2 \\ D_g u_1 & D_g u_2 \end{vmatrix}^2 \begin{vmatrix} v_1 & v_2 \\ D_g v_1 & D_g v_2 \end{vmatrix}^2.$$

(b) Generalize this result.

(c) Applications: For real x with $0 \le x \le 1$, positive integers s_j, and signs $\sigma_j \in \{1, -1\}$, let

$$\zeta(\sigma_1 s_1, \cdots, \sigma_k s_k; x) = \sum_{n_1 > \cdots > n_k > 0} x^{n_1} \prod_{j=1}^{k} n_j^{-s_j} \sigma_j^{n_j}.$$

For $0 \le x \le 1$ and complex z, let

$$\begin{aligned} Y_1(x, z) &= F(z, -z; 1; x), \\ Y_2(x, z) &= (1 - x)F(1 + z, 1 - z; 2; 1 - x), \\ G(z) &= \tfrac{1}{4}\{\psi(1 + iz) + \psi(1 - iz) - \psi(1 + z) - \psi(1 - z)\}, \end{aligned}$$

where F is the Gaussian hypergeometric function, and ψ is the logarithmic derivative of the Euler Gamma function: $\psi(z) = \Gamma'(z)/\Gamma(z)$. Then [36]

$$\sum_{n=0}^{\infty} (-1)^n z^{4n} 4^n \zeta(\{3,1\}_n; x) = Y_1(x,z) Y_1(x,iz),$$

and [72]

$$\sum_{n=0}^{\infty} (-1)^n z^{4n+2} 4^n \zeta(3,\{1,3\}_n; x) = G(z) Y_1(x,z) Y_1(x,iz)$$

$$- \frac{Y_1(x,iz) Y_2(x,z)}{4Y_1(1,z)} + \frac{Y_1(x,z) Y_2(x,iz)}{4Y_1(1,iz)}$$

define entire functions of z.

(d) Rederive that for positive integers n,

$$\zeta(\{3,1\}_n) = 4^{-n} \zeta(\{4\}^n) = \frac{2\pi^{4n}}{(4n+2)!}.$$

Additionally, show [72] that

$$\begin{aligned}
\zeta(3,\{1,3\}_n) &= 4^{-n} \sum_{k=0}^{n} \zeta(4k+3) \zeta(\{4\}_{n-k}) \\
&= \sum_{k=0}^{n} \frac{2\pi^{4k}}{(4k+2)!} \left(-\frac{1}{4}\right)^{n-k} \zeta(4n-4k+3),
\end{aligned}$$

and

$$\begin{aligned}
\zeta(2,\{1,3\}_n) &= 4^{-n} \sum_{k=0}^{n} (-1)^k \zeta(\{4\}_{n-k}) \left\{(4k+1)\zeta(4k+2)\right. \\
&\quad \left. - 4\sum_{j=1}^{k} \zeta(4j-1)\zeta(4k-4j+3)\right\}.
\end{aligned}$$

(e) For complex z, set

$$A(z) = \sum_{n=0}^{\infty} z^n \zeta(\{-1\}_n) = \prod_{j=1}^{\infty} \left(1 + \frac{(-1)^j z}{j}\right)$$

$$= \frac{\Gamma(1/2)}{\Gamma(1+z/2)\Gamma(1/2-z/2)}.$$

Let t and z be related by $z = (1+i)t/2$, and set $s = (1+x)/2$, where $0 \le x \le 1$. Define $U(s,z) = Y_1(s,z) - zY_2(s,z)$, where

Y_1 and Y_2 are the Gaussian hypergeometric functions previously defined. Then [72],

$$\sum_{n=0}^{\infty} \left[t^{2n}\zeta(\{-1,1\}_n;x) + t^{2n+1}\zeta(-1,\{1,-1\}_n;x) \right] = \frac{U(s,-z)U(s,iz)}{A(-z)A(iz)}$$

defines an entire function of z.

(f) Conclude that for all complex t,

$$\sum_{n=0}^{\infty} \left[t^{2n}\zeta(\{-1,1\}_n) + t^{2n+1}\zeta(-1,\{1,-1\}_n) \right] =$$

$$A\left(\frac{t}{1-i}\right) A\left(\frac{t}{1+i}\right) \quad (3.79)$$

and if $z = (1+i)t/2$, then

$$1 + \sum_{n=0}^{\infty} \left[t^{2n+1}\zeta(-1,\{-1,1\}_n) + t^{2n+2}\zeta(-1,-1,\{1,-1\}_n) \right]$$
$$= \tfrac{1}{2}(1+i)zA(z)A(-iz)\{\pi\csc(\pi z) - i\pi\operatorname{csch}(\pi z) + 4G(z)\}.$$
$$(3.80)$$

(g) Deduce explicit formulas [72] for the alternating unit Euler sums appearing as coefficients in (3.79) and (3.80).

(h) Find additional applications of these ideas to multiple polylogarithms or other special functions.

16. **A multizeta identity.** Consider the power series

$$J(x) = \sum_{n_1 > n_2 > 0} \frac{x^{n_1}}{n_1^2 n_2}.$$

(a) Show for $0 \le x \le 1$ that

$$J(x) = \int_0^x \frac{\log^2(1-t)}{2t}\, dt = \zeta(3) + \frac{1}{2}\log^2(1-x)\log(x)$$
$$+ \quad \log(1-x)\operatorname{Li}_2(x) - \operatorname{Li}_3(x),$$

(b) and that

$$J(-x) = -J(x) + \frac{1}{4}J(x^2) + J\left(\frac{2x}{x+1}\right) - \frac{1}{8}J\left(\frac{4x}{(x+1)^2}\right).$$
$$(3.81)$$

(c) Deduce that $J(1) = 8\,J(-1)$.

(d) Evaluate $J(1/2)$.

This functional equation was found, once the ingredients were determined by inspection, by evaluating (3.81) (actually, a version of it with undetermined coefficients) at a random point and then using LLL. Another successful strategy is to evaluate each J function at enough specific values of x to enable one to solve linear equations for the unknown coefficients. If $L(x)$ and $R(x)$ denote the left- and right-hand sides of (3.81), respectively, then computer manipulations (for $0 < x < 1$) show that $dL/dx = dR/dx$: Mechanically differentiating both sides and simplifying reduces the difference between the two expressions to zero. Now this completes a proof of $\zeta(2,1) = 8\zeta(\overline{2},1) = \zeta(3)$ (see (3.62)). Even the next case of (3.62) has only been established indirectly.

17. **Torus knots and zeta values.**   For integers $p, q > 1$, the $p - q$ torus knot is the knot that transpires when string is wound p times around one way on the torus, while being wound q times in the other direction. Figure 3.1 (see Color Plate IV) shows the $2 - 5$ and $5 - 2$ torus knots in three dimensions. Despite looking very different, this pair is mathematically the same knot (the torus is the product of two circles, and we just exchange generators). There is, along the lines of the discussion in Section 2.6 of [43], a connection between quantum field theory and multizeta values on one hand and between quantum field theory and knot theory on the other.

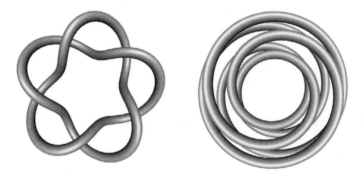

Figure 3.1. How does one identify two knots?

This has an especially interesting consequence for torus knots. Indeed, the $2 - (2n + 1)$ and $(2n + 1) - 2$ knots are indirectly, but

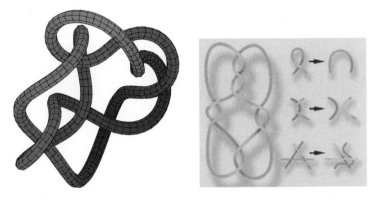

Figure 3.2. The knot 8_8 and Reidemeister moves.

tightly, coupled with $\zeta(2n + 1)$, for each $n \geq 1$. This is intriguing since the standard knot invariants of Alexander or Jones attach a polynomial (*algebraic*) quantity to a given knot. It would be very interesting to see a direct and natural identification. By contrast, the unknot is identified with π, and the $2 - (2n)$ knots are identified trivially with links via π^{2n} (Euler yet again!).

Figure 3.3. The knots 10_{161} and 10_{162}.

18. More knotty problems.

Figure 3.2 shows the 8_8 knot (from a standard catalogue in KnotPlot) and the famous Reidemeister moves, which are used to rearrange knots.

The knots in Figure 3.3 (see Color Plate III) were listed as separate knots in knot tables. These are 10_{161} and 10_{162} in [190], which notes

Figure 3.4. The knot equivalent to both 10_{161} and 10_{162}.

that in 1974 they were shown by Perko to be equivalent [178]. The knots are still listed as separate in some knot tables, including the recent book by Kawauchi [150].

A lengthy sequence of images showing the equivalence is at the URL http://www.cecm.sfu.ca/~scharein/projects/perko, with the nice experimental mathematics connection that these deformations were performed entirely automatically using the KnotPlot tool, available at the URL http://www.colab.sfu.ca/KnotPlot. Indeed, both may be deformed to the knot in Figure 3.4.

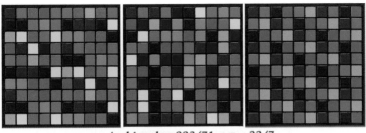

Archimedes: 223/71 < π < 22/7

Plate I. A pictorial proof of Archimedes' inequality. (See Section 1.1, Figure 1.1.)

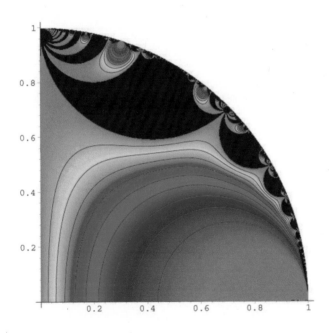

Plate II. The modulus of the modular function k. (See Chapter 4 Item 20, Figure 4.2.)

Plate III. The knots 10_{161} and 10_{162}. (See Chapter 3 Item 18, Figure 3.3.)

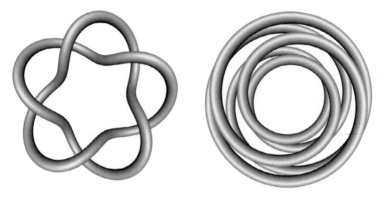

Plate IV. How does one identify two knots? (Chapter 3 Item 17. Figure 3.1.)

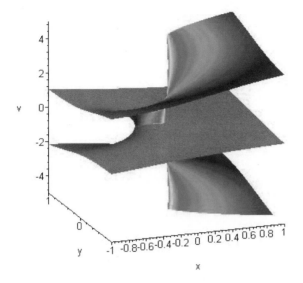

Plate V. The Riemann surface for the Lambert function. (See Section 6.6, Figure 6.3.)

Plate VI. Julia set associated with Newton solutions of $x^3 - 1 = 0$. (Section 7.3, Figure 7.1.)

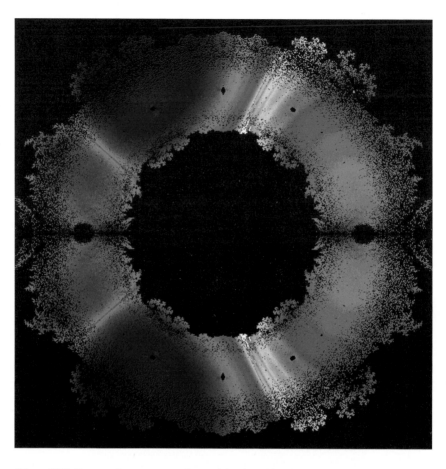

Plate VII. Roots of zero-one polynomials (See Chapter 5, Item 24, Figure 7.1.)

4 | Partitions and Powers

I'll be glad if I have succeeded in impressing the idea that it is not only pleasant to read at times the works of the old mathematical authors, but this may occasionally be of use for the actual advancement of science.

Constantin Carathéodory, speaking to an MAA meeting, 1936

In this chapter, we address the theory of additive partitions and the theory of representations as sums of squares, both from an experimental perspective. Each has a distinguished history. We will show that computational techniques can accelerate both solution and understanding of these problems. What's more, these techniques have a number of interesting applications, including, for instance, Madelung's constant in physical chemistry.

4.1 Partition Functions

The number of *additive partitions* of n, $p(n)$, is formally generated by

$$P(q) = 1 + \sum_{n \geq 1} p(n)q^n = \prod_{n \geq 1} (1 - q^n)^{-1}. \qquad (4.1)$$

One ignores "0" and permutations. Thus $p(5) = 7$ since

$$
\begin{aligned}
5 &= 4 + 1 = 3 + 2 = 3 + 1 + 1 = 2 + 2 + 1 \qquad (4.2)\\
&= 2 + 1 + 1 + 1 = 1 + 1 + 1 + 1 + 1.
\end{aligned}
$$

Additive partitions are less tractable than multiplicative ones as there is no analogue of unique prime factorization nor the corresponding structure.

Formula (4.1) is easily seen by expanding $(1 - q^n)^{-1}$ and comparing coefficients. It is relatively easy to deduce that $2^{\sqrt{n}} < p(n) < e^{\pi\sqrt{2n/3}}$ for $n > 3$ (see [172]), and that the series is absolutely convergent for $|q| < 1$. We return to the analytic behavior of this series below.

Figure 4.1. A Ferrer diagram

Partitions provide a wonderful example of why Keith Devlin calls mathematics "the science of patterns" [109]. Many geometric representations exist. For example, the partition $5 = 4 + 1$ can be represented as a point at $(0,0)$ and four points at $(0,1),(1,1),(2,1),(3,1)$. Read with axis reversed, this identifies $1 + 4$ with $2 + 1 + 1 + 1$ and so on. See Figure 4.1, which identifies $1 + 1 + 1 + 2 + 3 + 4$ and $6 + 3 + 2 + 1$. Such techniques provide alternate ways to prove results such as *the number of partitions of n with all parts odd is the number of partitions of n into distinct parts*, (see Exercise 1).

A modern computational temperament leads to:

Question: How hard is $p(n)$ to compute—in 1900 (for MacMahon the "father of combinatorial analysis") or in 2000 (for *Maple* or *Mathematica*)?

Answer: The computation of $p(200) = 3972999029388$ took MacMahon months and intelligence. Now, however, we can use the most naive approach: Computing 200 terms of the series for the inverse product in (4.1) instantly produces the result using either *Mathematica* or *Maple*. Obtaining the result $p(500) = 2300165032574323995027$ is not much more difficult, using the *Maple* code

```
>   N:=500; coeff(series(1/product(1-q^n,n=1..N+1),q,N+1),q,N);
```

$$2300165032574323995027$$

$\square$

4.1.1   Euler's Pentagonal Number Theorem

In early versions of *Maple*, computing $P(q)$ was quite slow, while taking the series for the reciprocal of the series for $Q(q) = \prod_{n \geq 1}(1 - q^n)$ was quite manageable? Why? Clearly the series for Q must have special properties. Indeed

$$Q(q) = 1 - q - q^2 + q^5 + q^7 - q^{12} - q^{15} + q^{22} + q^{26} - q^{35} - q^{40} + q^{51} + q^{57}$$
$$- q^{70} - q^{77} + q^{92} + O\left(q^{100}\right). \tag{4.3}$$

If we do not immediately recognize these *pentagonal numbers* $((3n \pm 1)n/2)$, Sloane's online Encyclopedia of Integer Sequences, available on the Internet at http://www.research.att.com/~njas/sequences, again comes to the rescue, with abundant references to boot.

So, algorithmic analysis predicts *Euler's pentagonal number theorem*:

$$\prod_{n \geq 1}(1 - q^n) \quad = \quad \sum_{n=-\infty}^{\infty} (-1)^n q^{(3n+1)n/2}. \tag{4.4}$$

One would be less prone to look at Q on the way to P today when the computation is very snappy.

With this success under our belt, we might well ask what about powers of Q? We obtain

$$Q^2(q) = 1 - 2q - q^2 + 2q^3 + q^4 + 2q^5 - 2q^6 - 2q^8 - 2q^9 + q^{10} + 2q^{13}$$
$$+ 3q^{14} - 2q^{15} + 2q^{16} - 2q^{19} - 2q^{20} - 2q^{23} - q^{24} + 2q^{26}$$
$$+ 2q^{27} - 2q^{28} + 2q^{29} + q^{30} + 2q^{31} + 2q^{33} - 2q^{34} - 2q^{35}$$
$$+ 2q^{36} - 2q^{38} - 4q^{40} + q^{44} - 2q^{45} + 2q^{48} + O\left(q^{50}\right),$$

which is not nearly as lacunary; but

$$Q^3(q) \quad = \quad 1 - 3q + 5q^3 - 7q^6 + 9q^{10} - 11q^{15} + 13q^{21} - 15q^{28}$$
$$+17q^{36} - 19q^{45} + O\left(q^{51}\right), \tag{4.5}$$

which has exposed another famous result—a special form of *Jacobi's triple product*. The general form is

$$\prod_{n \geq 1}(1 + xq^{2n-1})(1 + x^{-1}q^{2n-1})(1 - q^{2n}) \quad = \quad \sum_{n=-\infty}^{\infty} x^n q^{n^2}. \tag{4.6}$$

Then the formula implicit in (4.5) is

$$\prod_{n \geq 1}(1 - q^n)^3 \quad = \quad \sum_{m=0}^{\infty}(-1)^m(2m + 1)q^{m(m+1)/2}, \tag{4.7}$$

which may be obtained on replacing q by $q^{1/2}$ and x by $-wq^{1/2}$ in (4.6), differentiating with respect to w, and then letting $w \to 1$ from below.

If we write $P(q)Q(q) = 1$ in terms of the Cauchy convolution, we obtain

$$\sum_{k \leq n} q_k \, p(n-k) = \delta_n, \tag{4.8}$$

where q_k is the coefficient of q^k in (4.3), and δ_n is the Kronecker function which is 1 when $n = 0$ and 0 otherwise. It is a nice exercise to make this into Euler's explicit recursion for $p(n)$, which only needs to compute $O(\sqrt{n})$ smaller values of $p(k)$. One can similarly develop somewhat more efficient formulas, by relying on information such as (4.5).

4.1.2   Modular Properties of Partitions

Ramanujan used MacMahon's table of $p(n), 1 \leq n \leq 200$ to intuit remarkable and deep congruences such as

$$
\begin{aligned}
p(5n+4) &\equiv 0 \bmod 5 \\
p(7n+5) &\equiv 0 \bmod 7 \\
p(11n+6) &\equiv 0 \bmod 11,
\end{aligned}
$$

from data like

$$
\begin{aligned}
P(q) = {}& 1 + q + 2\,q^2 + 3\,q^3 + 5\,q^4 + 7\,q^5 + 11\,q^6 + 15\,q^7 + 22\,q^8 \\
& + 30\,q^9 + 42\,q^{10} + 56\,q^{11} + 77\,q^{12} + 101\,q^{13} + 135\,q^{14} + 176\,q^{15} \\
& + 231\,q^{16} + 297\,q^{17} + 385\,q^{18} + 490\,q^{19} + 627\,q^{20}b + 792\,q^{21}b \\
& + 1002\,q^{22} + 1255\,q^{23} + \cdots .
\end{aligned}
$$

If one generates more terms of $p(n)$ and displays them in an appropriately sized matrix, this becomes much clearer:

$$
\begin{bmatrix}
1 & 1 & 2 & 3 & 5 \\
7 & 11 & 15 & 22 & 30 \\
42 & 56 & 77 & 101 & 135 \\
176 & 231 & 297 & 385 & 490 \\
627 & 792 & 1002 & 1255 & 1575
\end{bmatrix}
\tag{4.9}
$$

shows clearly the congruence $p(5n+4) \equiv 0 \bmod 5$ in the last column.

Correspondingly,

$$
\begin{bmatrix}
1 & 1 & 2 & 3 & 5 & 7 & 11 \\
15 & 22 & 30 & 42 & 56 & 77 & 101 \\
135 & 176 & 231 & 297 & 385 & 490 & 627 \\
792 & 1002 & 1255 & 1575 & 1958 & 2436 & 3010 \\
3718 & 4565 & 5604 & 6842 & 8349 & 10143 & 12310
\end{bmatrix}
\tag{4.10}
$$

shows clearly the congruence $p(7n + 5) \equiv 0 \mod 7$ in the second to last column.

Driven entirely by limited experimental data, Ramanujan conjectured an audacious set of correct modular identities, and not surprisingly over-generalized! He conjectured that if $d = 5^a 7^b 11^c$ and $24n \equiv 1 \mod d$, then $p(n) \equiv 0 \mod d$. This is equivalent to the same conjectures for d a power of $5, 7, 11$. This holds for $a, b, c < 3$ and for all powers of 5 and 11, but fails for 7^3, as $p(243) = 133978259344888$, since $133978259344888 \equiv 245 \mod 343$ quickly shows in the 21st century, while $243 \cdot 24 = 17 \cdot 7^3 + 1$. Such modular identities (see [137, 44]) and their extensions remain an active source of research today. The simplest case is described in Exercise 2.

4.1.3   The "Exact" Formula for the Partition Function

One of the signal achievements of early twentieth century analysis was Hardy and Ramanujan's precise asymptotic for $p(n)$ [83]. It is based in part on an analysis of the *Dedekind η-function* $\eta(q) = e^{\pi i z / 12} \prod_{n \geq 1} (1 - e^{2\pi i n z})$. The function η is closely related to $Q(q)$, and $\theta_3(q)$ discussed in the next section, and satisfies a modular equation like (4.22). Their asymptotic is

$$
p(n) = \frac{e^{K \lambda_n}}{4\sqrt{3}\lambda_n^2} \left(1 + O\left(\frac{1}{\sqrt{n}}\right) \right),
\tag{4.11}
$$

where $K = \pi\sqrt{2/3}$ and $\lambda_n = \sqrt{n - 1/24}$.

This was subsequently refined by Rademacher to

$$
p(n) = \frac{1}{\pi\sqrt{2}} \sum_{k=1}^{\infty} \alpha_k(n)\sqrt{k} \frac{d}{dx} \left[\frac{\sinh\left(\frac{\pi}{k}\sqrt{\frac{2}{3}\left(x - \frac{1}{24}\right)}\right)}{\sqrt{\left(x - \frac{1}{24}\right)}} \right]_{x=n},
\tag{4.12}
$$

where

$$
\alpha_k(n) = \sum_{(h,k)=1}^{k} \omega_{h,k} e^{-2\pi i n h / k},
$$

and $\omega_{h,k} = \exp(\pi i \tau_{h,k})$ with

$$\tau_{h,k} = \sum_{m=1}^{k-1} \left(\frac{m}{k} - \left\lfloor \frac{m}{k} \right\rfloor - \frac{1}{2} \right) \left(\frac{hm}{k} - \left\lfloor \frac{hm}{k} \right\rfloor - \frac{1}{2} \right).$$

If order $\sqrt{n}$ terms are appropriately used, the nearest integer is $p(n)$.
    A mere five terms of this expansion provides

$$p(200) \approx 3972999029387.86108$$

and six terms yields $p(500) \approx 2300165032574323995027.196661$. As we have
seen, the underlying asymptotic is

$$p(n) \sim \frac{1}{4n\sqrt{3}} e^{\pi \sqrt{2n/3}}.$$

Later Erdős made an "elementary" derivation of the Hardy-Ramanujan
formula (4.11). A recent discussion of this formula is given by Almkvist
and Wilf in [8]. It is interesting to speculate how much corresponding
beautiful mathematics is not done when computation becomes too easy—
both *Maple* and *Mathematica* have good built-in partition functions.

4.2   Singular Values

The Jacobian theta functions are a very rich source mine for experimentation—both as a tool to learning classical theory and to discover new
phenomena. Further details of what follows are given fully in [44]. For our
purposes, we consider only the three classical θ-functions:

$$\theta_3(q) \;=\; \sum_{n=-\infty}^{\infty} q^{n^2}, \tag{4.13}$$

$$\theta_4(q) \;=\; \sum_{n=-\infty}^{\infty} (-1)^n q^{n^2},$$

$$\theta_2(q) \;=\; \sum_{n=-\infty}^{\infty} q^{(n+1/2)^2},$$

for $|q| \leq 1$. Note that θ_3^2 is the generating function for the number of ways
of writing a number as a sum of two squares, counting order and sign.
Similarly, θ_2^2 counts sums of two odd squares.
    A beautiful result of Jacobi's is

$$\theta_3^4(q) = \theta_2^4(q) + \theta_4^4(q). \tag{4.14}$$

If we write $k = \theta_2^2/\theta_3^2$ and $k' = \theta_4^2/\theta_3^2$, we note that $k^2 + (k')^2 = 1$. It transpires that

$$(i) \quad \theta_3^2(q^2) = \frac{\theta_4^2(q) + \theta_3^2(q)}{2} \qquad (ii) \quad \theta_4^2(q^2) = \theta_4(q)\,\theta_3(q). \qquad (4.15)$$

Now (4.14) and (4.15) can be proved in many ways and can be "verified" symbolically in many more. For example, Jacobi's triple product (4.6) with $x = \pm 1$ becomes product representations for $\theta_3(q)$ and $\theta_4(q)$, respectively. Multiplying these together yields (ii).

Even without access to the triple product, there is a simple algorithm (see [9, 10]) for converting a sum $A(q) = 1 + \sum_{n \geq 1} a_n q^n$ to a product $\prod_{n \geq 1} (1 - q^n)^{-b_n}$, even preserving rationality of coefficients. To get an idea why, differentiating logarithmically and expanding the denominator of the right side leads to

$$\frac{q\,A'(q)}{A(q)} = \sum_{n \geq 1} \frac{n\,b_n q^n}{1 - q^n} = \sum_{n,k \geq 1} n\,b_n\,q^{nk} = \sum_{n \geq 1} B_n^* q^n, \qquad (4.16)$$

where we set $a_0 = 1$ and

$$B_n^* = \sum_{d|n} d\,b_d. \qquad (4.17)$$

Thus, $q\,A'(q) = A(q) \sum_{n > 0} B_n^* q^n$, which is equivalent to the convolution

$$n\,a_n = \sum_{k=1}^{n} B_k^*\,a_{n-k} = \sum_{k=0}^{n-1} B_{n-k}^*\,a_k. \qquad (4.18)$$

This clearly determines $\{a_k\}$ given $\{b_m\}$ and the converse obtains from the Möbius inversion formula of (4.35).

Example 4.1. Counting rooted trees.

The number T_n of rooted trees with n branches is given by a formula due to Arthur Cayley (1821–1895)

$$T(x) = 1 + \sum_{k=1}^{\infty} T_{k+1} x^k = \prod_{n=1}^{\infty} (1 - x^k)^{-T_k}. \qquad (4.19)$$

As we remarked in Section 1.6 of [43], the product and the sum share their coefficients. The recursion (4.18) for T_n becomes

$$T_{n+1} = \frac{1}{n} \sum_{k=1}^{n} T_k^*\,T_{n+1-k}, \qquad T_n^* = \sum_{d|n} d\,T_d, \qquad (4.20)$$

which starts

$$T_1 = 1, T_2 = 1, T_3 = 2, T_4 = 4, T_5 = 9, T_6 = 20, T_7 = 48, \cdots . \qquad \square$$

Applied to $1 + 2\,q + 2\,q^4 + 2\,q^9 + 2\,q^{16}$, the algorithm produces

$$\frac{\left(1 - q^2\right)^3 \left(1 - q^4\right) \left(1 - q^6\right)^3 \left(1 - q^8\right) \left(1 - q^{10}\right)^3 \left(1 - q^{12}\right) \left(1 - q^{14}\right)^3 \left(1 - q^{16}\right)}{\left(1 - q\right)^2 \left(1 - q^3\right)^2 \left(1 - q^5\right)^2 \left(1 - q^7\right)^2 \left(1 - q^9\right)^2 \left(1 - q^{11}\right)^2 \left(1 - q^{13}\right)^2 \left(1 - q^{15}\right)^2},$$

from which a form of the product for $\theta_3(q)$ can be read off. By contrast, $1 + 3\,q + 3\,q^4 + 3\,q^9 + 3\,q^{16}$ produces

$$\frac{\left(1 - q^2\right)^6 \left(1 - q^4\right)^{15} \left(1 - q^6\right)^{97} \left(1 - q^8\right)^{573} \left(1 - q^{10}\right)^{3867} \left(1 - q^{12}\right)^{26446}}{\left(1 - q\right)^3 \left(1 - q^3\right)^8 \left(1 - q^5\right)^{39} \left(1 - q^7\right)^{231} \left(1 - q^9\right)^{1485} \left(1 - q^{11}\right)^{10056}}$$
$$\cdot \frac{\left(1 - q^{14}\right)^{187761} \left(1 - q^{16}\right)^{1356198}}{\left(1 - q^{13}\right)^{70305} \left(1 - q^{15}\right)^{503384}},$$

which is pretty good evidence that no natural product exists. We will establish (4.15) (i) in the next section.

Also, one notes that it follows from (4.15) that θ_3^2 and θ_4^2 *parametrize* the AGM and that

$$AG(\theta_3^2(q), \theta_4^2(q)) = AG(\theta_3^2(q^2), \theta_4^2(q^2)) = AG(\theta_3^2(q^4), \theta_4^2(q^4)) = \qquad (4.21)$$
$$\cdots AG(\theta_3^2(q^{2^n}), \theta_4^2(q^{2^n})) \cdots = AG(\theta_3^2(0), \theta_4^2(0)) = AG(1, 1) = 1,$$

since the iteration's limit is unchanged if one starts at the first or the second stage of the iteration, and since the AGM is continuous. Another marvellous fact that follows from *Poisson summation* is that

$$k(e^{-\pi s}) = k'(e^{-\pi/s}), \qquad (4.22)$$

for $s > 0$. In particular, $k(e^{-\pi}) = \sqrt{1/2}$. Then (4.22) and (4.21), in conjunction with the already explored relationship between elliptic integrals and the AGM (Section 5.6 of [43]), show that with the above definition of k,

$$K(k(q)) = \frac{\pi}{2}\,\theta_3^2(q). \qquad (4.23)$$

Now the classical theory of *modular equations* asserts that there is a algebraic relationship between $k = k(q)$ and $l = k(q^N)$ for each positive integer N. For example, the quadratic equation, implicit in (4.15), is $l' = 2\sqrt{k'}/(1 + k')$, while the *cubic modular equation* may be written as

$$\theta_3(q)\theta_3(q^3) = \theta_4(q)\theta_4(q^3) + \theta_2(q)\theta_2(q^3), \qquad (4.24)$$

or equivalently,

$$\sqrt{kl} + \sqrt{k'l'} = 1. \qquad (4.25)$$

Similarly, for $N = 7$ the equation is

$$\sqrt{\theta_3(q)\theta_3(q^7)} = \sqrt{\theta_4(q)\theta_4(q^7)} + \sqrt{\theta_2(q)\theta_2(q^7)}, \qquad (4.26)$$

or equivalently

$$\sqrt[4]{kl} + \sqrt[4]{k'l'} = 1. \qquad (4.27)$$

The existence of such modular equations means that there is an algebraic relationship between $K(k)$ and $K(l)$, and in particular that $k_N = k(e^{-\pi\sqrt{N}})$ is a (solvable) algebraic number, called the N-th *singular value*. It is also the case that two *invariants* used by Ramanujan reduce the degrees of these quantities. He used

$$G_N = (2kk')^{-1/12}, \qquad g_N = (2k/k'^2)^{-1/12}, \qquad (4.28)$$

and it transpires that G_N is better for odd N and g_N for even N.

From the equations above, since $k' = l, l' = k$ in this case, we may read off the values $G_1 = g_2 = 1$, $G_3 = 2^{1/12}$, and $G_7 = 2^{1/4}$; with a little more work, we may obtain $g_4 = 2^{1/8}, g_6^6 = \sqrt{2} + 1, g_8^8 = (\sqrt{2} + 1)/2$, and $G_9^6 = (2+\sqrt{3})$. From these evaluations, in turn, we may easily determine k_N for $N = 1, 2, 3, 4, 6, 7, 8, 9$. Had we supplied the quintic modular equation we could determine $G_5^{12} = \sqrt{5}+2, g_{10}^2 = (\sqrt{5}+1)/2, G_{15}^3 = 2^{3/4}(\sqrt{5}+1)/2$, and $G_{25} = (\sqrt{5} + 1)/2$.

Each of these has a reworking as an infinite series evaluation. Thus,

$$\theta_3(e^{-\pi}) = \sqrt[4]{2}\,\theta_4(e^{-\pi}) = \sqrt[4]{2}\,\theta_2(e^{-\pi}). \qquad (4.29)$$

But this is not the main point of this section. We have sketched that (modular) functions such as

$$N \mapsto \frac{\theta_2(q)}{\theta_3(q)}, \qquad q = e^{-\pi\sqrt{N}}, \qquad (4.30)$$

are guaranteed to have algebraic values; and by their nature, they are very rapidly computable to high precision. Thus, they provide excellent test beds for (i) recovering *minimal polynomials* from numerical data, and (ii) for simplifying the radicals so obtained.

For example, working to 15 places, the "MinimalPolynomial" feature of *Maple*, which uses lattice basis reduction, returns $x^2 - 2x - 1$ for g_{22}^2, returns

$x^2 - 12x - 1$ for G_{37}^4, and returns $1 + 10x + 23x^2 - 10x^3 + x^4$ for G_{58}^2, leading to three of the cleanest singular values. Correspondingly, G_{11}^4 solves the cubic $x^3 - 4x^2 + 4x - 1 = 0$, $g_{12}^4 = 2^{1/6}(\sqrt{3} + 1)$, $G_{13}^4 = (\sqrt{13} + 1)/2$, and g_{14}^2 yields the polynomial $x^4 - 2x^3 + 4x^2 - x + 1$, which gives $g_{14}^2 + g_{14}^{-2} = \sqrt{2} + 1$. Also, $G_{17}^{12} + G_{17}^{-12} = 40 + 10\sqrt{17}$. In each case, the root or radical obtained using a low-precision "hunt" can be checked almost instantly to many hundreds or thousands of digits precision.

For instance, we may discover that $x = G_{47}^4/2$ is a root of the solvable irreducible quintic $x^5 - 10x^4 + 9x^3 - 4x^2 - 1 = 0$. In cases of degree less than ten, *Maple* can provide the Galois group (in this case, the dihedral group D_5) and also the enormous radical:

$$5x = 10 + \left[\frac{273625}{4} + \frac{66025}{4}\sqrt{5} - \frac{53885}{772}\sqrt{45355 - 16826\sqrt{5}} \right.$$

$$\left. - \frac{1847377065}{772}\frac{\sqrt{5}}{\sqrt{45355 - 16826\sqrt{5}}} \right]^{1/5}$$

$$+ \frac{6285 + \frac{1265}{2}\sqrt{5} - \frac{2255}{386}\sqrt{45355 - 16826\sqrt{5}} - \frac{20331965}{193}\frac{\sqrt{5}}{\sqrt{45355 - 16826\sqrt{5}}}}{\left(\frac{273625}{4} + \frac{66025}{4}\sqrt{5} - \frac{53885}{772}\sqrt{45355 - 16826\sqrt{5}} - \frac{1847377065}{772}\frac{\sqrt{5}}{\sqrt{45355 - 16826\sqrt{5}}} \right)^{3/5}}$$

$$+ \frac{\frac{1375}{2} + 35\sqrt{5} + \frac{11}{386}\sqrt{45355 - 16826\sqrt{5}} - \frac{1767012}{193}\frac{\sqrt{5}}{\sqrt{45355 - 16826\sqrt{5}}}}{\left(\frac{273625}{4} + \frac{66025}{4}\sqrt{5} - \frac{53885}{772}\sqrt{45355 - 16826\sqrt{5}} - \frac{1847377065}{772}\frac{\sqrt{5}}{\sqrt{45355 - 16826\sqrt{5}}} \right)^{2/5}}$$

$$+ \frac{\frac{155}{2} + \frac{13}{2}\sqrt{5}}{\sqrt[5]{\frac{273625}{4} + \frac{66025}{4}\sqrt{5} - \frac{53885}{772}\sqrt{45355 - 16826\sqrt{5}} - \frac{1847377065}{772}\frac{\sqrt{5}}{\sqrt{45355 - 16826\sqrt{5}}}}},$$

which repeated massaging reduces to

$$x = 2 + \sqrt[5]{\frac{2189}{100} + \frac{2641}{500}\sqrt{5} - \frac{1}{2500}\sqrt{1436961550 + 641957866\sqrt{5}}}$$

$$+ \sqrt[5]{\frac{2189}{100} - \frac{2641}{500}\sqrt{5} - \frac{1}{2500}\sqrt{1436961550 - 641957866\sqrt{5}}}$$

$$+ \sqrt[5]{\frac{2189}{100} - \frac{2641}{500}\sqrt{5} + \frac{1}{2500}\sqrt{1436961550 - 641957866\sqrt{5}}}$$

$$+ \sqrt[5]{\frac{2189}{100} + \frac{2641}{500}\sqrt{5} + \frac{1}{2500}\sqrt{1436961550 + 641957866\sqrt{5}}}.$$

Likewise, Ramanujan's celebrated singular value, sent in his letter to Hardy, is

$$k_{210} = (\sqrt{2} - 1)^2(2 - \sqrt{3})(\sqrt{7} - \sqrt{6})^2(8 - \sqrt{63}) \qquad (4.31)$$
$$\times \ (\sqrt{10} - 3)^2(4 - \sqrt{15})^2(\sqrt{15} - \sqrt{14})(6 - \sqrt{35}) \,.$$

Indeed, k_{330} and k_{462} have a similar form involving fundamental solutions to Pell's equation (units of real quadratic fields).

Finally, we note that for small N the elliptic integral $K(k_N)$ is correspondingly susceptible to evaluation in terms of Gamma functions. Thus, to go along with our previous evaluation of $K(k_1)$, we have

$$K(k_3) = \frac{3^{1/4}\Gamma(\frac{1}{3})^3}{2^{7/3}\,\pi} \quad \text{and} \quad K(k_7) = \frac{\Gamma(\frac{1}{7})\,\Gamma(\frac{2}{7})\,\Gamma(\frac{4}{7})}{4\,\pi\,\sqrt[4]{7}}. \qquad (4.32)$$

In each case, there is a neater expression in terms of the β-function waiting to be disentombed.

4.3 Crystal Sums and Madelung's Constant

We have seen the power of converting series to products and making other changes of representation. We now introduce Lambert series, which are representations of the form

$$\sum_{n=1}^{\infty} f(n)\frac{x^n}{1 - x^n} = \sum_{n=1}^{\infty} F(n)x^n, \qquad (4.33)$$

where

$$F(n) = \sum_{d|n} f(d), \qquad (4.34)$$

summed over all positive divisors of n, due originally to Laguerre. The identity (4.33) is established by using the binomial theorem and gathering up terms, much as with the partition function above.

Thus, for $f(n) \equiv 1$, we have $F(n) = \tau(n) = \sigma_0(n)$, the number of divisors of n, while $f(n) = n^k(k \neq 0)$ yields $\sigma_k(n)$, the k-th power sum of the divisors. Recall that the Möbius function is defined by $\mu(1) = 1$, $\mu(n) = (-1)^m$ if n is the product of m distinct prime factors in n, and zero otherwise. Then the *Möbius inversion theorem* says that

$$\sum_{d|n} F(d)\,\mu(n/d) = f(n), \qquad (4.35)$$

for *any* arithmetic function f. This is an analogue of Cauchy convolution.

4.3.1   Sums of Squares

Let us observe that

$$\theta_3^m(q) = 1 + \sum_{n \geq 1} r_m(n) q^n, \tag{4.36}$$

where θ_3 is defined by (4.13), and where $r_m(n)$ counts the number of ways of writing $n = \sum_{k=1}^m n_k^2$, again distinguishing order and sign of the integers used.

It is easy to compute a significant number of terms by merely expanding truncations of the series on the right-hand side of (4.36). This is quite effective for small even numbers of squares.

Example 4.2. Two squares.

The first 60 terms of $r_2(n)/4$ are

$$1, 1, 0, 1, 2, 0, 0, 1, 1, 2, 0, 0, 2, 0, 0, 1, 2, 1, 0, 2, 0, 0, 0, 0, 3, 2, 0, 0, 2, 0,$$
$$0, 1, 0, 2, 0, 1, 2, 0, 0, 2, 2, 0, 0, 0, 2, 0, 0, 0, 1, 3, 0, 2, 2, 0, 0, 0, 0, 2, 0, 0,$$

which does not immediately show any clear pattern. However, applying (4.35) to the first 30 terms yields

$$1, 0, -1, 0, 1, 0, -1, 0, 1, 0, -1, 0, 1, 0, -1, 0, 1, 0, -1, 0, 1, 0, -1, 0, 1, 0,$$
$$-1, 0, 1, 0,$$

and the formula is immediately evident. It is

$$r_2(n) = 4 \left(d_1(n) - d_3(n) \right), \tag{4.37}$$

where d_k is the number of divisors of n congruent to k modulo four. Equivalently,

$$\theta_3^2(q) - 1 = 4 \sum_{n \geq 0} (-1)^n \frac{q^{2n+1}}{1 - q^{2n+1}}. \tag{4.38}$$

$\square$

Example 4.3. Four squares.

The series grows much faster ($r_2(n)$ is $O(n^\delta)$ for any $\delta > 0$) and the first 20 terms of $r_4(n)/8$ are

$$1, 3, 4, 3, 6, 12, 8, 3, 13, 18, 12, 12, 14, 24, 24, 3, 18, 39, 20, 18,$$

while Möbius inversion produces

$$1, 2, 3, 0, 5, 6, 7, 0, 9, 10, 11, 0, 13, 14, 15, 0, 17, 18, 19, 0,$$

from which it is obvious that

$$r_4(n) \;=\; 8 \sum_{d|n, 4\nmid d} d, \tag{4.39}$$

and a nice corollary is that since $1|n$, $r_4(n)$ is always positive (Lagrange's famous theorem). □

Example 4.4. Six squares.

Möbius inversion produces

$$12, 48, 148, 192, 300, 336, 948, 768, 716, 1200, 2388, 1344, 2028, 2256, 3700,$$
$$3072, 3468, 3120, 7188, 4800. \tag{4.40}$$

There is clearly structure here, but what? We leave this as a challenge and turn to eight squares. □

Example 4.5. Eight squares.

Möbius inversion now produces

$$11, 6, 27, 64, 125, 162, 343, 512, 729, 750, 1331, 1728, 2197, 2058, 3375, 4096. \tag{4.41}$$

Again, there is clearly some structure here, but what? If we apply inversion to $(-1)^d r_8(d)$ (this is using θ_4 instead of θ_3), we are rewarded with

$$-1, 8, -27, 64, -125, 216, -343, 512, -729, 1000, -1331, 1728, -2197, 2744,$$
$$-3375, 4096.$$

Thus,

$$\theta_3^8(-q) = \theta_4^8(q) = 1 + 16 \sum_{n \geq 1} \frac{(-1)^n n^3 q^n}{1 - q^n}. \tag{4.42}$$

□

We end this subsection by deriving (4.15) (i) as promised. Indeed, we note that this is equivalent to

$$\sum_{n>0} r_2(n)q^{2n} = \sum_{n>0} \frac{(1+(-1)^n)}{2} r_2(n)q^n.$$

This follows immediately from $r_2(2n) = r_2(n)$, given that (4.37) shows $r_2(n)$ only depends on the odd part of n. Of course, we have not proven any of these representations, but only uncovered them.

4.3.2 Multidimensional Sums

Consider the sums

$$\mathcal{M}_2(s) \quad = \quad \sum_{\substack{m,n \in \mathbb{Z} \\ (m,n) \neq 0}} \frac{(-1)^{m+n}}{(m^2+n^2)^{s/2}}, \tag{4.43}$$

$$\mathcal{M}_3(s) \quad = \quad \sum_{\substack{(m,n,p) \in \mathbb{Z} \\ (m,n,p) \neq 0}} \frac{(-1)^{m+n+p}}{(m^2+n^2+p^2)^{s/2}}, \tag{4.44}$$

and higher dimensional versions $\mathcal{M}_N(s)$ defined analogously. In the future, we write $\sum'$ to denote that poles of the summatory are left out. We are primarily interested in the value $M_3(1)$ which is called *Madelung's constant* for sodium chloride, as it is an attempt to count the potential at the origin if alternating charges are placed at all other points of an integer cubic lattice. The physical chemistry literature generally treats (4.43) and (4.44) as well-defined objects. Mathematically, this is far from so. Since these numbers are often computed, something more must be said (see [32]).

Example 4.6. Convergence over increasing circles, squares and diamonds.

1. *Squares.* The sum $b_2^N(s) = \sum_{n=1}^{N} \sum_{m=1}^{N} (n^2 + m^2)^{-s/2}$ converges to an analytic function $b_2(s)$, for $\mathrm{Re}(s) > 0$.

2. *Circles.* The sum $\sum_{n^2+m^2 \le N} (n^2 + m^2)^{-s/2} = \sum_{k \le N} r_2(k)/k^{s/2}$ converges to an analytic function $r_2(s)$, for $\mathrm{Re}(s) > 1/3$, but fails to converge somewhere above $1/4$. This relies on the fact that the average order of $r_2(n)$ is quite well understood, [138]. Where the limit exists, it must agree with that for squares by uniqueness of analytic continuation.

3. *Diamonds.* Consider adding up over increasing diamonds: $|n| + |m| = N$. Then the contribution of each shell is $\sum_{m=0}^{N} 1/\sqrt{N^2 - 2Nm + 2m^2}$ and the limit is the Riemann integral $\int_0^1 1/\sqrt{1 - 2t + 2t^2}\, dt = \sqrt{2}\log\left(1 + \sqrt{2}\right)$. So the terms of the series do not even go to zero.

Thus, the order of summation matters even for these "natural sums." In fact, if we always add these sums over increasing hypercubes, they converge to an analytic limit for $\mathrm{Re}(s) > 0$. So we shall take this as the default meaning of the sum [44, 31, 32]. We also note that these sums converge very slowly, so that direct summation methods are to be avoided. We shall return to the evaluation of $\mathcal{M}_3(1)$ in the next section.

The normalized *Mellin transform* (see also Item 3a at the end of this chapter) is a special form of the Laplace transform, which makes the link between theta functions and zeta functions, as we saw in Chapter 3. In this setting, we recall that

$$\mu_s(f) = \frac{1}{\Gamma(s)} \int_0^\infty f(t) t^{s-1}\, dt. \tag{4.45}$$

Then it is easy to check that $\mu_s(e^{-tn}) = n^{-s}$, and also that

$$\mathcal{M}_2(2s) = \mu_s(\theta_4^2(e^{-t}) - 1), \tag{4.46}$$

and, using (4.38), the fact that

$$\theta_4^2(q) - 1 = \theta_3^2(-q) - 1 = 4 \sum_{n>0} (-1)^n (-q)^{2n+1}/(1 - (-q)^{2n+1})$$

$$= 4 \sum_{n,m\geq 1} (-1)^{n+m-1} q^{m(2n-1)}$$

implies that

$$\mathcal{M}_2(2s) = 4 \sum_{n,m\geq 1} \frac{(-1)^{n+m-1}}{m(2n-1)^s} \tag{4.47}$$

$$= 4 \sum_{n\geq 0} \frac{(-1)^n}{(2n+1)^s} \sum_{m\geq 1} \frac{(-1)^m}{m^s} = -4\,\alpha(s)\beta(s),$$

where as before $\alpha(s) = \sum_{m\geq 1}(-1)^{m+1}/m^s$, the alternating zeta function, and $\beta(s) = \sum_{n\geq 0}(-1)^n/(2n+1)^s$ is the Catalan zeta function ($\beta(2) = G$ is Catalan's constant).

Similar arguments based on (4.39) lead to

$$\mathcal{M}_4(2s) = -8\,\alpha(s)\alpha(s-1), \tag{4.48}$$

and to a corresponding formula for $M_8(s)$. In each of these cases, the values are easily computed from the analytic continuations of the underlying zeta functions as given in Chapter 2. In particular, $\mathcal{M}_2(1) = -1.61554262671282472386\ldots$, and $\mathcal{M}_4(1) = -1.83939908404504706623\ldots$ Moreover, various closed forms exist, such as $\mathcal{M}_4(2) = -4\log(2)$.

We complete this subsection by listing some other formulae. Define $\mathcal{L}_N(2s) = \sum_{n=1}^{\infty} r_N(n)/n^s$. The corresponding $\mathcal{L}_N(2s)$ are known for $N = 2, 4, 6,$ and 8:

$$
\begin{aligned}
\mathcal{L}_2(2s) &= 4\zeta(s)\beta(s), \\
\mathcal{L}_4(2s) &= 8(1 - 4^{1-s})\zeta(s)\zeta(s-1), \\
\mathcal{L}_6(2s) &= 16\zeta(s-2)L_{-4}(s) - 4\zeta(s)\beta(s-2), \\
\mathcal{L}_8(2s) &= 16(1 - 2^{1-s} + 4^{2-s})\zeta(s)\zeta(s-3).
\end{aligned}
$$

From $\mathcal{L}_6(2s)$, one can reverse the steps we employed and discover the formula for $r_6(n)$ left open in (4.40).

Many more formulae are discussed in [44, 38] and references therein. For example, as discovered by Zucker, Glasser, and Robertson, we have similar closed forms for L-series based on the quadratic form $x^2 + 2Py^2$. We let $r_{2,2P}(n)$ be the number of representations of $n = m^2 + 2P\,k^2$ and let $\mathcal{L}_{2,2P}(2s) = \sum_{(n,m)\neq 0}(m^2 + 2P\,n^2)^{-s} = \sum_{n>0} r_{2,2P}(n)n^{-s}$. Then

$$
\mathcal{L}_{2,2P}(2s) = 2^{1-t} \sum_{\mu|P} \mathcal{L}_{\epsilon_\mu \mu}(s)\mathcal{L}_{-8P\epsilon_\mu/\mu}(s)
$$

for the *type two* integers

$$
P = 1, 3, 5, 11, 15, 21, 29, 35, 39, 51, 65, 95, 105, 165, 231,
$$

which we will meet again in Section 5.2. Here $\epsilon_\mu = \left(\frac{-1}{\mu}\right)$ and $L_\mu(s) = \sum_{n\geq 1}\left(\frac{\mu}{n}\right)n^{-s}$, where $\left(\frac{\mu}{n}\right)$ is the Legendre-Jacobi symbol. Thus $L_1(s) = \zeta(s)$ and $L_{-4}(s) = \beta(s)$. See also Section 3.3.

4.3.3   Madelung's Constant

Odd squares are notoriously less amenable to closed forms. Following Hardy, Bateman in [21] gives the following formula for $r_3(n)$. Let

$$
\chi_2(n) = \begin{cases} 0 & \text{if } 4^{-a}n \equiv 7 \pmod{8} \\ 2^{-a} & \text{if } 4^{-a}n \equiv 3 \pmod{8} \\ 3 \cdot 2^{1-a} & \text{if } 4^{-a}n \equiv 1, 2, 5, 6 \pmod{8}, \end{cases}
$$

where a is the highest power of 4 dividing n. Then

$$r_3(n) = \frac{16\sqrt{n}}{\pi} L_{-4n}(1) \chi_2(n)$$

$$\times \prod_{p^2|n} \left(\frac{p^{-\tau}-1}{p^{-1}-1} + p^{-\tau}\left(1 - \frac{1}{p}\left(\frac{-p^{-2\tau}n}{p}\right)\right)\right)^{-1}, \quad (4.49)$$

where $\tau = \tau_p$ is the highest power of p^2 dividing n.

The corresponding formula for $\mathcal{M}_3(s)$ or for $\mathcal{L}_3(s)$ is thus not tractable. We turn to Bessel functions and let K_s be the *modified Bessel function of the second kind*. Then

$$\mathcal{L}_3(2s) = \frac{6\pi}{s}\zeta(2s-2) + \frac{12\pi^{s+1}}{\Gamma(s+1)} \sum_{m=1}^{\infty} r_2(m) m^{s/2} \sum_{n=1}^{\infty} \frac{1}{n^{s-2}} K_s(2\pi n\sqrt{m}).$$

$$(4.50)$$

The second term of (4.50) can be rewritten as

$$\frac{12\pi^{s+1}}{\Gamma(s+1)} \sum_{k>0} k^{\frac{s}{2}} K_s(2\pi\sqrt{k}) \sum_{n^2|k} \frac{r_2(k/n^2)}{n^{2s-2}}.$$

Moreover, these Bessel functions are elementary when s is a half-integer. Most nicely, for "jellium," which is the Wigner sum analogue of Madelung's constant which arises when one considers bathing a positively charged cubic crystal in a continuous background charge, we have

$$\mathcal{L}_3(1) = -\pi + 3\pi \sum_{m>0} r_2(m) \operatorname{cosech}^2(\pi\sqrt{m}),$$

and the exponential rate of convergence is apparent. Exactly analogous is the most accessible expansion for Madelung's constant due to Benson and proved in [44]

$$\mathcal{M}_3(1) = -12\pi \sum_{m,n\geq 0} \operatorname{sech}^2\left(\frac{\pi}{2}((2m+1)^2 + (2n+1)^2)^{1/2}\right), \quad (4.51)$$

in which again the convergence is exponential. Summing for $m, n \leq 3$ produces $-1.747564594\ldots$, correct to 8 digits. If we write $o_3(n)$ for the number of ways of writing n as a sum of two odd squares, this becomes

$$\mathcal{M}_3(1) = -3\pi \sum_{m>0} o_3(m) \operatorname{sech}^2\left(\frac{\pi}{2}\sqrt{m}\right). \quad (4.52)$$

There is a corresponding formula for $\mathcal{M}_3(s)$.

There is also a beautiful formula for θ_2^3 due to Andrews (given with a typographical error in [44]):

$$\theta_2^3(q) = 8 \sum_{n=0}^{\infty} \sum_{j=0}^{2\,n} \left(\frac{1 + q^{4\,n+2}}{1 - q^{4\,n+2}} \right) q^{(2\,n+1)^2 - (j+1/2)^2}. \tag{4.53}$$

From (4.53), the reader will be able to derive almost immediately Gauss's result that *every number is the sum of three triangular numbers*, and is challenged to apply (4.53) to the study of $\mathcal{M}_3(1)$.

Another related class of physically meaningful integrals are the *logarithmic Watson integrals*, L_d which arise in the study of polymers and are studied in [146]:

$$L_d = \frac{1}{\pi^d} \int_0^{\pi} \cdots \int_0^{\pi} \log(d - \sum_1^d \cos(s_k))\, ds_1 \cdots ds_d. \tag{4.54}$$

For $d = 1, 2$, these reduce to $L_1 = (1/\pi) \int_0^{\pi} \log(1 - \cos(t))\, dt = -\log 2$ and $L_2 = (1/\pi^2) \int_0^{\pi} \int_0^{\pi} \log(2 - \cos(t) - \cos(s))\, dt\, ds = 4\beta(2)/\pi - \log(2)$. The evaluation of L_2 is equivalent to

$$\sum_{n=0}^{\infty} \frac{\binom{2\,n}{n}^2}{4^{2\,n+1}\,(2\,n+1)} = \frac{\beta(2)}{\pi}.$$

No closed form is known for $d > 2$. The prior evaluations and numerical exploration are facilitated by the lovely one-dimensional representation

$$L_d = \int_0^{\infty} \frac{e^{-t} - e^{-dt} I_0\,(t)^d}{t}\, dt, \tag{4.55}$$

where I_0 is a Bessel function of the first kind. A number of additional results related to Madelung's constant can be found in a series of papers by Richard Crandall [97, 100, 98, 99].

4.4   Some Fibonacci Sums

Theta functions turn up in quite unexpected places as we now show. The *Fibonacci sequence*, namely

$$1, 1, 2, 3, 5, 8, 13, 21, 34, 55, 89, 144, \cdots ,$$

takes its name from its first appearance in print, which seems to have been in the famous book *Liber Abaci*, published by Leonardo Fibonacci (also known as Leonardo of Pisa) in 1202. He asked:

How many pairs of rabbits can be produced from a single pair in a year if every month each pair begets a new pair which from the second month on becomes productive?

Lest one thinks the problem is imprecise, Fibonacci describes the solution in the text and in the margin. There one finds written vertically

Parium 1 Primus 2 Secundus 3 Tercius 5 Quartus 8 Quintus 13 Sestus 21 Septimus 34 Octauus 55 Nonus 89 Decimus 144 Undecimus 233 Duodecimus 377.

We leave it to the reader to decide that this indeed leads to the Fibonacci sequence, but we do note that "the proof is left as an exercise" seems to have occurred first in *De Triangulis Omnimodis* by Regiomontanus, written in 1464 (but published in 1533). He is quoted as saying, "This is seen to be the converse of the preceding. Moreover, it has a straightforward proof, as did the preceding. Whereupon I leave it to you for homework."

Among its many other contributions such as popularizing Hindu-Arabic notation in the west, *Liber Abaci* contains methods for extracting cube roots, for solving quadratics, and the lovely identity $(a^2 + b^2)(c^2 + d^2) = (ac \pm bd)^2 + (ad \mp bc)^2$, which shows the product of sums of two squares is such a sum.

The Fibonacci sequence occurs in many contexts both serious and quirky. For example, 144 is the only Fibonacci square. A moment's inspection shows that it is generated by

$$F_0 = 1, \qquad F_1 = 1, \qquad F_{n+1} = F_n + F_{n-1}. \tag{4.56}$$

It grows quickly (like rabbits) and is monotonic. In particular, $F_{n+2} > 2F_n$. If we look computationally at F_{n+1}/F_n, for $n = 10, 20, 30, 40$, we obtain the numerical values 1.61818181818, 1.61803399852, 1.61803398875, 1.61803398875, which either the human eye or a constant recognition facility reveals to be the *Golden Mean* $\phi = (\sqrt{5} + 1)/2$, to the precision used.

Indeed, the standard theory of two term linear recurrence relations leads to

$$F_n = \frac{1}{\sqrt{5}} \left(\frac{1 + \sqrt{5}}{2} \right)^n - \left(\frac{1 - \sqrt{5}}{2} \right)^n, \tag{4.57}$$

where $g = (1 - \sqrt{5})/2$ is the other root of $x^2 = x + 1$.

It is easy to check that the sequence in (4.57) satisfies the recursion in (4.56), and has the correct initial conditions. Since $|g| < 1$, it is also easy to see that $F_{n+1}/F_n \to \phi$, as claimed, and to deduce many other identities such as $F_{n+1} F_{n-1} = F_n^2 + (-1)^n$ for $n \geq 2$.

There is a slightly less well known companion *Lucas sequence*, named after the French number theorist Edouard Lucas (1842–1891):

$$L_0 = 2, \qquad L_1 = 1, \qquad L_{n+1} = L_n + L_{n-1}, \tag{4.58}$$

which is correspondingly solved by

$$L_n = \left(\frac{\sqrt{5}+1}{2}\right)^n + \left(\frac{1-\sqrt{5}}{2}\right)^n. \tag{4.59}$$

As both Fibonacci and Lucas sequences are built of geometric sequences, it is clear that we can easily evaluate sums like $\sum_{n=1}^{N} F_n^k$, for positive integer k. What happens for negative integers is more interesting.

A preparatory lemma is useful ([44]):

Lemma 4.7. *For $0 < \beta < \alpha$ with $\alpha\beta = 1$,*

$$\sum_{n=1}^{\infty} \frac{1}{\alpha^n + \beta^n} = \sum_{n=1}^{\infty} \frac{\beta^n}{1 + \beta^{2n}} = \theta_3^2(\beta), \tag{4.60}$$

$$\sum_{n=0}^{\infty} \frac{1}{\alpha^{2n+1} + \beta^{2n+1}} = \sum_{n=0}^{\infty} \frac{\beta^{2n+1}}{1 + \beta^{2n+1}} = \frac{1}{4} \theta_2^2(\beta^2). \tag{4.61}$$

Proof. The proof of the first formula is a consequence of (4.38), discovered in our discussion of sums of squares. This relies on confirming that

$$\sum_{n=1}^{\infty} \frac{\beta^n}{1 + \beta^{2n}} = \sum_{n=0}^{\infty} (-1)^n \frac{\beta^{2n+1}}{1 - \beta^{2n+1}}. \tag{4.62}$$

(Try expanding both sides as double sums.)

The second formula then follows by applying the first to α^2 and β^2, and then subtracting that result from the first to obtain $(\theta_3^2(\beta) - \theta_3^2(\beta^2))/4$, which equals $\theta_2^2(\beta^2)/4$.   □

Two immediate consequences are

$$\sum_{n=0}^{\infty} \frac{1}{F_{2n+1}} = \frac{\sqrt{5}}{4} \theta_2^2 \left(\frac{3-\sqrt{5}}{2}\right) \tag{4.63}$$

$$\sum_{n=1}^{\infty} \frac{1}{L_{2n}} = \frac{1}{4} \theta_3^2 \left(\frac{3-\sqrt{5}}{2}\right) + \frac{1}{4}. \tag{4.64}$$

Two somewhat more elaborate derivations, (see [44], Section 3.7), lead to

$$\sum_{n=1}^{\infty} \frac{1}{F_n^2} = \frac{5}{24}\left(\theta_2^4\left(\frac{3-\sqrt{5}}{2}\right) - \theta_4^4\left(\frac{3-\sqrt{5}}{2}\right) + 1\right) \quad (4.65)$$

$$\sum_{n=1}^{\infty} \frac{1}{L_n^2} = \frac{1}{8}\left(\theta_3^4\left(\frac{3-\sqrt{5}}{2}\right) - 1\right). \quad (4.66)$$

Since it is known that the classical theta functions are transcendental for algebraic values $q, 0 < |q| < 1$, we discover the far-from-obvious result that the left-hand side of each of (4.63), (4.64), (4.66) is a transcendental number, as probably is (4.65).

Moreover, since both the initial sums and especially the theta functions are easy to compute numerically, we can hunt for other such identities using integer relation methods. In this way, we find:

$$\sum_{n=1}^{\infty} \frac{(-1)^n}{F_n^2} = \frac{5}{48}\left(2 - \theta_2^4\left(\frac{3-\sqrt{5}}{2}\right) - 2\theta_4^4\left(\frac{3-\sqrt{5}}{2}\right)\right), \quad (4.67)$$

and a host of more recondite identities.

By contrast, a remarkable elementary identity is

$$\sum_{n=0}^{\infty} \frac{1}{F_{2n+1} + F_{2k-1}} = \frac{(2k-1)\sqrt{5}}{2\,F_{2k-1}}, \quad (4.68)$$

for $k = 1, 2, 3, \cdots$. So while $\sum_{n=0}^{\infty} F_{2n+1}^{-1}$ is transcendental, $\sum_{n=0}^{\infty}(F_{2n+1} + 1)^{-1} = \sqrt{5}/2$. If we compute the corresponding continued fractions of the two sums, we obtain the quite different results

$$[1, 1, 4, 1, 2, 3, 6, 2, 1, 3, 1, 189, 1, 3, 12] \quad \text{and} \quad [1, 8, 2, 8, 2, 8, 2, 8, 2, 8]$$

in partial confirmation.

4.5   A Characteristic Polynomial Triumph

We illustrate the possibilities of computing with symbolic characteristic polynomials, with an example arising in partial factorizations relating to double Euler sums (see Section 3.5). The rationale for looking at these matrices was discussed in the previous chapter. Consider $n \times n$ matrices A, B, C

$$A_{kj} = (-1)^{k+1}\binom{2n-j}{2n-k}, \qquad B_{kj} = (-1)^{k+1}\binom{2n-j}{k-1},$$

$$C_{kj} = (-1)^{k+1} \binom{j-1}{k-1}$$

$(k, j = 1, \cdots, n)$, and a composite matrix

$$M = A + B - C.$$

We aim to prove M is invertible, indeed that

$$M^{-1} = \frac{M+I}{2}.$$

The key is discovering the following theorem:

Theorem 4.8.

$$A^2 = C^2 = I, \qquad BC = A, \qquad B^2 = CA. \tag{4.69}$$

It follows that $B^3 = BCA = AA = I$, and that the *group generated by A, B and C is the symmetric group S_3.* Once (4.69) is discovered, combinatorial proofs are quite routine, either for a human or a computer, as we now show. It will help to look at Lemma 3.7.

Proof.

$$(A^2)_{kj} = (-1)^{k+1} \sum_{i=1}^{n} (-1)^{i+1} \binom{2n-i}{2n-k}\binom{2n-j}{2n-i}$$

$$= (-1)^{k+1} \sum_{i=n+1}^{2n} (-1)^{i} \binom{i-1}{2n-k}\binom{2n-j}{i-1} = (-1)^{k+1}(-1)^{j+1}\delta_{kj}.$$

$$(C^2)_{kj} = (-1)^{k+1} \sum_{i=1}^{n} (-1)^{i+1} \binom{i-1}{k-1}\binom{j-1}{i-1} = (-1)^{k+1}(-1)^{j+1}\delta_{kj}.$$

$$(BC)_{kj} = (-1)^{k+1} \sum_{i=1}^{n} (-1)^{i+1} \binom{2n-i}{k-1}\binom{j-1}{i-1} = (-1)^{k+1}\binom{2n-j}{2n-k}.$$

It follows also that $AC = BC^2 = B$, and similarly $AB = AAC = C$. To prove the third identity, we proceed as follows:

$$(B^2)_{kj} - (CA)_{kj} = (-1)^{k+1} \sum_{i=1}^{n} (-1)^{i+1} \binom{2n-i}{k-1}\binom{2n-j}{i-1}$$

$$- (-1)^{k+1} \sum_{i=1}^{n} (-1)^{i+1} \binom{i-1}{k-1}\binom{2n-j}{2n-i}$$

$$= (-1)^{k+1} \sum_{i=1}^{n} (-1)^{i+1} \binom{2n-i}{k-1}\binom{2n-j}{i-1}$$

$$+ (-1)^{k+1} \sum_{i=n+1}^{2n} (-1)^{i+1} \binom{2n-i}{k-1}\binom{2n-j}{i-1}$$

$$= (-1)^{k+1} \binom{j-1}{2n-k} = 0.$$

Then $B^3 = BCA = A^2 = I$ follows from the other identities. Finally, $CB = AB^2 = ACA = BA$, and the group is indeed S_3. Additionally, one now easily shows $M^2 + M = 2I$ as formal algebra, using (4.69) and its consequences, since $M = A + B - C$.   □

The truth is that after unsuccessfully peering at various instances of M, the authors of ([50]) decided to look at instances of "$minpoly(M, x)$" and then, emboldened, tried "$minpoly(B, x)$" in *Maple*, when the minimal polynomial for all $n < 8$ of M was the same quadratic $t^2 + t - 2$. By contrast, random $n \times n$ matrices have full degree minimal polynomials, as is guaranteed by the Cayley-Hamilton theorem.

By chance, a much weaker related fact appeared as *American Mathematical Monthly* Problem 01735 in 1999.

Problem. If L_n is the n-by-n matrix with i, j-entry equal to $\binom{i-1}{j-1}$, then $L_n^2 \equiv I_n$ mod 2, where I_n is the n-by-n identity matrix. Show that if R_n is the n-by-n matrix with the i, j-entry equal to $\binom{i-1}{n-j}$, then $R_n^3 \equiv I_n$ mod 2.

Solution: Let A, B, C be the $n \times n$ matrices with i, j-entries given by

$$A_{ij} = (-1)^j \binom{n-i}{n-j}, \quad B_{ij} = (-1)^j \binom{i-1}{n-j}, \quad C_{ij} = (-1)^j \binom{i-1}{j-1}.$$

Since $L_n \equiv C$ mod 2 and $R_n \equiv B$ mod 2, it suffices to prove that $C^2 = I_n$ and $B^3 = -I_n$, which is entirely analogous to the proof of (4.69) given above, and also $A^2 = I_n$. By contrast, the minimum polynomial of L_n is $t \mapsto (t-1)^n$ and that for R_n is less elegant.   □

In a related analysis, for $n > 3$, however, the corresponding $(n-1) \times (n-1)$ matrix $\tilde{M} = \tilde{A} + \tilde{B} - \tilde{C}$, with

$$\tilde{A}_{kj} = (-1)^{k+1} \binom{2n-j-1}{2n-k-1}, \quad \tilde{B}_{kj} = (-1)^{k+1} \binom{2n-j-1}{k-1},$$

$$\tilde{C}_{kj} = (-1)^{k+1} \binom{j-1}{k-1}, \quad \text{for } j, k = 1, \cdots, n-1,$$

arose, and has minimal polynomial $\tilde{M}^3 + 2\tilde{M}^2 - 3\tilde{M} = 0$.

This may be proved in much the same way as in the previous case. It follows from analysis of the trace of $\tilde{M}$ and of $\tilde{M}^2$ that the number of null eigenvalues is $\lfloor (n-1)/3 \rfloor$, and, since the minimal polynomial has no repeated roots, that the dimension of the null space is $\lfloor (n-1)/3 \rfloor$.

The characteristic or the minimal polynomial, like partial fractions, is an object brought fully to life by computation. In much the same way Jordan Forms and other normal forms can be productively used to study singular values—in the matrix sense!

4.6   Commentary and Additional Examples

1. **Partitions with all parts odd.** Prove, analytically and combinatorially, that the number of partitions of n with all parts odd equals the number of partitions of n into distinct parts.

2. **The partition function of 5n+4 is divisible by 5.**

   Proof Sketch. With Q as in the text above, we obtain

   $$
   \begin{aligned}
   qQ^4(q) &= q\,Q(q)q^3(q) && (4.70) \\
   &= \sum_{m\geq 0}\sum_{n=-\infty}^{\infty}(-1)^{n+m}(2m+1)q^{1+(3n+1)n/2+m(m+1)/2},
   \end{aligned}
   $$

   from the triple product and pentagonal number theorems. Now consider when k is a multiple of 5, and discover this can only happen if $2m+1$ is divisible by 5 as is the coefficient of q^{5m+5} in $qQ^4(q)$. Then by the binomial theorem,

   $$(1-q)^{-5} \equiv (1-q^5)^{-1} \mod 5,$$

   and so the coefficient of the same term in $qQ(q^5)/Q(q)$ is divisible by 5. Finally,

   $$q + \sum_{n>1}p(n-1)q^n = qQ^{-1}(q) = \frac{qQ(q^5)}{Q(q)}\prod_{m=1}^{\infty}\sum_{n=0}^{\infty}q^{5mn},$$

   as claimed.

3. **A combinatorial determinant problem.** Find the determinant of

$$
\begin{bmatrix}
\binom{n}{p} & \binom{n}{p+1} & \binom{n}{p+2} \\
\binom{n+1}{p} & \binom{n+1}{p+1} & \binom{n+1}{p+2} \\
\binom{n+2}{p} & \binom{n+2}{p+1} & \binom{n+2}{p+2}
\end{bmatrix}
$$

$$
\begin{bmatrix}
\binom{n}{p} & \binom{n}{p+1} & \binom{n}{p+2} & \binom{n}{p+3} \\
\binom{n+1}{p} & \binom{n+1}{p+1} & \binom{n+1}{p+2} & \binom{n+1}{p+3} \\
\binom{n+2}{p} & \binom{n+2}{p+1} & \binom{n+2}{p+2} & \binom{n+2}{p+3} \\
\binom{n+3}{p} & \binom{n+3}{p+1} & \binom{n+3}{p+2} & \binom{n+3}{p+3}
\end{bmatrix}
$$

and its q-dimensional extension as a function of n, p, q. (Taken from [132].)

Solution: The pattern is clear from the first few cases on simplifying in *Maple* or *Mathematica*.

4. **A sum-of-powers determinant.** Find the determinant of

$$
\begin{bmatrix}
\sum_{k=0}^{1} k^4 & \sum_{k=0}^{1} k^4 & \sum_{k=0}^{1} k^4 & \sum_{k=0}^{1} k^4 \\
\sum_{k=0}^{1} k^4 & \sum_{k=0}^{2} k^4 & \sum_{k=0}^{2} k^4 & \sum_{k=0}^{2} k^4 \\
\sum_{k=0}^{1} k^4 & \sum_{k=0}^{2} k^4 & \sum_{k=0}^{3} k^4 & \sum_{k=0}^{3} k^4 \\
\sum_{k=0}^{1} k^4 & \sum_{k=0}^{2} k^4 & \sum_{k=0}^{3} k^4 & \sum_{k=0}^{4} k^4
\end{bmatrix}
$$

and its q-dimensional extension. (Taken from [132].)

Solution: The first few instances of this sequence are

$$1, 4, 216, 331776, 24883200000, 139314069504000000,$$

which can be quickly identified as $(q!)^q$, using the Sloane online sequence recognition tool. This fact can be proved by taking cofactors on the last row, and observing that only the final two entries have nonzero cofactors with value $(q-1)!^{q-1}$.

5. **Putnam problem 1995–B3.** For each positive integer with n^2 digits write the digits as a square matrix in order row by row. Thus 2354 becomes

$$
\begin{bmatrix}
2 & 3 \\
5 & 4
\end{bmatrix}.
$$

Find, as a function of n, the sum of all $9 \cdot 10^{n^2-1}$ such determinants, which arise on assuming that leading digits are non-zero.

Hint: With the help of a symbolic math program, observe that almost all sample matrices of this form have zero determinant. Then use multilinearity of the determinant to reduce the problem to computing the determinant of just one $n \times n$ matrix.

Answer: For $n = 1$ the answer is 45. For $n = 2$, the matrix may be taken to be

$$\begin{bmatrix} 450 & 405 \\ 450 & 450 \end{bmatrix},$$

with determinant 20250. For $n > 2$, the value is zero.

6. **Crandall's integral representation for Madelung's constant.** The following identity is both beautiful and effective—though less effective for computational purposes than Benson's formula. For example, 60 digits of $\mathcal{M}_3(1)$ can be obtained in seconds in *Maple* or *Mathematica* using Benson's identity, while using the numerical quadrature tools of Section 7.4 to compute the integral to the same 60 digits takes roughly one hour runtime on a 2003-era computer. Richard Crandall's formula is derived in [97] from the Andrews formula for θ_2^3. It is

$$\mathcal{M}_3(1) = -\frac{2}{\pi} \int_0^1 r\, dr \int_{-\pi}^{\pi} \frac{1 + 2/(1 + r^{2(1-\sin\theta)})}{(1 + r^{1+\cos\theta})(1 + r^{1-\cos\theta})}\, d\theta$$

$$= -1.7475645946332\ldots \tag{4.71}$$

7. **Repeated exponential integrals.** Show that

(a)

$$\sqrt{\frac{2}{\pi}} \int_0^\infty \frac{e^{-x} \sin(ax)}{\sqrt{x}}\, dx = \frac{\sqrt{\sqrt{1+a^2}-1}}{\sqrt{1+a^2}},$$

(b)

$$\sqrt{\frac{2}{\pi}} \int_0^\infty \int_0^\infty \frac{e^{-x-y} \sin(ax+by)}{\sqrt{x+y}}\, dx\, dy = \frac{\frac{\sqrt{1+\sqrt{1+b^2}}}{\sqrt{1+b^2}} - \frac{\sqrt{1+\sqrt{1+a^2}}}{\sqrt{1+a^2}}}{a - b},$$

and evaluate

(c)

$$\sqrt{\frac{2}{\pi}} \int_0^\infty \int_0^\infty \int_0^\infty \frac{e^{-x-y-z} \sin(ax + by + cz)}{\sqrt{x + y + z}} \, dx \, dy \, dz,$$

for real coefficients a, b and c.

(d) Generalize the results above, by dimension and to cosines.

(e) Evaluate

$$\int_0^\infty \cdots \int_0^\infty \frac{e^{-\left(\sum_{k=1}^n x_k\right)}}{\sqrt{\sum_{k=1}^n x_k}} \, dx_1 \cdots dx_n,$$

for $n = 1, 2, \cdots$. This shows the prior integrals are absolutely convergent.

8. **Andrews' convolution.** Prove or disprove that (4.18) preserves integrality of coefficients in both directions.

Hint: To determine whether $\{b_n\}$ is integer if and only if $\{a_k\}$ is, expand $\prod_{k>0} \left(1 - q^k\right)^{-b_k}$ in one direction, by the binomial theorem, and note that the coefficients are integers when the values of b_k are integers. In the other direction, observe that $b_1 = a_1$ and inductively consider

$$\prod_{k=1}^n \left(1 - q^k\right)^{b_k} \left(1 + \sum_{m=\ell}^\infty a_m q^m\right) = \prod_{n+1}^\infty \left(1 - q^k\right)^{-b_k}.$$

This is the basis for an efficient algorithm but, in a modern computational package, (4.18) is very easy to program and likely to be more efficient, especially as one will rarely want more that a few hundred terms of the product.

9. **Berkeley problem 6.13.15.** Determine the final digit of $23^{23^{23^{23}}}$. *Answer:* The last digit is a "7."

Hint: Maple or *Mathematica* can verify that $23^{23} \equiv 7 \bmod 4$ and $23^{23^{23}} \equiv 7 \bmod 4$. To prove this observed trend, work modulo four and observe that as $\phi(10) = 4$, $3^r \equiv 3^s \bmod 10$ when $r \equiv s \bmod 4$. Then use $3^{23^{23}} \equiv -1 \bmod 4$.

10. **A series with binomial coefficients.** Prove that for all $n \geq 0$,

$$\sum_{k=0}^n \frac{1}{\binom{n}{k}} = (n+1) \sum_{k=0}^n \frac{1}{(n-k+1) 2^k}.$$

Hint: Consider computing, in two different ways, the electrical resistance between two points distance $n + 1$ apart in the n-dimensional unit cube, if every edge has unit resistance.

11. **A binomial coefficient inequality.** (From [140, pg. 137]). Show inductively for $n > 1$ that

$$\frac{4^n}{n+1} < \binom{2n}{n} < 4^n,$$

and for $n > 6$ that

$$\left(\frac{n}{3}\right)^n < n! < \left(\frac{n}{2}\right)^n.$$

12. **An n-th root inequality.** (From [140, pg. 162]). Show that, for all nonnegative numbers a_i,

$$\sqrt[n]{(1 + a_1)(1 + a_2)\cdots(1 + a_n)} \geq 1 + \sqrt[n]{a_1 a_2 \cdots a_n}.$$

13. **A polygon problem.** Count (i) the number of ways a polygon with $n+2$ sides can be cut into n triangles; (ii) the number of ways in which parentheses can be placed in a sequence of numbers to be multiplied, two at a time; and (iii) the number of paths of length $2n$ through an n-by-n grid that do not rise above the main diagonal (*Dijk paths*).

Hint: In each case the sequence starts

$$1, 2, 5, 14, 42, 132, 429, 1430, 4862.$$

The "gfun" package returns the ordinary generating function $4\left(1 + \sqrt{1 - 4x}\right)^{-2}$ and the recursion $(4n+6)u(n) = (n+3)u(n+1)$, which gives rise to the Catalan numbers $(1/(n+1))\binom{2n}{n}$ named after Eugéne Charles Catalan (1814–1894).

14. **A cubic theta function identity.** If we define

$$
\begin{aligned}
a(q) &= \sum_{m,n\in\mathbb{Z}} q^{m^2+mn+n^2} & b(q) &= \sum_{m,n\in\mathbb{Z}} \omega^{n-m} q^{m^2+mn+n^2} \\
c(q) &= \sum_{m,n\in\mathbb{Z}} q^{(n+1/3)^2+(n+1/3)(m+1/3)+(m+1/3)^2},
\end{aligned}
$$

where $\omega = \exp(2\pi i/3)$, then we have a remarkable cubic identity parallel to Jacobi's quartic identity:

$$a^3 = b^3 + c^3, \tag{4.72}$$

and a lovely parameterization of the $_2F_1$ hypergeometric function [49]:

$$F\left(\frac{1}{3}, \frac{2}{3}, 1; \frac{c^3}{a^3}\right) = a, \tag{4.73}$$

which we will meet in another guise in (6.16).

(a) Choosing $q = \exp(-2\pi\sqrt{N/3})$ for rational N, it can be shown that $s_N = c/a$ is an algebraic number expressible by radicals; see [49]. If N is a positive integer, then s_N is the N-th *cubic singular value*. As above, what can we discover computationally about s_N? For example, can we determine radical formulae for the higher order cubic singular values? The following helps the computations. It is known that, in terms of the classical theta functions,

$$\begin{aligned} a(q) &= \theta_3(q)\theta_3(q^3) + \theta_2(q)\theta_2(q^3) \\ b(q) &= (3a(q^3) - a(q))/2 \qquad c(q) = (a(q^{1/3}) - a(q))/2. \end{aligned}$$

The lacunarity of these series allows for very rapid computation.

(b) Compute the product formula for a—it is not very pretty.

$$a(q) = \frac{\left(1-q^2\right)^{21}\left(1-q^4\right)^{345}\left(1-q^6\right)^{8906}\left(1-q^8\right)^{250257}\left(1-q^{10}\right)^{7538421}}{\left(1-q\right)^6\left(1-q^3\right)^{76}\left(1-q^5\right)^{1734}\left(1-q^7\right)^{46662}\left(1-q^9\right)^{1365388}}$$
$$\cdots.$$

(c) While a does not have a nice product, one should persevere

$$\begin{aligned} b(q) &= (1-q)^3\left(1-q^2\right)^3\left(1-q^3\right)^2\left(1-q^4\right)^3\left(1-q^5\right)^3\left(1-q^6\right)^2 \\ &\times \left(1-q^7\right)^3\left(1-q^8\right)^3\left(1-q^9\right)^2\left(1-q^{10}\right)^3\left(1-q^{11}\right)^3 \\ &\times \left(1-q^{12}\right)^2\left(1-q^{13}\right)^3\left(1-q^{14}\right)^3\left(1-q^{15}\right)^2 \cdots. \end{aligned}$$

This turns out to be the key in providing a computer-guided, but very intuitive proof, given in [49].

15. **Triangles inscribed in a sphere.** Show that $3\sqrt{3}/4$ is the maximum area for triangles inscribed in a unit sphere, and is attained only by equilateral triangles inscribed in a great circle of the sphere.

16. **Nests of radicals.** Identify the limits of the following infinite nested radicals and establish a rigorous sense in which the evaluations are justified.

(a)

$$\sqrt{1+2\sqrt{1+3\sqrt{1+4\sqrt{1+5\cdots}}}}$$

(b)

$$\sqrt{6+2\sqrt{7+6\sqrt{2+\sqrt{9+5\cdots}}}}$$

(c)

$$\sqrt[3]{4+\sqrt[3]{10+9\sqrt[3]{16+25\sqrt[3]{22+\cdots}}}}$$

(d)

$$\sqrt[3]{a+\sqrt[3]{a+\sqrt[3]{a+\cdots}}}$$

for $a=\sqrt{5/3}$.

(e)

$$\sqrt[p]{a+\sqrt[p]{a+\sqrt[p]{a+\cdots}}}$$

for $p>1, a>0$.

(f)

$$\sqrt{2-\sqrt{2-\sqrt{2+\sqrt{2-\sqrt{2-\sqrt{2+\sqrt{2-\sqrt{2-\cdots}}}}}}}}.$$

Hint: Find a functional equation for a *large* class of functions so that iteration in that class solves the functional equation uniquely. Many such equations were evaluated informally by Ramanujan. More details are given in [35].

Answers: (a) 3; (b) 4; (c) 2; (d) $\frac{2}{\sqrt{3}}\left(\sqrt[3]{\frac{1}{2}\sqrt{5}+\frac{1}{2}}+\sqrt[3]{\frac{1}{2}\sqrt{5}-\frac{1}{2}}\right)$;

(e) the positive root of $x^p = x+a$;

(f) the positive root of $x^3 + x^2 - 2x = 1$.

17. **Some unconditional sums.** Evaluate

(a)

$$\sum_{k=1}^{\infty}\sum_{l=1}^{\infty}\frac{1}{(4l-1)^{2k}} \qquad \left(=\frac{1}{4}\log(2)\right)$$

(b)

$$\sum_{k=1}^{\infty}\sum_{l=1}^{\infty}\frac{1}{(4l-1)^{2k+1}} \qquad \left(=\frac{1}{8}\pi-\frac{1}{2}\log(2)\right)$$

(c)

$$\sum_{k=1}^{\infty}\sum_{l=1}^{\infty}\frac{1}{(4l-2)^{2k}} \qquad \left(=\frac{1}{8}\pi\right)$$

18. **Some conditional sums.** Evaluate

(a)

$$\sum_{n=1}^{\infty}\sum_{m=1}^{\infty}\frac{(-1)^{m+n}}{(m+n)^{s}} \qquad (=\alpha(s)-\alpha(s-1))$$

(where $\alpha(s)=(1-2^{1-s})\,\zeta(s)$), and thence justify the conditional evaluation $\sum_{n=1}^{\infty}\sum_{m=1}^{\infty}(-1)^{m+n}/(m+n)=\log(2)-1/2$.

(b)

$$\sum_{n=1}^{\infty}\sum_{m=1}^{\infty}(-1)^{m+n}\frac{mn}{(m+n)^{2}} \qquad \left(=\frac{1}{6}\log(2)-\frac{1}{24}\right)$$

(c)

$$\sum_{n=1}^{\infty}\sum_{m=1}^{\infty}\frac{m^{2}-n^{2}}{(m^{2}+n^{2})^{2}} \qquad \left(=\frac{\pi}{4}\right).$$

(Note that exchanging the order changes the sign of the answer.)

19. **A matrix problem.** Let

$$M = \begin{bmatrix} p & q & 1-p-q \\ 1-p-q & p & q \\ q & 1-p-q & p \end{bmatrix}.$$

Determine the behavior of M^{n} when $p+q<1, p>0, q>0$. *Answer:* $3\,M^{n}\to E$, the matrix with all entries 1.

20. **Theta and self-similarity.** For $k=2$ or 4, plot the set

$$\mathcal{K}_{k}(\alpha)=\left\{|q|<1:\left|\frac{\theta_{k}}{\theta_{3}}(q)\right|>\alpha, q\in\mathbf{C}\right\}$$

for various values of $\alpha>0$. For appropriate k, this is shown in Figure 4.2 (see Color Plate II) for $\alpha=1$. Recall that $(\theta_{2}^{4}/\theta_{3}^{4})\,(e^{-\pi s})=(\theta_{4}^{4}/\theta_{3}^{4})\,(e^{-\pi/s})$, for $\mathrm{Re}(s)>0$. Use the complex AGM iteration [44]

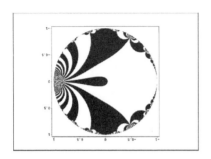 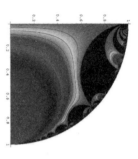

Figure 4.2. Where $|\theta_2/\theta_3| > 1$ and where $|\theta_4/\theta_3| > 1$ (first quadrant)

in θ-form to explore the structures and relations suggested in Figure 4.2. More details on these and like images are given in [102].

David Mumford and his colleagues' book *Indra's Pearls* [170] offers a wealth of important and visually enticing material. In the gloss to their book, they write:

> It is the story of our computer aided explorations of a family of unusually symmetrical shapes, which arise when two spiral motions of a very special kind are allowed to interact. These shapes display intricate "fractal" complexity on every scale from very large to very small. Their visualisation forms part of a century-old dream conceived by the great German geometer Felix Klein.

21. **Putnam problem 1994–B4.** Let d_n be the greatest common divisor of the entries of $A^n - I$ where

$$A = \begin{bmatrix} 3 & 2 \\ 4 & 3 \end{bmatrix}.$$

Show that $d_n \to \infty$ with n.

Hint: Observe numerically, then prove by induction, that A^n has determinant 1 and is of the form

$$\begin{bmatrix} a_n & b_n \\ 2b_n & a_n \end{bmatrix}.$$

Hence, $(a_n-1)|2b_n^2$. Then write A^n explicitly via the Cayley-Hamilton theorem, which tells us that $A^{n+1} = 6\,A^n - A^{n-1}$.

22. **Putnam problem 1999–B5.** Evaluate the determinant of $I + A_n$ where A_n is the $n \times n$ matrix with entries $\cos((j+k)\pi/n)$.

Answer: $I + A_n$ has determinant $1 - n^2/4$. *Hint:* The determinant equals $\prod_{k=1}^{n}(1+\lambda_k)$ where λ_k ranges over the eigenvalues of A_n. One may discover numerically that A_n has eigenvalue zero with multiplicity $n-2$ and remaining eigenvalues $\pm n/2$. Let $v^{(m)}$ denote the vector with $v_k^{(m)} = \exp(ikm\,2\pi/n)$. One may also be led to discover that the eigenvectors are $v^{(0)}, v^{(2)}, v^{(3)}, \cdots, v^{(n-2)}, v^{(1)} \pm v^{(n-1)}$. This is then easy to formally confirm.

23. **Fibonacci and Lucas numbers in terms of hyperbolic functions.** Show that

$$F_n = \frac{2}{\sqrt{5}}\, i^{-n} \sinh(n\theta) \qquad \text{and} \qquad L_n = 2\, i^{-n} \cosh(n\theta),$$

where

$$\theta = \log\left(\frac{\sqrt{5}+1}{2}\right) + i\,\frac{\pi}{2}.$$

Many Fibonacci formulas are then easy to obtain from the addition formulas for sinh and cosh—for example consider $5\,F_n^2 - L_n^2$. (See [110], which should be consulted whenever one "discovers" a result in classical number theory.)

24. **Berkeley problem 7.5.25.** Let M_n^2 be the $n \times n$ tridiagonal matrix with $a_{ij} = 1$ if $|i - j| = 1$ and all other entries zero. (a) Find the determinant of M_n^2, and (b) show that the eigenvalues are symmetric around the origin.

Answer: (a) $\det(M_n^2) = \cos(n\,\pi/2)$. (b) Compute the characteristic polynomial inductively and observe that it contains only odd (respectively even) powers for n odd (respectively even).

25. **Gersgorin circles.** Let $A = \{a_{ij}\}$ be a real $n \times n$ matrix. (a) Show that if

$$|a_{ii}| > \sum_{\substack{t=1 \\ t \neq i}}^{n} |a_{it}|,$$

then A is nonsingular. (b) Deduce that each eigenvalue of A lies in one of the discs

$$|z - a_{ii}| \leq \sum_{\substack{t=1 \\ t \neq i}}^{n} |a_{it}|.$$

Hint: (a) Consider the largest coordinate in absolute value of an element, x, in the null space of A.

26. **Putnam problem 1996–B4.** For a square matrix, A define $\sin(A)$ via the power series

$$\sin(A) = \sum_{n=0}^{\infty} \frac{(-1)^n}{(2n+1)!} A^{2n+1}. \tag{4.74}$$

Prove or disprove that

$$\begin{bmatrix} 1 & 1996 \\ 0 & 1 \end{bmatrix}$$

is in the range of (4.74). *Answer*: It is not. Which matrices are?

Hint: Program the above in a symbolic math program, and observe what the range looks like for various input 2×2 matrices. The result can be obtained by considering the normal form of any A with

$$\sin(A) = \begin{bmatrix} 1 & 1996 \\ 0 & 1 \end{bmatrix},$$

which shows A cannot be diagonalizable. Thus, A must have equal eigenvalues and so is conjugate to a matrix of the form

$$B = \begin{bmatrix} a & b \\ 0 & a \end{bmatrix}.$$

Now determine that

$$\sin(B) = \begin{bmatrix} \sin(x) & y\cos(x) \\ 0 & \sin(x) \end{bmatrix}.$$

Since $\det(\sin(A)) = \det(\sin(B))$, deduce that $|\sin(x)| = 1$ and so $\cos(x) = 0$.

27. **Formula for $\zeta(2,1)$.** Obtain the formula for $\zeta(2,1)$ using the Cauchy-Lindelöf theorem (Theorem 6.10) applied to $\pi\cot(\pi z)\Psi(-z)$ and $1/z^2$.

28. **A log-trig integral.** Show that the following identity holds (due to Victor Adamchik):

$$-\frac{1}{4\pi^2} \int_0^{2\pi} \int_0^{2\pi} \ln\left(3 - [\cos(x) + \cos(y) + \cos(x+y)]\right) dx\,dy$$
$$= \frac{\pi}{\sqrt{3}} + \ln(2) - \frac{\Psi'\left(\frac{1}{6}\right)}{2\sqrt{3}\,\pi}.$$

29. **The Schur functions.** We consider the symmetric functions, Λ, that is, the polynomials in indeterminates x_r, for r in N. (See [161].)

(a) Let $\lambda = (l_1, l_2, \cdots, l_n)$ be a partition of length n. Write $l(\lambda)$ for the length of λ. The corresponding *Schur function* is defined by

$$s_\lambda = \det\left(x_i^{l_j+n-j}\right) / \det\left(x_i^{n-j}\right)_{1 \le i,j \le n}.$$

(b) Show that the *elementary symmetric functions*, $e_0 = 1$ and $e_r = \sum_{l_1 < l_2 < \cdots < l_r} x_{l_1} x_{l_2} \cdots x_{l_r}$, for $r \ge 1$, have an ordinary generating function

$$E(t) = \sum_{r=0}^{n} e_r t^r = \prod_{i=1}^{n} (1 + tx_i).$$

Correspondingly, the *complete symmetric functions*, h_r, have the generating function

$$H(t) = \sum_{r \ge 0} h_r t^r = \prod_{i \ge 1} (1 - tx_i)^{-1}.$$

(c) For $r \ge 1$, the formal r-th *power sum* is defined by $p_r = \sum x_i^r$. Show that the ordinary generating function

$$P(t) = \sum_{r \ge 1} p_r t^{r-1} = \sum_{i \ge 1} \frac{x_i}{1 - x_i t}$$

satisfies $P(t) = E'(t)/E(t)$ and $P(-t) = H'(t)/H(t)$.

(d) Show that the determinant expression for s_λ in terms of the h_r is

$$s_\lambda = \det\left(h_{l_i+i-j}\right),$$

for $1 \le i, j \le n$, and for any $n \ge l(\lambda)$. Find a similar formula in terms of the e_r.

(e) Conclude that every symmetric function is uniquely expressible as a polynomial in the elementary symmetric, the complete symmetric functions or in the Schur functions. That is, each class is a Z-basis for the ring Λ.

(f) Since the h_r and the e_r are each algebraically independent, it is possible to substitute as one wishes. For example

$$H(t) = \prod_{i \ge 0} \frac{1 - bq^i t}{1 - aq^i t}.$$

leads to

$$h_r = \prod_{i=1}^{r} \left(a - bq^{i-1}\right) / \left(1 - q^i\right), e_r = \prod_{i=1}^{r} \left(aq^{i-1} - 1\right) / \left(1 - q^i\right)$$

and

$$p_r = (a^r - b^r)/(1 - q^r).$$

(g) The *natural partial ordering* orders partitions by $\lambda \geq \mu$ if $\sum \lambda_i \lambda_i \geq \sum^i \mu_i$ for all i. Show that for two partitions of n, $\lambda \geq \mu$ if and only if there is a $n \times n$ doubly stochastic matrix S with $\mu = S\lambda$. Represent this result graphically.

30. **Schur functions and Young tableaux.** Another often more useful representation is the *Young diagram*, in which the points of Ferrer's diagram are replaced by squares. The *conjugate partition*, λ', arises on exchanging rows and columns of the diagram. Shading squares allows one to usefully represent adding one partition to another, and much else. We now sketch a combinatorial approach to Schur functions following [202].

A tableaux of *shape* λ is an array $T = (T_{ij})$ of positive integers such that $1 \leq i \leq \ell(\lambda), 1 \leq j \leq \lambda_i$ that is nondecreasing in each row and strictly increasing in each column. We define the *type* $\alpha(T) = (\alpha_1, \alpha_2, \cdots)$ to be the number of occurrences of i in T. Thus, Figure 4.3 shows a *(semistandard) Young tableau* $\widehat{T}$ of shape $(6, 5, 3, 3)$ and of type $(3, 1, 1, 4, 4, 1, 1, 0, 2)$. A *standard tableau* has each row and column strictly increasing.

A natural combinatorial definition of the Schur functions is to define

$$s_\lambda(x) = \sum_T x^T,$$

summed over all monomials $x^T = x^{\alpha_1(T)} x^{\alpha_2(T)} \cdots$ of shape λ. For the tableau illustrated, $x^{\widehat{T}} = x_1^3 x_2 x_3 x_4^4 x_5^4 x_6 x_7 x_9^2$.

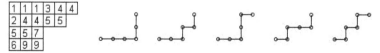

Figure 4.3. A Young tableau along with all lattice paths of shape (3,2)

(a) Write out $s_{(2,1)}$ explicitly.

(b) Let f^λ denote the number of standard tableaux of shape λ. Show that $f^{(3,2)} = 5$ and then show that f^λ counts the number of lattice paths from 0 to λ in $\mathbf{R}^{\ell(\lambda)}$ so that each step is a unit vector staying in the cone $x_1 \geq x_2 \geq \cdots \geq x_l \geq 0$. These paths for $\lambda = (3,2)$ are drawn in Figure 4.3.

(c) Prove that this definition of Schur function coincides with the definition of Exercise 29.

John Stembridge's *Maple* package, http://www.math.lsa.umich.edu/~jrs computes with Schur functions and much else.

31. **Repeated exponentiation.** Recursions like $x_1 = t > 0$ and $x_n = t^{x_{n-1}}$ for $n > 0$ have been subjected to considerable scrutiny.

(a) Study the existence and behavior of

$$\overline{x}_\infty = \lim_{n\to\infty} x_{2n}, \qquad \underline{x}_\infty = \lim_{n\to\infty} x_{2n+1},$$

for $0 \leq x \leq 1$. Note that, in Figure 4.4, the even approximations on $[0,1]$ decrease while the odd ones increase. Thus the limits are taken uniformly by Dini's theorem, which asserts that the monotone limit of continuous functions on a compact set is uniform if and only if the limit is continuous. (Compare Exercises 3 and 10 of Chapter 1.)

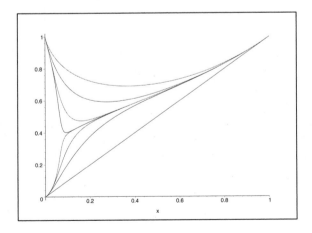

Figure 4.4. Approximations to $x^{x^{x^{x^{\cdot^{\cdot^{\cdot}}}}}}$.

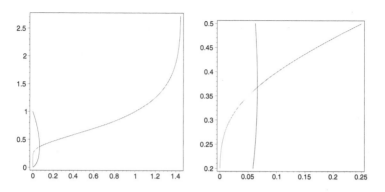

Figure 4.5. Solutions to $x^{x^{x^{x^{\cdots}}}}$.

Hint: Show that (i) the solution to $t^x = x$ is

$$t \mapsto -W\left(-\log t\right)/\log t$$

and that (ii) $\overline{x}_\infty(t)$ and $\underline{x}_\infty(t)$ are the two solutions to $t^{t^x} = x$, which bifurcate at $\hat{b} = \exp(-\exp(1)) \approx 0.06598803584$. In terms of the real branches of the Lambert W function, they are portions of $\exp\left((W_k(t\log t))/t\right)$ for $k = 0, -1$ on $[0, \exp(-1)]$ and on $[\exp(-1), 1]$ respectively. The shared component is

$$-W\left(-\log t\right)/\log t$$

on $[\hat{b}, 1]$ (see Figure 4.5). What happens for $t > 1$?

(The righthand asymptote in Figure 4.5 is at approximately 1.44466786100976613366.) This is discussed in [88].

(b) Estimate $\int_0^1 \overline{x}_\infty(t)\, dt$ and $\int_0^1 \underline{x}_\infty(t)\, dt$.

(c) How many distinct meanings may be assigned to the n-fold exponentiation $x *_n^{x} = x^{x^{x^{\cdots x}}}$?

(d) Show that

$$\int_0^1 (x^x)^x\, dx = 1 + \frac{1}{2}\sum_{n=1}^{\infty}\frac{1}{2^n}\sum_{k=1}^{n}\frac{(-1)^{k+1}}{(k+1)^{n+1}}\binom{n}{k} \qquad (4.75)$$

$$\int_0^1 x^{(x^x)}\, dx = \frac{1}{2} + \sum_{n=1}^{\infty}(-1)^n\sum_{k=1}^{n}\frac{(n-k)^k}{(k+1)^{n+1}}\binom{n}{k}, \qquad (4.76)$$

where the sums are absolutely convergent. Try to further refine both these sums. A useful source for classical combinatorial identities is [188].

(e) Investigate $\int_0^1 x *_n^{\frac{x}{x}} dx$ for more general n.

Proof of (4.76). Note that $0 < x < x^{x^x} < 1$ when $0 < x < 1$, and hence, by dominated convergence, that

$$\lim_{\varepsilon \to 0+} \int_0^1 x^\varepsilon x^{x^x} dx = \int_0^1 x^{x^x} dx.$$

In what follows, suppose that $\varepsilon > 0$. We will prove that

$$\int_0^1 x^\varepsilon x^{x^x} dx = \sum_{n=0}^\infty \sum_{k=0}^\infty (-1)^{k+n} \frac{n^k}{(k+1+\varepsilon)^{n+k+1}} \binom{n+k}{k}, \quad (4.77)$$

where the double sum is absolutely convergent. We have

$$\int_0^1 x^\varepsilon x^{x^x} dx = \int_0^1 x^\varepsilon e^{x^x \log x} dx = \sum_{n=0}^\infty \frac{1}{n!} \int_0^1 x^\varepsilon (x^x \log x)^n dx$$

$$= \sum_{n=0}^\infty \frac{1}{n!} \sum_{k=0}^\infty \int_0^1 x^\varepsilon \log^n x \frac{(xn \log x)^k}{k!} dx$$

$$= \sum_{n=0}^\infty \sum_{k=0}^\infty \frac{n^k}{n!k!} \int_0^1 x^{k+\varepsilon} \log^{n+k} x \, dx.$$

Observe next that

$$\sum_{n=0}^\infty \sum_{k=0}^\infty \frac{n^k}{n!k!} \int_0^1 x^{k+\varepsilon} \left| \log^{n+k} x \right| dx$$

$$= \sum_{n=0}^\infty \sum_{k=0}^\infty \frac{n^k}{n!k!} \int_0^1 x^{k+\varepsilon} \log^{n+k} \left(\frac{1}{x}\right) dx$$

$$= \sum_{n=0}^\infty \sum_{k=0}^\infty \frac{n^k}{n!k!} \int_1^\infty t^{-k-\varepsilon-2} \log^{n+k} t \, dt$$

$$= \int_1^\infty \frac{t^{t^{\frac{1}{t}}}}{t^{2+\varepsilon}} dt < \infty, \quad (4.78)$$

since $t^{t^{1/t}} \sim t$ as $t \to \infty$. It follows from (4.78) that the changes in order of sums and integrals in (4.78) are valid; and hence, on making the change of variable $(k + 1 + \varepsilon) \log x = -u$ in the final integral in (4.78), that

$$\int_0^1 x^\varepsilon x^{x^x}\, dx = \sum_{n=0}^{\infty} \sum_{k=0}^{\infty} \frac{(-1)^{n+k} n^k}{n! k! (k+1+\varepsilon)^{n+k+1}} \int_0^\infty u^{n+k} e^{-u}\, du,$$

which establishes (4.77). We have shown that

$$\int_0^1 x^x\, dx = \lim_{\varepsilon \to 0+} \sum_{n=0}^{\infty} \sum_{k=0}^{\infty} (-1)^{k+n} \frac{n^k}{(k+1+\varepsilon)^{n+k+1}} \binom{n+k}{k}.$$

We cannot automatically replace the limit by

$$\sum_{n=0}^{\infty} \sum_{k=0}^{\infty} (-1)^{k+n} \frac{n^k}{(k+1)^{n+k+1}} \binom{n+k}{k},$$

for this double sum is not absolutely convergent (as is shown by (4.78) with $\varepsilon = 0$), but on treating the terms with $n = 0$ and $n > 0$ separately (as above), one does obtain (4.76).

32. **Fibonacci squares.** Decide whether or not 144 is the only square Fibonacci number. Relatedly, $1729 = 12^3 + 1^3 = 10^3 + 9^3$ is the smallest integer with two distinct representations as a sum of two cubes. What is the next instance?

33. **Fibonacci.** A more precise description of Fibonacci's question mentioned in Section 3.4 is taken from *Scritti di Leonardo Pisano*, pubblicati da Baldassare Boncompagni, Roma, 1857 (Vol. 1), Liber Abaci, chapter XII. This corresponds to folio pages 123 verso and 124 recto of the manuscript *Conversi Soppressi C.1. nr 2616, Codice Magliabechiano* (Biblioteca di Firenze):

> Quot paria coniculorum in uno anno ex uno pario germinentur.

> Quidam posuit unum par cuniculorum in quodam loco, qui erat undique pariete circondatus, ut sciret, quot ex eo paria germinarentur in uno anno: cum natura eorum sit per singulum mensem aliud par germinare; et in secundo mense ab eorum natiuitate germinant. Quia suprascriptum par in primo mense germinat, duplicabis ipsum, erunt paria duo in uno mense. Ex quibus unum, scilicet primum, in secundo mense geminat; et sic sunt in secundo mense paria 3; ex quibus, in uno mense duo pregnantur; et ...

L.E. Sigler translates this as [197]:

How Many Pairs of Rabbits Are Created by One Pair in One Year.

A certain man had one pair of rabbits together in a certain enclosed place, and one wishes to know how many are created from the pair in one year when it is the nature of them in a single month to bear another pair, and in the second month those born to bear also. Because the above written pair in the first month bore, you will double it; there will be two pairs in one month. One of these, namely the first, bears in the second month, and thus there are in the second month three pairs; of these in one month two are pregnant, and ...

Note that it is assumed rabbits are immortal! Alternatively, assume pairs breed after one month and die after two reproductions.

5 | Primes and Polynomials

I hope that ... I have communicated a certain impression of the immense beauty of the prime numbers and the endless surprises which they have in store for us.

Don Zagier, "The First 50 Million Prime Numbers," 1997

Many computational number theory problems involve careful combinatorial search techniques. To be successful, the search space must be reasonably sized. Often significant mathematical insight is essential to trim the search space, and thus is typically more helpful than raw computational power in large, challenging problems.

5.1 Giuga's Prime Number Conjecture

Giuga's prime number conjecture was formulated in 1950 by G. Giuga [124].

Conjecture 5.1. (Giuga's conjecture). *A number $n \in \mathbb{N}$, $n > 1$, is a prime if and only if*

$$s_n = \sum_{k=1}^{n-1} k^{n-1} \equiv n - 1 \ (mod \ n).$$

Whenever n is a prime, then s_n must be congruent to $n - 1$. This is a consequence of Fermat's little theorem: *If p is a prime, then $k^{p-1} \equiv 1 \mod p$) for all $k = 1, \cdots, p-1$.* Thus the question is if there are composite numbers n with $s_n \equiv n - 1 \ (mod \ n)$. It is known that such a number n (a counterexample) must have at least 14,000 decimal digits, if it exists at all. This is a huge number! It is clear that this bound has not been computed by checking every composite number n up to $n = 10^{14,000}$. Instead, Giuga's problem has been converted to a combinatorial search problem, which we shall now describe in detail.

5.1.1   Computation of Exclusion Bounds

The following reformulation of the problem, already given by Giuga, is essential.

Theorem 5.2. *Let $n \in \mathbb{N}$, $n \geq 1$, be given. Then $s_n \equiv n - 1 \bmod n$ if and only if for every prime divisor p of n the relations $(p - 1) \mid (n/p - 1)$ and $p \mid (n/p - 1)$ hold.*

Proof.   As a preliminary consideration, write $n = p \cdot q$ and note that $s_n \equiv q \cdot \sum_{k=1}^{n} k^{n-1} \equiv -q \pmod{p}$ if $(p - 1) \mid (n - 1)$ by Fermat's little theorem, and $s_n \equiv 0 \pmod{p}$ if $(p - 1) \nmid (n - 1)$ since the multiplicative group mod p is cyclic.

Now assume that $s_n \equiv -1 \pmod{n}$. Then the preliminary consideration implies that $(p - 1) \mid (n - 1)$ and $s_n \equiv -1 \equiv -q \pmod{p}$ since $s_n \equiv 0 \pmod{p}$ is impossible. Thus also $(p - 1) \mid (q - 1)$ and $p \mid (q - 1)$.

On the other hand, assume that $(p - 1) \mid (q - 1)$ and $p \mid (q - 1)$. Then also $(p - 1) \mid (n - 1)$, and thus $s_n \equiv -q \pmod{p}$, by the preliminary consideration. The second assumption $q \equiv 1 \pmod{p}$ implies that $s_n \equiv -1 \pmod{p}$, thus $p \mid (s_n + 1)$. This holds for every prime divisor p of n, and since n must be square-free (as $p \mid (q - 1)$ implies $p^2 \nmid n$), we have that $n \mid (s_n + 1)$, thus $s_n \equiv -1 \pmod{n}$.   □

Composite squarefree numbers n that satisfy $(p - 1) \mid (n - 1)$ (or equivalently $(p - 1) \mid (n/p - 1)$), for each prime divisor p of n, are called *Carmichael numbers*. Interestingly, they are pseudoprimes in the sense that Fermat's little theorem is satisfied by them in the following form: The Carmichael numbers are precisely the composite numbers n for which $n \mid (k^n - k)$, for every $k \in \mathbb{N}$. This is Korselt's criterion (1899). The smallest Carmichael numbers are $561, 1105$, and 1729. It has only recently been proved by Alfors, Granville, and Pomerance [5] that there are infinitely many Carmichael numbers. Note that if p is a prime divisor of a Carmichael number n, then for no $k \in \mathbb{N}$ can $kp + 1$ be a prime divisor of n; otherwise, we would have $kp \mid (n - 1)$, a contradiction to $p \mid n$. In particular, every Carmichael number is odd.

Correspondingly, we call a composite integer n with $p \mid (n/p - 1)$ for every prime divisor p of n a *Giuga number*. As seen in the proof of Theorem 5.2, every Giuga number is square-free. The smallest examples are $30, 858$, and 1722; at the present time, only 13 Giuga numbers are known. The largest Giuga number found so far has 97 digits and 10 prime factors (one has 35 digits). It is not known if there are infinitely many Giuga numbers. Nor is it known if there is an odd Giuga number or if any n can be Giuga and Carmichael at the same time.

The following characterization of Giuga numbers is very helpful for finding such numbers. Moreover, it is needed in the computation of exclusion bounds for counterexamples.

Theorem 5.3. *A composite $n \in \mathbf{N}$ with $n > 1$ is a Giuga number if and only if it is squarefree and satisfies*

$$\sum_{p|n} \frac{1}{p} - \prod_{p|n} \frac{1}{p} \in \mathbf{N} \qquad (5.1)$$

where the sum and the product go over all prime divisors p of n.

Proof. Write $n = p_1 \cdots p_m$ and $q_i = n/p_i$. Then n is a Giuga number iff $p_i \mid (q_i - 1)$ for all i iff $p_i \mid (q_1 + \cdots + q_m - 1)$ for all i iff $n \mid (q_1 + \cdots + q_m - 1)$ iff $(q_1 + \cdots + q_m - 1)/n = \sum \frac{1}{p} - \prod \frac{1}{p} \in \mathbf{N}$. $\square$

By Theorem 5.3, the sum over the reciprocals of the prime divisors of a Giuga number must be greater than one. Thus an odd Giuga number must have at least 9 distinct prime factors. Since the product of the 9 smallest odd primes has 10 decimal digits, this already proves, without much computation, and certainly without checking billions of numbers, that a counterexample to Giuga's conjecture must be greater than 10^9. This proof does not even take into account that any counterexample must also be a Carmichael number. Since this condition further restricts the possible prime factors in any counterexample, its systematic use will increase the lower bound significantly. This idea is precisely what led to the computation of the lower bound of $10^{14,000}$, and thus we shall now explain it in more detail. From now on, denote by q_k the k-th *odd* prime; hence, $q_1 = 3$, $q_2 = 5$, $q_3 = 7$, $\cdots$. Denote further $Q = \{3, 5, 7, 11, \cdots\}$ and $Q_k = \{q_1, q_2, \cdots, q_{k-1}\}$.

The precise criterion that is used for these computations follows, as we have seen above, from the Carmichael- and Giuga-ness of any counterexample: *If a number $n = p_1 \cdots p_m$ with $m > 1$ prime factors p_i is a counterexample to Giuga's conjecture (i.e., satisfies $s_n \equiv n - 1 \mod n$), then $p_i \neq p_j$ and $p_i \not\equiv 1 \mod p_j$ for $i \neq j$, and $\sum_{i=1}^{m} 1/p_i > 1$.*

Definition 5.4. *A set of primes is* normal *if it contains no primes p, q such that p divides $q - 1$.*

Thus if 3 is in a normal set of primes, then $7, 13, 19, \cdots$ cannot be in the set, and if 5 belongs, then $11, 31, 41, \cdots$ are excluded. The prime factors of any counterexample form a normal set.

Denote by $\mathcal{N}_k$ the system of all normal subsets of Q_k. For example, $\mathcal{N}_1 = \{\{\}\}$ and $\mathcal{N}_2 = \{\{\}, \{3\}\}$. For fixed k and $N \in \mathcal{N}_k$, denote by $T_k(N)$ the subset of Q determined by the following algorithm: (1) Start with $T = N$ and $j = k$; (2) while $\sum_{p \in T} \frac{1}{p} \leq 1$ do: if $N \cup \{q_j\}$ is normal then $T = T \cup \{q_j\}$; fi; $j = j + 1$; od; (3) return $T_k(N) = T$. By Exercise 4 at the end of this Chapter, this algorithm always terminates and produces a set $T_k(N)$.

It is clear that the sets $T_k(N)$ have the following properties:

(i) $N \subseteq T_k(N)$,

(ii) if $p \in T_k(N) \setminus N$, then $p \geq q_k$ and $N \cup \{p\}$ is normal,

(iii) $\sum_{p \in T_k(N)} \frac{1}{p} > 1$, but $\sum_{\substack{p \in T_k(N) \\ p \neq q}} \frac{1}{p} \leq 1$ for every prime $q \in T_k(N) \setminus N$.

We will be mainly interested in the number of elements of the sets $T_k(N)$ as well as their product; thus set $j_k(N) = |T_k(N)|$, $P_k(N) = \prod_{p \in T_k(N)} p$, and $j_k = \min\{r_k(N) : N \in \mathcal{N}_k\}$, $P_k = \min\{P_k(N) : N \in \mathcal{N}_k\}$. For example,

$$T_1(\{\}) = \{3, 5, 7, 11, 13, 17, 19, 23, 29\}, \quad j_1(\{\}) = 9 \quad \text{and } P_1(\{\}) > 10^9,$$

$$T_2(\{\}) = \{5, 7, 11, 13, 17, \cdots, 107, 109\}, \ j_2(\{\}) = 27 \quad \text{and } P_2(\{\}) > 10^{42},$$

$$T_2(\{3\}) = \{3, 5, 11, 17, \cdots, 317, 347\}, \quad j_2(\{3\}) = 36 \quad \text{and } P_2(\{3\}) > 10^{71}.$$

Thus, $j_1 = 9$, $P_1 > 10^9$, and $j_2 = 27$, $P_2 > 10^{42}$. We have already noted that, since the sum over the reciprocals of the prime factors of a counterexample n must exceed one, n must contain at least 9 prime factors and have at least 10 decimal digits. This is recaptured by the computation of j_1 and P_1 above. However, the computation of j_2 and P_2 now gives a better exclusion bound: Any counterexample n either does or doesn't contain the prime factor 3. If it doesn't, then for every j, the j-th largest element of $T_2(\{\})$ is a lower bound for the j-th largest prime factor of n. Therefore n must have at least 27 different prime factors and must be larger than 10^{42}. If n does contain the prime factor 3, then the elements of $T_2(\{3\})$ are lower bounds for the prime factors of n. Therefore, in this case n has at least 36 distinct prime factors and must be larger than 10^{71}.

In general, we can conclude that every counterexample n has at least j_k distinct prime factors and exceeds P_k, for every $k \in \mathbb{N}$. Both j_k and P_k are increasing in k, as we will see shortly. To find exclusion bounds for counterexamples, one therefore has to compute j_k (then $n \geq \prod_{j=1}^{j_k} q_j$) or P_k (then $n \geq P_k$), for values of k as high as possible. On the surface, this is a daunting task, since the number of normal sets, $|\mathcal{N}_k|$, probably grows exponentially in k. Matters can be accelerated slightly by noting that, since j_k and P_k are minimal values, we do not have to continue the computation of a certain $j_k(N)$ or $P_k(N)$ if their intermediate values already exceed a previously computed $j_k(\widetilde{N})$ or $P_k(\widetilde{N})$.

In this way, with 1950's technology, Giuga [124] computed $j_8 = 323$ and estimated $j_9 > 361$; this gives an exclusion bound of $\prod_{j=1}^{361} q_j > 10^{1039}$. Later, in 1985, E. Bedocchi [22] computed $j_9 = 554$; this gives a bound of $\prod_{j=1}^{554} q_j > 10^{1716}$. In 1994, two of the present authors, with two coauthors [29], computed by the same method $j_{19} = 825$, and thus a lower bound of 10^{2722}. But at this point the method seems exhausted; because of the exponential growth, not much more information can be expected from additional computations.

However, appearances are deceiving—the story does not end here. With one additional bit of insight, the exclusion bounds can be forced up significantly. This bit of insight is to note that the systems $\mathcal{N}_k$ have a tree structure. Consider a normal set $N \in \mathcal{N}_k$. Then this set has one or two successors in $\mathcal{N}_{k+1}$. The first successor is N itself; it is normal and thus contained in $\mathcal{N}_{k+1}$. The second successor may be the set $N' = N \cup \{q_k\}$; this set is contained in $\mathcal{N}_{k+1}$ iff it is normal. The insight now is that the j and P values always increase from N to its successors.

Theorem 5.5. *For each k, $j_{k+1}(N) \geq j_k(N)$, $j_{k+1}(N') \geq j_k(N)$ and $P_{k+1}(N) \geq P_k(N)$, $P_{k+1}(N') \geq P_k(N)$.*

Proof. If $\sum_{p \in N} > 1$, then the assertion is trivial. There is something to prove only when $N \subsetneq T_k(N)$. We have to distinguish two cases.

(1) If $N \cup \{q_k\}$ is normal, then N has the two successors N and N' in $\mathcal{N}_{k+1}$.

Regarding N, the prime q_k is included in $T_k(N)$, but not in $T_{k+1}(N)$. Other than that, we have $T_k(N) \setminus \{q_k\} \subseteq T_{k+1}(N)$, since the normality condition for including primes is the same for both sets. Because of property (iii) above, the prime q_k, missing in $T_{k+1}(N)$, must be compensated by at least one prime higher than any element of $T_k(N)$. Therefore $j_{k+1}(N) = |T_{k+1}(N)| \geq |T_k(N)| = j_k(N)$, and similarly $P_{k+1}(N) \geq P_k(N)$.

Regarding N', $T_k(N)$ can contain primes which are congruent to 1 modulo q_k, but $T_{k+1}(N')$ cannot. $T_k(N)$ minus these primes is a subset of $T_{k+1}(N')$. Therefore, each of these primes has to be compensated by at least one higher prime. Again, we get $j_{k+1}(N') \geq j_k(N)$ and $P_{k+1}(N') \geq P_k(N)$.

(2) If $N \cup \{q_k\}$ is not normal, then the only successor of N in $\mathcal{N}_{k+1}$ is N itself. Since q_k neither appears in $T_k(N)$ nor in $T_{k+1}(N)$, these sets are equal and we have $j_{k+1}(N) = j_k(N)$ and $P_{k+1}(N) = P_k(N)$. $\qquad \square$

The consequence from this theorem is that the computation of the numbers j_k and P_k can be done recursively: Assume that an upper bound u for, say, j_k is already known. Then it is not necessary to compute the j-value of any successor of a normal set with a j-value exceeding u. In other words, the tree of the $\mathcal{N}_k$ can be partially trimmed at an early level. Indeed, this significantly reduces the number of cases to be checked. Our runtimes seem to suggest that this reduces an exponential algorithm to a polynomial algorithm. We cannot prove this reduction, but, of course, we can happily run the computations.

How do we get upper bounds for j_k and P_k, needed for the algorithm? Since these values are minima, it is enough to compute $j_k(N)$ and $P_k(N)$ for one normal set $N \in \mathcal{N}_k$; these will then be upper bounds. Of course, the smaller these bounds are, the faster the algorithm will be, because phony branches of the tree will be trimmed at an earlier stage. Which normal sets give small j_k and P_k values? From his computations, Giuga noticed that sets L_k, defined below, always seem to have the minimal value. Although we cannot prove that these sets always give the minimum, we can at least use their values as good upper bounds. In our computations, we have never found a set with smaller values. These sets are given by

$$L_5 \;=\; \{5,7\} \quad \text{and} \quad L_{k+1} = \begin{cases} L_k \cup \{q_k\} & \text{if this set is normal,} \\ L_k & \text{otherwise.} \end{cases}$$

With this improved, recursive algorithm, we could increase the exclusion bounds significantly. In 1994, using *Maple* on a workstation, we could compute the values j_k up to $k = 100$. For values of k around 100, this took a few CPU hours for each k. We got $j_{100} = 3050$, leading to an exclusion bound of $10^{12,054}$. We then continued the computations in C++ (the recoding took two months), and could extend the range up to $k = 135$, with the result $j_{135} = 3459$ and an exclusion bound of $10^{13,886}$. This algorithm then crashed (irrevocably for linguistic reasons) in the Tokyo Computer Centre before doing any new work. We see here forcibly the dilemma of when to use high-level languages and when to opt for computational speed.

We did not compute the P_k-values in 1994, since the slightly higher exclusion bounds they could give us were more than offset by the additional cost to compute many products. This disadvantage only disappears for high k levels, and recently Holger Rauhut has computed P_{106} on a Sun workstation, leading to an exclusion bound of 10^{14164} (achieved by L_{106}, of course). This computation took about five days.

Note that the set $L_{27,692}$ is normal, has 8,135 elements, and satisfies

$$\sum_{q \in L_{27,692}} 1/q > 1.$$

Therefore, $j_k \leq 8135$ for all $k \geq 27,692$, and the method would be used up at this level. It can never lead to higher and higher exclusion bounds. With current technology, we are still far away from exhausting the algorithm.

5.1.2   Giuga Sequences

As we have seen, if there is a counterexample to Giuga's conjecture, then it is to be found among the Giuga numbers: Composite integers n with $p \mid (n/p - 1)$ for every prime divisor p of n; or, equivalently, square-free composite integers n with $\sum_{p|n} 1/p - \prod_{p|n} 1/p \in \mathbb{N}$.

The problem of finding Giuga numbers can be relaxed to a *combinatorial* problem by dropping the requirement that all factors must be prime. More precisely, we define a *Giuga sequence* to be a finite sequence of integers, $[n_1, \cdots, n_m]$, satisfying $n_j \mid (\prod_{i \neq j} n_i - 1)$. As in the prime case, it follows from this definition that the n_j in a Giuga sequence must be relatively prime; an equivalent definition is: *A sequence of integers* $[n_1, \cdots, n_m]$ *is a Giuga sequence if and only if the n_j are relatively prime and satisfy* $\sum 1/n_j - \prod 1/n_j \in \mathbb{N}$. An example of a nonprime Giuga sequence is

$$\frac{1}{2} + \frac{1}{3} + \frac{1}{7} + \frac{1}{83} + \frac{1}{5 \times 17} - \frac{1}{296310} = 1.$$

For all known examples, the "sum minus product" value is one; to reach any higher value, as we saw, the sequence would have to have at least 59 factors. To find all Giuga sequences of a given length, one could check in principle all sequences of this length whose elements are not too large (the sum over their reciprocals must be greater than one to be ruled out). However, the number of these grows exponentially; even for length seven there are too many to check them all.

Luckily, we have the following reasonably effective theorem which tells us how to find all Giuga sequences of length m with a given initial segment of length $m - 2$.

Theorem 5.6.

(a) Take an initial sequence of length $m - 2$, $[n_1, \cdots, n_{m-2}]$. Let

$$P = n_1 \cdots n_{m-2}, \quad S = 1/n_1 + \cdots + 1/n_{m-2}.$$

Fix an integer $v > S$ (this will be the sum minus product value). Take any integers a, b with $a \cdot b = P(P + S - v)$ and $b > a$. Let

$$n_{m-1} = (P + a)/P(v - S), \quad n_m = (P + b)/P(v - S).$$

Then

$$S + 1/n_{m-1} + 1/n_m - 1/Pn_{m-1}n_m = v.$$

The sequence $[n_1, \cdots, n_{m-1}, n_m]$ *is a Giuga sequence if and only if* n_{m-1} *is an integer.*

(b) Conversely, if $[n_1, \cdots, n_{m-1}, n_m]$ *is a Giuga sequence with sum minus product value* v, *and if we define*

$$a = n_{m-1}P(v - S) - P, \quad b = n_m P(v - S) - P$$

(with P *and* S *the product and the sum of the first* $m - 2$ *terms) then* a *and* b *are integers and* $a \cdot b = P(P + S - v)$.

To conclude the section on Giuga's conjecture, we note Agoh's conjecture (1995), which is equivalent:

$$nB_{n-1} \equiv -1 \pmod{n} \quad \text{if and only if } n \text{ is prime;}$$

here B_n is a *Bernoulli number*.

5.1.3 Lehmer's Problem

Conjecture 5.7. (Lehmer's conjecture [1932]). $\phi(n) \mid (n-1)$ *if and only if* n *is prime.*

Lehmer called this "A problem as hard as the existence of odd perfect numbers."

For Lehmer's conjecture, the set of prime factors of any counterexample n is a normal family. Lehmer's conjecture has now been verified for up to 14 prime factors of n. The related condition,

$$\phi(n) \mid n + 1,$$

is known to have eight solutions with up to six prime factors: $2, F_0, \cdots, F_4$ (the *Fermat primes*), and a rogue pair: 4919055 and 6992962672132095. Fermat primes are primes of the form $F_n = 2^{2^n} + 1$. As an early example of experimental error, the sequence starts $3, 5, 17, 257, 65537$, all of which are prime. On this inductive basis, Fermat conjectured all Fermat numbers were prime, despite the fact that $F_5 = 4294967297$ is divisible by 641, as Euler discovered.

Recently, this result was extended by one to seven prime factors—by dint of a heap of factorizations! But the next cases of Lehmer's two problems (15 and 8, respectively) are much too large for current methods and machines. The *curse of exponentiality* strikes again!

5.2 Disjoint Genera

The role of experimental pattern recognition is reiterated in the description of how the following theorem [55] was found, in response to a question posed by Richard Crandall in his analysis of the Madelung constant [100].

Theorem 5.8. *There are at most 19 positive integers not of the form of* $xy + yz + xz$ *with* $x, y, z \geq 1$. *The only non-square-free cases are 4 and 18. The first 16 square-free cases are*

$$1, 2, 6, 10, 22, 30, 42, 58, 70, 78, 102, 130, 190, \mathbf{210}, 330, 462, \qquad (5.2)$$

which correspond to "discriminants with one quadratic form per genus."

If the 19th exists, it is greater than 10^{11}, *which the* Generalized Riemann Hypothesis *excludes.*

These exceptions were found with a crude MATLAB program, which also showed there were no others less than $50,000$. One may then note that the largest three numbers correspond to three very special singular values ($k_{210}, k_{330}, k_{462}$ of Section 4.2). Further inspection showed that the square-free solutions in (5.2) corresponded precisely to quadratic forms $Q_{2P}(n, m) = n^2 + 2Pm^2$, with precisely one quadratic form per genus, see Section 4.3. In this manner was the theorem discovered.

After the research was nearly finished, the authors remembered to consult Sloane's online encyclopedia and learned that the theorem was true! This was based only on email communications, though there are now several published proofs. Moreover, if you now consult the encyclopedia, it will tell you the numbers are those with one form per genus, and give details! Had the authors consulted the database earlier, they would have considered Crandall's question answered. This would have saved time, but left the database and literature poorer.

5.3 Gröbner Bases and Metric Invariants

In this section, we introduce *Gröbner bases* and the Pedersen-Roy-Szpirglas real solution counting method. We apply these modern tools together with ideas from classical distance geometry (the Cayley-Menger determinant) to show how Petr Lisoněk settled two open problems in Euclidean geometry posed in the *American Mathematical Monthly* in 1999 [165].

By analyzing systems of algebraic equations satisfied by metric invariants of a tetrahedron, we shall conclude (i) in general, the four face areas, circumradius, and volume together do not uniquely determine a tetrahedron, and that (ii) there exist nonregular tetrahedra that are uniquely determined just by the four face areas and circumradius.

The development outlines the main steps of [159], which can be obtained (along with a *Maple* worksheet containing all computations) on the Internet at http://www.cecm.sfu.ca/~lisonek/tetrahedron.html. Executing all computations in *Maple* takes about 40 seconds CPU time on a 2003-era computer.

We should add here that the *Magma* tool is also very useful for performing this type of algebraic computation. Information on *Magma* is available at http://magma.maths.usyd.edu.au/magma.

5.3.1   Formulation of the Polynomial System

Let $\mathbf{R}^n$ denote n-dimensional Euclidean space, with the Euclidean distance between points v and w denoted by $d(v, w)$. Consider a general tetrahedron T in $\mathbf{R}^3$ and let v_i $(i = 1, 2, 3, 4)$ be the vertices of T. Clearly, T is determined uniquely (up to a rigid motion) if the lengths $d(v_i, v_j)$ of its six edges are given. We shall analyze whether T can be determined uniquely by sets of *metric invariants* other than that of all edge lengths.

Denote the *squared edge lengths* by

$$s_{i,j} = d(v_i, v_j)^2, \tag{5.3}$$

the volume by V, and the circumradius (i.e., radius of the circumscribed sphere) by R. We let the area of the face $v_i v_j v_k$ be denoted by A_l, where $\{i, j, k\} \cup \{l\} = \{1, 2, 3, 4\}$.

Let P_i $(1 \le i \le n)$ be n points in $\mathbf{R}^{n-1}$ and denote $d_{i,j} = d(P_i, P_j)$. The *Cayley-Menger determinant* ([27], §40) associated with the points P_i $(1 \le i \le n)$ is the determinant of the $(n + 1) \times (n + 1)$ matrix

$$D(P_1, \cdots, P_n) = \begin{vmatrix} 0 & 1 & 1 & \cdots & 1 \\ 1 & 0 & d_{1,2}^2 & \cdots & d_{1,n}^2 \\ 1 & d_{2,1}^2 & 0 & \cdots & d_{2,n}^2 \\ \vdots & \vdots & \vdots & \ddots & \vdots \\ 1 & d_{n,1}^2 & d_{n,2}^2 & \cdots & 0 \end{vmatrix}. \tag{5.4}$$

The $(n - 1)$-dimensional volume $\mathrm{Vol}(P_1, \cdots, P_n)$ of the convex hull of $P_1, \cdots, P_n$ satisfies the equality ([27], page 98)

$$\mathrm{Vol}(P_1, \cdots, P_n)^2 = \frac{(-1)^n}{2^{n-1}((n-1)!)^2} D(P_1, \cdots, P_n). \tag{5.5}$$

Hence, the facial areas and the volume of the tetrahedron can be expressed as low-degree polynomials in the squared edge lengths $s_{i,j}$. In the case of a triangle, T, (5.5) is a disguised version of the *Heron's formula*:

$$A(T) = \sqrt{s(s-a)(s-b)(s-c)}, \qquad (5.6)$$

where $s = (a+b+c)/2$ is the *semiperimeter* of the circle, and a, b, c are the lengths of the sides (Exercise 13).

Let C denote the center of the circumscribed sphere of our tetrahedron. The five points v_1, v_2, v_3, v_4, C lie in a three-dimensional space (a hyperplane of $\mathbb{R}^4$), and therefore $\mathrm{Vol}(v_1, v_2, v_3, v_4, C) = 0$. Hence, the algebraic equation for the circumradius R is

$$D(v_1, v_2, v_3, v_4, C) = 0, \qquad (5.7)$$

where $d(v_i, C) = R$ for $1 \le i \le 4$. Consider seven positive real numbers $a_{1,2}, a_{1,3}, \cdots, a_{3,4}, W$. Clearly, the statement "The volume of the tetrahedron with edge lengths $\sqrt{a_{1,2}}, \sqrt{a_{1,3}}, \cdots, \sqrt{a_{3,4}}$ is W" implies the statement "The point $(a_{1,2}, a_{1,3}, \cdots, a_{3,4}, W)$ is a zero of the polynomial $f_1(s_{1,2}, s_{1,3}, \cdots, s_{3,4}, V)$." Here the explicit form of f_1 can be easily extracted from (5.3–5.5) (see Exercise 13b at the end of this chapter).

The analogous statements for the other metric invariants of the tetrahedron (e.g., circumradius, four face areas) imply corresponding statements about zeroes of five other computable polynomials $f_2, \cdots, f_6$ with rational coefficients.

Since we transform a geometric problem into an algebraic one, we should make sure that we are working with concrete algebraic objects that have a meaning (interpretation) back in the original geometric domain. The essential question here is for which positive sextuples is $(s_{1,2}, s_{1,3}, \cdots, s_{3,4}) \in \mathbb{R}^6$ does there exist a tetrahedron whose squared edge lengths are the values $s_{i,j}$? It is shown in [27], §40 that this is the case exactly when all squared volumes (5.5) evaluate to nonnegative, which therefore is not only a necessary, but also a sufficient condition for the existence of the tetrahedron.

The important consequence is that whenever there exists a positive solution $(a_{1,2}, a_{1,3}, \cdots, a_{3,4}) \in \mathbb{R}^6$ to a system $\{f_1^* = 0, \cdots, f_6^* = 0\}$, where f_i^* is f_i with $V, R, A_1, \cdots, A_4$ replaced by positive real constants, then this solution does have a geometric interpretation and corresponds to a tetrahedron in $\mathbb{R}^3$, whose edge lengths are $\sqrt{a_{i,j}}$.

Since scaling does not affect the answer to our problem in any way, we typically allow ourselves to normalize the value of one of the variables, say R.

5.3.2   Gröbner Bases

Let $\mathbf{Q}[\mathbf{x_1}, \cdots, \mathbf{x_n}]$ denote the ring of n-variable polynomials with rational coefficients in indeterminates $x_1, \cdots, x_n$. Let $f_1, \cdots, f_s \in \mathbf{Q}[\mathbf{x_1}, \cdots, \mathbf{x_n}]$. The *ideal generated by* $f_1, \cdots, f_s$ is defined by

$$\langle f_1, \cdots, f_s \rangle = \left\{ \sum_{i=1}^{s} h_i f_i \ : \ h_i \in \mathbf{Q}[\mathbf{x_1}, \cdots, \mathbf{x_n}] \right\} .$$

The (affine) *variety* determined by $f_1, \cdots, f_s$ is $\mathbf{V}(f_1, \cdots, f_s) \subset C^n$ defined as

$$\mathbf{V}(f_1, \cdots, f_s) = \{ x \in C^n \ : \ f_1(x) = \cdots = f_s(x) = 0 \} .$$

If $I = \langle f_1, \cdots, f_s \rangle$, then we say that $\{f_1, \cdots, f_s\}$ is a *basis* of I. Clearly, if $\{f_1, \cdots, f_s\}$ and $\{g_1, \cdots, g_t\}$ are two bases of the same ideal, then $\mathbf{V}(f_1, \cdots, f_s) = \mathbf{V}(g_1, \cdots, g_t)$. That said, one of the bases may be much more suitable than the other for assessing properties of the variety (such as nonemptiness, cardinality, finiteness, dimension, characterizing all points in the variety, etc.).

Roughly speaking, *Gröbner bases* are those special bases of polynomial ideals that are especially suited for studying properties of affine varieties such as those listed above, and for answering questions about the ideal itself—such as determining whether or not a given polynomial belongs to the ideal. There are many different Gröbner bases for each polynomial ideal, all of which can be computed using Bruno *Buchberger's algorithm*. Which Gröbner basis is most suitable in a given situation depends on the question to be resolved for the given variety or ideal. The book [92] is a very accessible introduction to polynomial ideals, their varieties, and the algorithms operating on them.

5.4   A Sextuple of Metric Invariants

As noted above, the question as to whether a tetrahedron is uniquely determined by its volume, circumradius and face areas was posed as an open problem in [165]. In this section, we constructively answer this question negatively, by producing two (or more) tetrahedra that share the same volume, circumradius, and face areas.

Consider a tetrahedron T given by squared edge lengths. Equations for the volume, circumradius and face areas of T may be obtained on substituting the values of $s_{i,j}$ into the polynomials $f_1, \cdots, f_6$ introduced earlier. Conversely, by substituting the numerical values of $V, R, A_1, \cdots, A_4$

into the f_i, we set up a polynomial system in the squared edge lengths as unknowns—let us call this system $F^* = \{f_1^*, \cdots, f_6^*\}$.

Thus, we hope for distinct positive solutions to F^*. To determine these solutions we first find G, a *grevlex Gröbner basis* for F^*. In all numerical examples that Lisoněk and Israel [159] tried, they found that $\langle F^* \rangle$ was a zero-dimensional ideal. They then used the method of Pedersen, Roy and Szpirglas to count all real solutions of F^*. (See Chapter 2 of [91]; we adopt the terminology and notation introduced therein.)

Let $p_{i,j}$ be the generators for the univariate elimination ideals; that is, $\langle p_{i,j} \rangle = \mathrm{C}[s_{i,j}] \cap \langle F^* \rangle$. Using the theory in Chapter 2, Section 2 of [91], first one finds the $p_{i,j}$ from the grevlex Gröbner basis G by working in the algebra $A = \mathrm{C}[s_{1,2}, s_{1,3}, \cdots, s_{3,4}] / \langle G \rangle$. Secondly, one uses interval arithmetic (together with knowledge of the total number of real solutions) to isolate these real solutions of F^*. Thirdly, one selects the positive solutions.

It turns out to be easy to obtain examples in which several different tetrahedra share the same volume, circumradius and face areas, as is illustrated in the following numerical example. Some intermediate expressions are omitted because of their large size.

Example 5.9. Tetrahedral volume.

Consider the tetrahedron defined by

$$(s_{1,2}, s_{1,3}, s_{1,4}, s_{2,3}, s_{2,4}, s_{3,4}) = (1, 1, 2, 2, 1, 2).$$

We find that $V = \sqrt{1/48}$, $R = \sqrt{7/12}$, $A_1 = A_2 = \sqrt{7/16}$ and $A_3 = A_4 = 1/2$.

1. By substituting these six values in the polynomials $f_1, \cdots, f_6$, we obtain the set $F^* = \{f_1^*, \cdots, f_6^*\}$. Next, compute G, the Gröbner basis for F^* with respect to the grevlex ordering induced by $s_{1,2} > s_{1,3} > \cdots > s_{3,4}$, and the basis for the algebra $A = \mathrm{C}[s_{1,2}, s_{1,3}, \cdots, s_{3,4}] / \langle F^* \rangle$.

2. It turns out that the dimension of A is 8, and that a monomial basis of A is $B = (1, s_{1,2}, s_{2,3}, s_{2,4}, s_{2,4}^2, s_{1,2}s_{2,4}, s_{3,4}, s_{2,4}s_{3,4})$. Let h be the constant function 1 and construct the symmetric bilinear form S_1 as described in [91, page 65].

3. Compute the characteristic polynomial of M_1, the matrix of S_1 with respect to B. It turns out that there are six sign variations in the coefficient list of this characteristic polynomial, whence by Déscartes' rule of signs, the signature of S_1 is four, and there are four distinct real solutions to the system F^* by [91, Theorem 5.2].

4. Using the *Maple* procedure Groebner[univpoly] (which implements the Faugère-Gianni-Lazard-Mora basis conversion method), we use G to find the generators $p_{i,j}$ for the univariate elimination ideals $\langle p_{i,j} \rangle = C[s_{i,j}] \cap \langle F^* \rangle$. They turn out to be the following polynomials:

$$
\begin{aligned}
p_{1,2}(x) &= (x-1)a(x) \\
p_{1,3}(x) &= p_{1,4}(x) = p_{2,3}(x) = p_{2,4}(x) \\
&= (x-1)(x-2)b(x) \\
p_{3,4}(x) &= (x-2)c(x),
\end{aligned}
$$

where

$$
\begin{aligned}
a(x) &= 9x^3 - 123x^2 + 491x - 249 \\
b(x) &= 81x^6 - 1485x^5 + 10215x^4 - 25803x^3 + 22865x^2 \\
&\quad - 3303x + 162 \\
c(x) &= 9x^3 - 150x^2 + 1028x - 960.
\end{aligned}
$$

5. Using Sturm's theorem (the *Maple* procedure realroot), we determine the isolating intervals for all real roots of a, b, and c. It turns out that a and c have one real root each, namely $\alpha_1 \in [75/128, 19/32]$ and $\gamma_1 \in [35/32, 141/128]$, respectively, while b has two real roots, $\beta_1 \in [219/128, 55/32]$ and $\beta_2 \in [267/128, 67/32]$.

6. These isolating intervals are narrow enough to prove (using interval arithmetic) that only four boxes in $\mathbf{R}^6$ (Cartesian products of the separating intervals) can possibly contain a solution of F^*.

7. Since we know that there are exactly four real solutions, we have isolated all solutions of F^*. Their values $(s_{1,2}, s_{1,3}, s_{1,4}, s_{2,3}, s_{2,4}, s_{3,4})$ are $(1,2,1,1,2,2)$, $(1,1,2,2,1,2)$, $(\alpha_1, \beta_1, \beta_2, \beta_2, \beta_1, \gamma_1)$, and $(\alpha_1, \beta_2, \beta_1, \beta_1, \beta_2, \gamma_1)$.

8. As each $s_{i,j}$ value is positive for all four solutions, it follows (see the discussion at the end of Section 5.3.1) that we have constructed four different tetrahedra that share the same volume, circumradius and face areas. On inspecting the four solutions, we see that the four tetrahedra pair up as two pairs, with the elements of each pair related by interchanging the labels of the vertices v_1 and v_2. Thus, we have obtained two essentially different tetrahedra.   □

5.5 A Quintuple of Related Invariants

It was noted in [165], part (b), that the quintuple of values (A_1, A_2, A_3, A_4, R) uniquely determines any *regular* tetrahedron. In this case, of course $A_1 = A_2 = A_3 = A_4 = \sqrt{4/3}R^2$. Whether there exist *nonregular* tetrahedra uniquely determined by (A_1, A_2, A_3, A_4, R) was posed as an open problem in [165]. This question is answered affirmatively on showing:

Example 5.10. Every nondegenerate tetrahedron, having a face that is an equilateral triangle inscribed in a great circle of the circumscribed sphere, is determined uniquely by its four face areas and circumradius.

1. Assume without loss of generality that $R = 1$. It is easy to see that $3\sqrt{3}/4$ is the maximum area among all triangles inscribed in a unit sphere, and it is attained only by equilateral triangles inscribed in a great circle of the sphere (Exercise 15 of Chapter 4).

2. Therefore the shape of one of the faces of the tetrahedron is determined by requiring its area to be equal to $3\sqrt{3}/4$. Assume this face is $v_1v_2v_3$. Then $R = 1$, $s_{1,2} = s_{1,3} = s_{2,3} = 3$ and $A_4 = 3\sqrt{3}/4$.

3. Let F' be the set of polynomials $f_2, \cdots, f_6$ from Section 5.3.1 with the values from the previous sentence substituted. It takes only a few seconds to compute the Gröbner basis for the ideal generated by F' for the lexicographic ordering induced by $s_{3,4} > s_{2,4} > A_1 > A_3 > A_2 > s_{1,4}$. This basis contains (among others) the polynomials f^2 and g^2, where

$$
\begin{aligned}
f &= 81s_{1,4}^4 - 432s_{1,4}^3 + 288(A_2^2 + A_3^2)s_{1,4}^2 \\
&\quad + 256A_2^4 + 256A_3^4 - 512A_2^2A_3^2 \\
g &= 9s_{1,4}^2 - 54s_{1,4} - 16A_1^2 + 32A_2^2 + 32A_3^2 + 27.
\end{aligned}
$$

4. Further,

$$
f = Q \cdot g + S, \tag{5.8}
$$

where

$$
\begin{aligned}
Q &= 9s_{1,4}^2 + 6s_{1,4} + 16A_1^2 + 9 \\
S &= (960A_1^2 - 192A_2^2 - 192A_3^2 + 324)s_{1,4} \\
&\quad + 256(A_1^4 + A_2^4 + A_3^4) - 512(A_1^2A_2^2 + A_1^2A_3^2 + A_2^2A_3^2) \\
&\quad - 288(A_1^2 + A_2^2 + A_3^2) - 243.
\end{aligned}
$$

If both f and g vanish, then S must vanish by (5.8). The equation $S = 0$ determines $s_{1,4}$ uniquely if L, the coefficient at $s_{1,4}$ in S, is nonzero.

5. The discriminant of g as a quadratic polynomial in $s_{1,4}$ is

$$Z = 576A_1^2 - 1152A_2^2 - 1152A_3^2 + 1944.$$

Observe that $6 \cdot L = 5184A_1^2 + Z$.

6. If there exists a tetrahedron with face areas A_1, A_2, A_3, $3\sqrt{3}/4$, and circumradius 1, then g must have a real (positive) root, hence $Z \geq 0$ and consequently $L > 0$ by nondegeneracy, in particular $L \neq 0$. Therefore, $s_{1,4}$ is determined uniquely by $S = 0$.

7. By applying analogous arguments to $s_{2,4}$ and $s_{3,4}$ we prove that all $s_{j,4}$ are determined uniquely ($1 \leq j \leq 3$). Since also $s_{1,2} = s_{1,3} = s_{2,3} = 3$ are determined uniquely by $A_4 = 3\sqrt{3}/4$, we have established that the tetrahedron is determined uniquely. □

5.5.1 Some Open Questions on Invariants

An intriguing question is whether, for *every* set of positive real constants $V, R, A_1, \cdots, A_4$, there are *only finitely many* tetrahedra, all having these values as their respective metric invariants.

If the answer is affirmative, what is then the *maximal number* of such tetrahedra? For example, the values $(s_{1,2}, s_{1,3}, s_{1,4}, s_{2,3}, s_{2,4}, s_{3,4}) = (16, 25, 9, 9, 33, 54)$ yield a family of six such tetrahedra, but it is not known whether this is maximal.

The bounds obtained by applying general theorems from algebraic geometry seem much larger than the empirical results obtained in [159], by running the algorithm outlined above on a number of different examples.

In cases when the number of solutions to the system F^* is finite, one can apply *Bézout's theorem*, and so obtain $3 \cdot 4 \cdot 2^4 = 192$ as an upper bound on the number of solutions. A theorem of Milnor [167] gives 9,375 as an upper bound on the number of real solutions of F^* that are isolated points. Here, the central issue is whether the variety is always zero-dimensional (is a finite set of isolated points), or whether it can have (components of) positive dimension. In the zero-dimensional case, Bézout's theorem applies, while Milnor's result unconditionally bounds the number of isolated real points.

5.6   Sloane's Harmonic Designs

In this final section, we describe and illustrate an ambitious and successful experimental approach to finding spherical designs used by N. J. A. Sloane, R. H. Hardin, and P. Cara [135, 136, 198, 199].

A set of N points $\{P_1, \cdots, P_N\}$ on the unit sphere $\Omega_n = S^{n-1} = \{x = (x_1, \cdots, x_n) \in \mathbf{R}^n : x \cdot x = 1\}$ forms a *spherical t-design* if the identity

$$\int_{\Omega_n} f(x) d\mu(x) = \frac{1}{N} \sum_{i=1}^{N} f(P_i) \tag{5.9}$$

holds for all polynomials f of degree $\leq t$, where μ is uniform measure on the sphere normalized to have total measure one ([107, 125, 126, 127] and [86, §3.2]). A spherical t-design is also a t'-design for all $t' \leq t$. The largest t for which the points form a t-design is called the *strength* of the design.

It is known that if N is large enough, then a spherical t-design in Ω_n always exists [193]. The problem is to find the smallest value of N for a given strength and dimension, or equivalently to find the largest strength t that can be achieved with N points in Ω_n.

Not surprisingly, the main application is to numerical integration, but spherical designs also have applications to the design of experiments in statistics. Their study has involved many interesting questions in algebra and group theory. For searching for a spherical t-design, and for verifying that a set of points does form a spherical t-design, the following equivalent condition is more useful than the definition:

$P_1, \cdots, P_N$ forms a spherical t-design if and only if the polynomial identities

$$\frac{1}{N} \sum_{i=1}^{N} (P_i \cdot x)^{2s} = \left(\prod_{j=0}^{s-1} \frac{2j+1}{2j+n} \right) (x \cdot x)^s \tag{5.10}$$

and

$$\frac{1}{N} \sum_{i=1}^{N} (P_i \cdot x)^{2\bar{s}+1} = 0 \tag{5.11}$$

hold, where s and $\bar{s}$ are defined by $\{2s, 2\bar{s}+1\} = \{t-1, t\}$ ([126], [184, page 114], [185]).

The approach taken by Sloane et al. follows the following three stages.

1. *Experimentation.* Given the dimension n, a specified number of points N, and a desired value of the strength t, search for a set of points satisfying (5.10) and (5.11). If no solution seems to exist, decrease t and try again. This is repeated a large number of times

(with different starting configurations), until a collection of putative designs has been assembled. At this stage, these are only numerical approximations to the desired designs; that is, numerical coordinates for points which appear to satisfy (5.10) and (5.11), with an error less than 10^{-10}. The search algorithm used was a modification of the "pattern search" of Hooke and Jeeves [142], a distant cousin of the conjugate gradient method.

2. *Beautification.* They now attempt to show that there *is* a spherical t-design in the neighborhood of the numerical points, that is, to find algebraic expressions for the coordinates of points, such that (5.10) and (5.11) are satisfied exactly. This step involves a considerable amount of guesswork, guided by computations of the geometric structure of the computer-produced points, such as their apparent automorphism group.

The crucial step in the beautification process is to use knowledge of the automorphism group to reduce the number of unknowns in the design. If the points appear to fall into k orbits under the group, the number of unknowns is reduced from $N(n-1)$ (the number of degrees of freedom in the original design) to $k(n-1)$. With luck, they are now able to solve equations (5.10) and (5.11) exactly.

3. **Generalization.** Try to find infinite families of designs that generalize those found in Step 2.

One example will serve to illustrate the process, a four-dimensional spherical 6-design (in Ω_4) with 42 points. After beautification, this turned out to consist of six heptagons, each in a different plane, with a group of order 42 acting transitively on the 42 points. In other words, the computer-produced design was suggesting that they should choose six planes in R^4, i.e., six points in the Grassmann manifold $G(4,2)$, and draw a heptagon in each plane. Sloane and his colleagues found this an appealing idea, in view of the recent work on finding packings and designs in Grassmann manifolds (see [15, 77, 85, 194])! It immediately suggested several general constructions, one of which is the following.

Let $\Pi_1, \cdots, \Pi_6$ be the planes in four dimensions spanned by the rows of the following six matrices:

$$\begin{bmatrix} 1 & 0 & 0 & 0 \\ 0 & 1 & 0 & 0 \end{bmatrix}, \begin{bmatrix} 0 & 0 & 1 & 0 \\ 0 & 0 & 0 & 1 \end{bmatrix},$$

$$\begin{bmatrix} s & 0 & -h & h \\ 0 & s & -h & -h \end{bmatrix}, \begin{bmatrix} -h & h & s & 0 \\ -h & -h & 0 & s \end{bmatrix},$$

$$\begin{bmatrix} -s & 0 & -h & -h \\ 0 & -s & h & -h \end{bmatrix}, \begin{bmatrix} -h & -h & -s & 0 \\ h & -h & 0 & -s \end{bmatrix},$$

where $s = 1/\sqrt{2}$, $h = 1/2$.

Theorem 5.11. *Let p be an integer ≥ 3 and draw regular p-gons in each plane. The resulting $N = 6p$ points*

$$\{\cos(j\theta)u_i + \sin(j\theta)v_i : 0 \leq j < p,\ 1 \leq i \leq M\}, \tag{5.12}$$

where u_i and v_i span Π_i, form a $6p$-point spherical t-design with $t = \min\{p - 1, 7\}$.

This is an interesting result, since there are very few general constructions known for infinite families of spherical designs.

The theorem yields the following t-designs.

p	3	4	5	6	7	8	9	10	11	12	$\cdots$
N	18	24	30	36	42	48	54	60	66	72	$\cdots$
t	2	3	4	5	6	7	7	7	7	7	$\cdots$

The papers [136, 198, 199] contain many other examples.

5.7 Commentary and Additional Examples

1. **A prime generating number.** Let

$$\alpha = \sum_{m=1}^{\infty} \frac{p_m}{10^{2^m}} = 0.0203000500000007000000000000000110\ldots,$$

where p_m denotes the m-th prime. Show that

$$p_n = \lfloor 10^{2^n}\alpha \rfloor - 10^{2^{n-1}}\lfloor 10^{2^{n-1}}\alpha \rfloor,$$

which would be useful if one could find a method of computing α to the needed precision to obtain p_n without knowing p_n. This seems unlikely.

2. **Carmichael and Lucas-Carmichael numbers.** The Carmichael numbers are usually defined as those *pseudoprimes* such that $a^{n-1} \equiv 1 \bmod n$ for each a relatively prime to n. Korselt (1899) showed this coincides with the definition in the text: Namely n is square-free and $(p - 1)|(n - 1)$ whenever $p|n$. The smallest examples are

$561, 1105, 1729$, and a much larger one is $17 \cdot 37 \cdot 41 \cdot 131 \cdot 251 \cdot 571 \cdot 4159 = 2013745337604001$. We suggest you consult Sloane's table (or the online version) to learn about the sequence starting

$$399, 935, 2015, 2915, 4991, 5719, 7055.$$

3. **Odd perfect numbers.** It is known that there is no odd perfect number with seven or fewer prime factors. Show that any odd perfect number with eight prime factors must be divisible by $3, 5$, or 7. More generally, in 1888 Servais proved that the smallest prime factor cannot exceed the number of prime factors.

4. **Primes in arithmetic progression.** Let $\pi_{d,r}(x)$ denote the number of primes of the form $nd + r \le x$, with d and r relatively prime. Then Dirichlet proved the number of primes in each such progression is infinite. Indeed,

$$\pi_{d,r}(x) \sim \frac{1}{\phi(d)} \frac{x}{\log x}.$$

This is due to de la Vallée Poussin [186, page 149].

   (a) Use this to show that every normal sequence of primes can be extended so that the sum of its reciprocals exceeds one.

   (b) It is conjectured that one can find arbitrarily long arithmetic progressions, all of whose members are prime [186, pg. 153]. The longest known arithmetic sequence of primes is currently 22, starting with the prime 11410337850553, and continuing with common difference 4609098694200 [181]. The longest known sequence of *consecutive* primes in arithmetic progression is ten. It starts with the 93-digit prime

   $$100996972469714247637786655587969840329509324689190 0418$$
   $$0360341775890434170334 8882159067229719,$$

   and has difference 210.

5. **A fourth degree polynomial problem.** (From [140, page 87]). Let $\alpha_1, \alpha_2, \alpha_3, \alpha_4$ be the roots of the polynomial

$$P(x) = x^4 + px^3 + qx^2 + rx + 1.$$

Show that

$$(1 + \alpha_1^4)(1 + \alpha_2^4)(1 + \alpha_3^4)(1 + \alpha_4^4) = (p^2 + r^2)^2 + q^4 - 4pq^2r.$$

Hint: Consider $\prod_{k=1}^{4} P(e^{(2k-1)i\pi/4})$.

6. **Putnam problem 1991–B4.** Show that for p an odd prime,

$$\sum_{j=0}^{p} \binom{p}{j}\binom{p+j}{j} \equiv 2^p + 1 \quad \mathrm{mod}\ p^2.$$

7. **1-additive sequences.** Given (u, v), the 1-additive sequence generated by (u, v) is defined as $a_1 = u$, $a_2 = v$, and for $n \geq 3$, a_n is the least integer exceeding a_{n-1} and possessing a unique representation of the form $a_i + a_j$ for $i < j$. Many 1-additive sequences behave quite erratically. The sequence generated by $(2, 3)$, for example, defies any simple characterization. Finch conjectured, and later Schmerl and Spiegel proved, that the sequence generated by $(2, v)$ for odd $v \geq 5$ has precisely two even terms, so that the sequence of successive differences is eventually periodic. Finch has further conjectured, based on extensive computations, that for odd $v \geq 5$:

Conjecture 5.12. (Finch). *If $v \neq 2^m - 1$ for any $m \geq 3$, the sequence generated by $(4, v)$ has precisely three even terms: $4, 2v + 4$ and $4v + 4$. When $v = 2^m - 1$ for some $m \geq 3$, then the sequence generated by $(4, v)$ has precisely four even terms: $4, 2v + 4, 4v + 4$, and $4v^2 + 2v - 4$.*

Cassaigne and Finch have proven this conjecture for the case $v \equiv 1$ mod 4, but the question is open for $v \equiv 3$ mod 4. See [82] for further details.

8. **Amicable numbers.** Two numbers are *amicable* if, like 220 and 284, each is the sum of the others proper divisors, Thabit ibn Kurrah (ca. 850 CE) noted that if $n > 1$ and each of $p = 3 \cdot 2^{n-1} - 1, q = 3 \cdot 2^n - 1$, and $r = 9 \cdot 2^{2n-1} - 1$ are prime, then $2^n pq$ and $2^n r$ are amicable numbers. It was many years until this formula led to a second and third pair of amicable numbers! Fermat provided the pair 17,296 and 18,416 (n=4) in a letter to Mersenne in 1636. Computer searches have found all such numbers with 10 or fewer digits. It is unknown if there are infinitely many amicable pairs or any relatively prime pair. (Such a pair must be more than 25 digits long, and the product must be divisible by at least 22 distinct primes.)

9. **More on amicable numbers.** The smallest example of an amicable four-cycle is

$$n_1 = 2^2 \cdot 5 \cdot 17 \cdot 3719, \qquad n_3 = 2^2 \cdot 521 \cdot 829,$$
$$n_2 = 2^2 \cdot 5 \cdot 193 \cdot 401, \qquad n_4 = 2^5 \cdot 40787,$$

discovered by Henri Cohen in 1970, during an exhaustive search up to sixty thousand. Very recently, Blankenagel, Borho and vom Stein have obtained 50 new amicable four-cycles by a seed-and-complete method akin to the way in which Giuga sequences were generated.

10. **Aliquot sequences.** Consider the iteration $n \mapsto s(n) = \sigma(n) - n$, where σ is the divisor function. It is unknown whether this iteration must eventually become periodic, as is clearly the case for an amicable pair. There are five numbers less than 1000 whose status is unsettled ("the Lehmer five": 276, 552, 564, 660, and 966). Such Aliquot sequences can grow very rapidly before subsiding. One may consider iterating many other arithmetic functions with irregular growth (see [134]).

11. **Prime power problem.** Let $\Lambda(n) = \log(p)$ for $p = n^m$ a prime power and be zero otherwise. Show that

$$\sum_{n=1}^{\infty} \frac{\Lambda(n)}{n^s} = \frac{\zeta'(s)}{\zeta(s)},$$

and so that

$$\log(n) = \sum_{d|n} \Lambda(d).$$

12. **Putnam problem 1988–B1 (disjoint genera).** Show that every composite number can be expressed in the form $xy + yz + zx + 1$ for positive integers x, y, z. Compare what is true in the case of primes.

13. **Tetrahedral volume formulas.**

   (a) Verify that (5.6) is a specialization of (5.5).

   (b) Confirm that the geometric statement of Section 5.3.1 implies the algebraic one.

   (c) Show that a positive sextuple $(s_{1,2}, s_{1,3}, \cdots, s_{3,4})$ in $\mathbb{R}^6$ comes from a tetrahedron whose squared edge lengths are the values $s_{i,j}$ exactly when all the squared volumes in (5.5) are nonnegative.

14. **Putnam problem 1992–B5.** For each n, evaluate the determinant D_n of the $(n-1) \times (n-1)$ matrix of which the 5×5 case is

$$\begin{bmatrix} 3 & 1 & 1 & 1 & 1 \\ 1 & 4 & 1 & 1 & 1 \\ 1 & 1 & 5 & 1 & 1 \\ 1 & 1 & 1 & 6 & 1 \\ 1 & 1 & 1 & 1 & 7 \end{bmatrix}.$$

Generalize this result to the case of arbitrary entries $1 + a_k$ along the diagonal.

Hint: Note that $D_n = n \, D_{n-1} + (n-1)!$.

15. **Wilson's theorem.** Wilson's theorem is the assertion that if p is prime, then $(p-1)! \equiv -1 \bmod p$. (The Lagrange converse is also true; namely, the congruence is necessary for primality.) This can be proved using a group argument, as follows. Note first that only $1, -1$ are self-inverses modulo p, so that the product of *all other residues* consists of element-inverse pairs aa^{-1}. Hence, $\prod_{a=2}^{p-2} a \equiv 1 \bmod p$, and Wilson's theorem follows when you include $1, -1$ in the product. Such observations actually lead to certain computational advantages in the evaluation of very large factorials, as in [101].

16. **Prouhet-Tarry-Escott problem.** Given positive integers n and k, this Diophantine problem asks for nontrivial solutions to

$$\begin{aligned} \alpha_1 + \alpha_2 \cdots + \alpha_n &= \beta_1 + \beta_2 + \cdots \beta_n \\ \alpha_1^2 + \alpha_2^2 + \cdots + \alpha_n^2 &= \beta_1^2 + \beta_2^2 + \cdots \beta_n^2 \\ &\cdots \\ \alpha_1^k + \alpha_2^k + \cdots + \alpha_n^k &= \beta_1^k + \beta_2^k + \cdots \beta_n^k. \end{aligned}$$

A solution is abbreviated as $[\alpha] =_k [\beta]$. For example,

$$[-2, -1, 3] =_2 [2, 1, -3], \qquad [-5, -1, 2, 6] =_3 [-4, -2, 4, 5]$$

$$[-8, -7, 1, 5, 9] =_4 [8, 7, -1, -5, -9],$$

and such *ideal solutions* with $k = n - 1$ are known for $n < 12$. This can be equivalently rewritten as a question about $\pm$ polynomials dividing $(z-1)^n$ of minimal length (sum of the absolute values of the coefficients) [65]. A good deal is known about the problem computationally, and the main open questions include:

(a) Find a second inequivalent solution for $n = 12$ where only

$$[\pm 151, \pm 140, \pm 127, \pm 186, \pm 61, \pm 22] =_{11} [\pm 148, \pm 146, \pm 1271, \pm 94, \\ \pm 47, \pm 35]$$

is known. (Any other symmetric solution has some entry exceeding 1000 in absolute value.)

(b) Find ideal solutions with $n > 12$.

(c) Find ideal solutions for each n or find a n with no ideal solution (more likely).

17. **Zeta and arithmetic functions.** Show that

$$\zeta(s) \sum_{n=1}^{\infty} \frac{a_n}{n^s} = \sum_{n=1}^{\infty} \frac{b_n}{n^s}$$

if and only if

$$\sum_{n=1}^{\infty} a_n \frac{x^n}{1 - x^n} = \sum_{n=1}^{\infty} b_n x^n,$$

$$\alpha(s) \sum_{n=1}^{\infty} \frac{a_n}{n^s} = \sum_{n=1}^{\infty} \frac{b_n}{n^s}$$

if and only if

$$\sum_{n=1}^{\infty} a_n \frac{x^n}{1 + x^n} = \sum_{n=1}^{\infty} b_n x^n.$$

Let ϕ be Euler's totient function and denote $\sigma_k(n) = \sum_{d|n} d^k$. Write $\sigma = \sigma_1$ and $\tau = \sigma_0$ the number of divisors. Let λ be the number of prime factors of n counting multiplicity. Let $q(n) = |\mu(n)|$, where μ is the Möbius function, so that $q(n)$ is 1 when n is *quadratfrei* (square-free) and 0 otherwise. Deduce, using the above and facts such as Euler's product for ζ that for s large enough to assure convergence of the Dirichlet series, the following hold.

(a)

$$\sum_{n=1}^{\infty} \frac{\phi(n)}{n^s} = \frac{\zeta(s - 1)}{\zeta(s)}$$

$$\sum_{n=1}^{\infty} \frac{\mu(n)}{n^s} = \frac{1}{\zeta(s)}$$

(b)

$$\sum_{n=1}^{\infty} \frac{\tau(n)}{n^s} = \zeta^2(s),$$

and, more generally,

(c)

$$\sum_{n=1}^{\infty} \frac{\sigma_k(n)}{n^s} = \zeta(s-k)\zeta(s)$$

(d)

$$\sum_{n=1}^{\infty} \frac{\lambda(n)}{n^s} = \frac{\zeta(2s)}{\zeta(s)}$$

(e)

$$\sum_{n=1}^{\infty} \frac{\tau^2(n)}{n^s} = \frac{\zeta^4(s)}{\zeta(2s)}$$

(f)

$$\sum_{n=1}^{\infty} \frac{q(n)}{n^s} = \frac{\zeta(s)}{\zeta(2s)},$$

and, more generally, discover what d_k and q_k must count if

(g)

$$\sum_{n=1}^{\infty} \frac{q_k(n)}{n^s} = \frac{\zeta(s)}{\zeta(ks)}$$

and

(h)

$$\sum_{n=1}^{\infty} \frac{d_k(n)}{n^s} = \zeta^k(s).$$

If $\sigma_k^*(n) = \sum_{d|n}(-1)^k d^k$, consider similar formulae in terms of α. Note that in each case, it is very easy to check the claimed identities numerically for reasonable large s.

18. **Hurwitz's results on three and five squares.** If n is a square, then

$$r_3(n) = 6 \prod_p \left[\frac{p^{\lambda_p/2+1} - 1}{p - 1} - (-1)^{(p-1)/2} \frac{p^{\lambda_p/2} - 1}{p - 1} \right],$$

and

$$r_5(n) = 10 \frac{2^{3\lambda/2+3} - 1}{2^3 - 1} \prod_p \left[\frac{p^{3\lambda_p/2+1} - 1}{p^3 - 1} - p \frac{p^{3\lambda_p/2} - 1}{p^3 - 1} \right],$$

where p ranges over the odd prime factors of n and λ_p is its multiplicity. This and much more is discussed in [87].

19. **Berkeley problem 6.11.33.** Let n be a positive integer and let P be a given polynomial of degree n. Explicitly compute a nontrivial polynomial Q of the form

$$Q(x) = \sum_{i=0}^{n} a_i x^{2^i}$$

containing P as a factor. *Hint*: Using a computer algebra system, perform the Euclidean algorithm to write

$$x^{2^i} = Q_i(x)P(x) + R_i(x)$$

for each $0 \leq i \leq n$, where degree $(R_i) \leq n-1$. Since $\{R_0, R_1, \cdots, R_n\}$ are dependent, there are a_i, not all zero, with $\sum_{i=0}^{n} a_i R_i = 0$.

20. **A polynomial problem.** (From [140, pg. 86]). Find all real numbers $|r| < 2$ such that $x^{14} + rx^7 + 1$ divides $x^{154} - rx^{77} + 1$.

21. **Putnam problem 1989–A3.** Show all roots of

$$11 z^{10} + 10 iz^9 + 10 iz - 11 = 0$$

lie on the unit circle. *Hint*: This can be solved explicitly using *Maple*, *Mathematica*, or a custom-written root-finding program that employs Newton iterations (see Chapter 7).

22. **Berkeley problem 6.11.23.** Prove that $P(x) = 1 + x + \cdots + x^{p-2} + x^{p-1}$ is irreducible over $\mathbf{Q}$ when p is prime. Is this true more generally? *Hint*: Using a computer algebra system, compute $P(y + 1)$ and use Eisenstein's criterion: *It suffices to find a prime q which divides all but the leading coefficient while q^2 does not divide the constant coefficient.*

23. **The Hilbert matrix.** The n-dimensional Hilbert matrix $H(n)$ is the banded Hankel (constant on-off diagonals) matrix whose entry is

$H_{i,j} = 1/(i + j - 1)$. Hence,

$$H(3) = \begin{bmatrix} 1 & 1/2 & 1/3 \\ 1/2 & 1/3 & 1/4 \\ 1/3 & 1/4 & 1/5 \end{bmatrix}.$$

(a) Let $d_n = \det^{-1}(H(n))$. Then the sequence starts

$$1, 12, 2160, 6048000, 266716800000, 186313420339200000.$$

Determine the closed form for this sequence, say using Sloane's encyclopedia, and prove it. Similarly, consider $K(n)$ with $K_{i,j} = 1/(i+j)$.

(b) Discover the structure of $H^{-1}(n)$. For example,

$$H^{-1}(3) = \begin{bmatrix} 9 & -36 & 30 \\ -36 & 192 & -180 \\ 30 & -180 & 180 \end{bmatrix}.$$

(c) Hilbert matrices are very poorly conditioned, and so make good tests of numerical routines. Let $H_D(N)$ denote the floating point evaluation of $H(N)$ using D digits. Compare $\det(H(15))$ and $\det(H_D(15))$ for $D = 10^k, k = 1, \cdots, 5$.

(d) The corresponding infinite Hilbert matrix

$$H = \left(\frac{1}{i+j-1} \right)_{i,j=1}^{\infty}$$

induces a bounded linear operator on the square summable sequences, whose operator norm is no greater than π.

(e) Indeed, $\|H\| = \pi$.

A lovely paper on many aspects of the Hilbert matrix is [84].

Answers:
(a) The value is $d_n = \prod_{k=1}^{n-1} (2k+1) \binom{2k}{k}^2$.
(b) The general term of the $n \times n$ inverse is

$$(-1)^{i+j}(i+j-1) \binom{i+j-2}{i-1}^2 \binom{n+i-1}{n-j} \binom{n+j-1}{n-i}.$$

To prove this, it may help to use the determinant formula

$$\det\left(\left[\frac{1}{x_i + y_j}\right]\right) = \frac{\prod_{i>j}(x_i - x_j)(y_i - y_j)}{\prod_{i,j}(x_i + y_j)},$$

and Cramer's rule.

(c) This dramatically highlights the difference between computing symbolically and numerically.

(d) Let $\Lambda = (\lambda_{i+j})$ be the doubly infinite matrix, where $\lambda_0 = 0$ and $\lambda_k = 1/k$ otherwise for $k \in \mathbf{Z}$. Note that one may write

$$\Lambda = \begin{bmatrix} -H & L \\ L^* & H \end{bmatrix},$$

so that Λ is a dilation of H. It follows that $\|H\| \le \|\Lambda\| = \sqrt{\|\Lambda^2\|}$, since Λ is symmetric. Now direct calculation shows $\Lambda^2 = (a_{i+j})$ is a nonnegative doubly infinite Hankel matrix with $a_0 = 2\,\zeta(2)$ and $a_k = 2/k^2$ otherwise. As each row sums to $\sum_{i\in\mathbf{Z}} a_i = 6\,\zeta(2) = \pi^2$, we are done once we observe that $\Lambda^2 = \sum_{k\in\mathbf{Z}} a_k J_k$, where the operator $J(k)_{i,j} = \delta_{i-j}(k)$ is zero except when $|i - j| = k$, and 1 in that case. Thus,

$$\|\Lambda^2\| \le \sum_k |a_k|\,\|J(k)\| = \sum_k |a_k| = \pi^2$$

—Jensen and Euler strike again!

(e) Let $T(n)$ be the Hankel matrix $(\tau_{i+j})_{i,j=1}^n$ with $\tau_k = 1/k$ for $k \le n + 1$ and zero otherwise. Observe that $H(n) \ge T(n) \ge 0$ coordinate-wise and so $\|H\| \ge \|H(n)\| \ge \|T(n)\|$. It thus suffices to show $\liminf_n \|T(n)\| \ge \pi$.

To see this, use $\|T(n)\| \ge \langle T(n)v_n, v_n\rangle/\|v_n\|^2$ for $v_n = (1, 1/\sqrt{2}, \cdots, 1/\sqrt{n})$. Then with $\alpha_k = \sum_{j=1}^{k-1} (j(k-j))^{-1/2}$ and $H_n = \sum_{k=1}^n 1/k$,

$$\|H\| \ge \liminf_n \frac{1}{H_n} \sum_{k=1}^n \alpha_k/k = \liminf_k \alpha_k$$

$$= \liminf_k \frac{1}{k} \sum_{j=1}^{k-1} \left(\frac{j}{k}\right)^{-1/2} \left(1 - \frac{j}{k}\right)^{-1/2},$$

which is a Riemann sum for $\int_0^1 x^{-1/2}(1-x)^{-1/2}\,dx = \beta(1/2, 1/2) = \pi$.

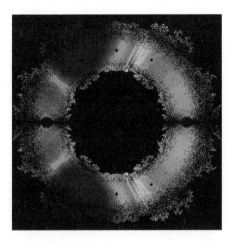

Figure 5.1. Zeros of zero-one polynomials.

24. **The holes in the argument.** Figure 5.1 (see Color Plate VII) plots all roots of polynomials, $\mathcal{B}_N$, with coefficients in $\{0, 1, -1\}$ up to degree $N = 18$. The graphically displayed information would be very hard to digest numerically. The zeroes in Figure 5.1 (also in the Color Supplement) are colored by their local density normalized to the range of densities; from red for low density to yellow for high density. The fractal structure and the holes around the roots are of different shapes and precise locations. This and more is described in [66]. For example, when α is a Pisot number

$$C_1 \frac{1}{\alpha^N} \leq \min_{\alpha \neq \beta \in \mathcal{B}_N} |\alpha - \beta| \leq C_2 \frac{1}{\alpha^N},$$

for constants $C_1, C_2 > 0$. Similarly, for α a d-th root of unity,

$$C_3 \frac{1}{N^{(1+\phi(d)/2)(k+1)}} \leq \min_{\alpha \neq \beta \in \mathcal{B}_N} |\alpha - \beta| \leq C_4 \frac{1}{N^{(1+\phi(d)/2)(k+1)}},$$

for constants $C_1, C_2 > 0$. Unexplained phenomena can be seen in images at the URL http://www.cecm.sfu.ca/personal/loki/Projects/Roots/Book.

25. **Rudin-Shapiro polynomials.** These are defined recursively by $P_0(z) = Q_0(z) = 1$ and

$$P_{n+1}(z) = P_n(z) + z^{2^n} Q_n(z),$$
$$Q_{n+1}(z) = P_n(z) - z^{2^n} Q_n(z).$$

(a) P_n and Q_n are even of degree $2^n - 1$ and have only ± 1 coefficients.

(b) For $|z| = 1$

$$|P_{n+1}(z)|^2 + |P_{n+2}(z)|^2 = 2\left(|P_n(z)|^2 + |P_{n+1}(z)|^2\right)$$

and so $|P_n(z)|, |Q_n(z)| \le 2^{(n+1)/2} \le \sqrt{2}\,(2^n - 1)$.

In consequence, these are examples of ± 1 polynomials p_n that grow no faster than $\sqrt{2}$ degree(p_n) on the unit disk. A famous problem of Littlewood [65] is to find polynomials p_n of degree n with

$$c_1\sqrt{n} \le |p_n(z)| \le c_2\sqrt{n},$$

for all n.

26. **Hyperbolic polynomials and self-concordant barriers for hyperbolic means.** Following the development in [158], we shall construct two classes of self-concordant barrier functions on natural convex sets. Let $p : \mathrm{R}^n \mapsto \mathrm{R}$ be a polynomial on n variables, *homogeneous* of degree m, that is, $p(tx) = t^m p(x)$ for every $t \in \mathrm{R}$ and $x \in \mathrm{R}^n$. We say that $p(x)$ is *hyperbolic* in the direction $d \in \mathbf{R}^n$ if the polynomial $t \mapsto p(x + td)$ has m real roots for every x. Denote the negative of these roots, which depend on x and d, by $t_i(x, d)$, $i = 1, \cdots, m$. The following are three important examples of hyperbolic polynomials.

(a) $p_1 : x \in \mathrm{R}^n \mapsto \prod_{i=1}^n x_i$ with respect to the direction $d = (1, ..., 1)$.

(b) $p_2 : x \in \mathrm{R}^n \mapsto x_1^2 - \sum_{i=2}^n x_i^2$ with respect to the direction $d = (1, 0, ..., 0)$.

(c) Let $\mathcal{S}^n$ be the space of $n \times n$ symmetric matrices and recall that it is isomorphic to $\mathrm{R}^{n(n+1)/2}$. The polynomial in the entries of the symmetric matrix given by $p_3 : X \in \mathcal{S}^n \mapsto \det(X)$ is hyperbolic with respect to the direction $d = I$ (the $n \times n$ identity matrix).

Define the sets

$$C(p, d) = \{x \in \mathrm{R}^n : p(x + td) \ne 0, \forall t \ge 0\},$$
$$C_a(p, d) = \{x \in \mathrm{C} : p(x) > a\}, \text{ for } a \ge 0.$$

It is known (see [119]) that $C(p, d)$ is an open, convex cone; that p is hyperbolic in the direction of any vector in $C(p, d)$ (also for any

$c \in C(p, d)$, $C(p, c) = C(p, d)$); and that $p(x)^{1/m}$ is a concave function on $C(p, d)$ and zero on the boundary. (See Figure 5.2.) On the other hand, let Q be a convex set in R, and F a real valued function defined on Q. We denote the boundary of Q by ∂Q, and denote the k-th directional derivative of F in the direction h evaluated at zero by $D^k F(x)[h, \cdots, h]$. We say that F is a θ-self-concordant barrier on Q if the following three conditions are satisfied:

(i) $|DF(x)[h]| \leq \sqrt{\theta}(D^2 F(x)[h, h])^{1/2}$

(ii) $|D^3 F(x)[h, h, h]| \leq 2(D^2 F(x)[h, h])^{3/2}$, and

(iii) $F(x^r) \to \infty$ for any sequence $x^r \to x \in \partial Q$.

(a) Following the steps given below, prove that:

    i. The function $F(x) = -\log(p(x))$ is an m-self-concordant barrier on the set $C(p, d)$ (see [133]).

    ii. For any $a \geq 0$, the function $F(x) = -m \log(p(x) - a)$ is an m^2-self-concordant barrier on the set $C_a(p, d)$ (see [158]).

Self concordance plays a central role in interior point methods of modern optimization.

(b) Suggested steps on the path:

    i. The barrier condition (iii) should be clear from the mentioned properties of hyperbolic polynomials on the set $C(p, d)$.

    ii. Show that the following representation holds

$$p(x + r\,d) = p(d) \prod_{i=1}^{m}(r + t_i(x, d)).$$

    The part that requires thought is to determine the constant $p(d)$ on the right-hand side. (Keep in mind that $\{t_i(x, d)\}_{i=1}^{m}$ are the negatives of the roots of the hyperbolic polynomial.)

    iii. Use the above factorization and the homogeneity of p to show that for any $x \in C(p, d)$ and any $h \in \mathbf{R}^n$, we have

$$p(x + r\,h) = p(x) \prod_{i=1}^{m}(1 + r\,t_i(h, x)).$$

    iv. Use this representation to obtain the derivatives of $r \mapsto p(x + r\,h)$ for any $x \in C(p, d)$ and $h \in \mathbf{R}^n$:

$$\frac{d}{dr}p(x + r\,h) = p(x + r\,h) \sum_{i=1}^{m} \frac{t_i(h, x)}{1 + r\,t_i(h, x)}.$$

  v. Calculate the directional derivatives of $F(x) = -\log(p(x))$
     and prove the self-concordant inequalities hold with $\theta = m$.

 vi. For the second part, consider first the case $a = 0$. Next, use
     a linear substitution to argue without loss of generality that
     $a = 1$.

vii. Set $t_i = t_i(h, x)$, and define $\alpha = p(x) - 1$,

$$C_1 = \sum_{i=1}^{m} t_i, C_2 = \sum_{i=1}^{m} t_i^2, C_3 = \sum_{i=1}^{m} t_i^3.$$

   Show that for $F(x) = -m \log(p(x) - 1)$, we have

$$DF(x)[h] = -m \frac{\alpha + 1}{\alpha} C_1,$$

$$D^2 F(x)[h, h] = m \frac{\alpha + 1}{\alpha^2} C_1^2 + m \frac{\alpha + 1}{\alpha} C_2, \text{ and}$$

$$D^3 F(x)[h, h, h] = -m \frac{(\alpha + 1)(\alpha + 2)}{\alpha^3} C_1^3 - 3m \frac{(\alpha + 1)}{\alpha^2} C_1 C_2$$
$$- 2m \frac{\alpha + 1}{\alpha} C_3.$$

viii. Use the Cauchy-Schwarz inequality to show that the first
      inequality in the definition of self-concordancy holds, with
      $\theta = m^2$.

 ix. To show the second inequality in the definition, square both
     sides and group terms with respect to powers of α. Notice
     that the inequality is homogeneous of degree one with re-
     spect to the vector $(t_1, \cdots, t_m)$, thus without loss $C_1 = \pm 1$.
     Notice also that $\alpha > 0$ since $x \in C_a(p, d)$, and $a = 1$. Now,
     show that all coefficients in front of the different powers of
     α are positive.

(c) Verify the three examples of hyperbolic polynomials given at the
    beginning of this item.

(d) Try to prove the inequalities in a computer algebra system.

27. **The Lax conjecture is true.** An elegant 1958 conjecture of Lax
    concerning hyperbolic polynomials has recently been settled by Lewis,
    Parrilo, and Ramana [157]:

Theorem 5.13. *A polynomial p in three real variables is hyperbolic
of degree d with respect to $e = (1, 0, 0)$, with $p(e) = 1$, if and only if*

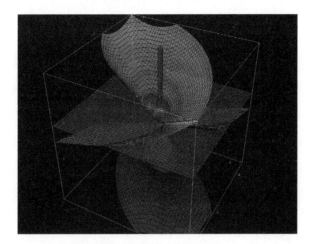

Figure 5.2. A hyperbolicity cone.

there are symmetric $d \times d$ matrices B, C such that

$$p(x, y, z) = \det(xI + yB + zC).$$

Deduce what characterizes the corresponding hyperbolic polynomials in two variables.

Figure 5.2 shows a third-order hyperbolic cone. It was drawn by Pablo Parrilo in a fine personal computer drawing package called *DPGraph* (for "dynamic photorealistic graphing"). The cone is actually the part on the top of the figure; all the rest is the "hidden." There is a horizontal plane at $z = 1$, and on that plane a nice set with the "rigid convexity" property that contains the point $(0, 0)$ near the center. The vertical line is in the direction $e = (0, 0, 1)$ of hyperbolicity.

28. **Inverse problems.** Inverse problems provide a host of challenges for the computationally and experimentally minded. Examples, both explicit and implicit, have occurred throughout our volumes (e.g., maximum entropy problems, the self-concordant barriers above, the nuclear magnetic resonance example, and while exploring JPEG compression or the Watson integrals).

29. **Nonnegative polynomials.** The nonnegative polynomials of degree d, $P_+^d(I)$, on an interval I, form an interesting convex cone in the

vector space of all polynomials $P^d(I)$. Recall that an *extreme ray* of a convex cone K is an element with the property that $e = p+q, p, q \in K$ implies q and p are nonnegative multiples of e.

(a) Characterize the extreme rays of $P_+^d([0, \infty))$, using the fundamental theorem of algebra. Deduce that every polynomial p that is nonnegative on R^+ can be expressed as

$$p(t) = t \sum_I p_i^2(t) + \sum_J p_j^2(t),$$

where p_i, p_j are finitely many real polynomials with nonnegative roots.

(b) Deduce that every polynomial p that is nonnegative on R can be expressed as

$$p(t) = \sum_I p_i^2(t),$$

where p_i are polynomials with real roots. Prove also that when p has even degree, it is the square of at most two polynomials.

(c) Characterize the extreme rays of $P_+^d(I)$, for a finite interval.

(d) *Hilbert's theorem.* A nonnegative polynomial in three variables, homogeneous of degree four, is expressible as a sum of squares of quadratic forms (in the three variables). The hypotheses are needed as the nonnegative polynomials below show.

    i. $x^4y^2 + x^2y^4 + z^6 - 3x^2y^2z^2$ (Motzkin);

    ii. $w^4 + x^2y^2 + y^2z^2 + 2z^2x^2 - 4xyzw$ (Choi and Lam).

Hint: A boundary polynomial must have a zero in the interval. An extreme polynomial must be of full degree with all roots in the interval. In each case, every polynomial in the cone is in the closed conical hull of the extreme directions. See [20] for details.

30. **The Ehrhardt polynomial of a polytope.** An *integer polytope* is the convex hull of finitely many points with integer coefficients. The following result neatly links convexity, lattices, and polynomials.

Theorem 5.14. **(Ehrhardt polynomial).** *Let P be an integer polytope in R^d. There exists a univariate polynomial of degree at most d, p_P, such that*

$$p_P(k) = \left| (kP) \cap \mathrm{Z}^d \right|,$$

for all nonnegative integers k.

Let P be a given integer polytope.

(a) Show that dim $P = $ degree p_P.

(b) For integer polygons, use Pick's theorem to establish the existence of the Ehrhardt polynomial, and show that it has nonnegative coefficients.

(c) Show that

$$p_P(-k) = (-1)^{\dim P} \left| \text{relint} \, (kP) \cap Z^d \right|,$$

for all positive integers k. Here "relint" denotes the relative interior of P in its affine span (that is, the inside).

(d) Show that the constant term of p_P is 1 and the coefficient of t^d is the volume of P (which could be zero). The coefficient of t^{d-1} is half of a "surface area."

(e) Fix a positive integer m. Consider the Ehrhardt polynomial, p_m, of the tetrahedron with vertices $(0,0,0), (1,0,0), (0,1,0)$, and $(1,1,m)$. Show that

$$p_m(k) = \frac{m}{6}k^3 + k^2 + \frac{12 - m}{6}k + 1.$$

This is described in Barvinok's fine recent book [20].

31. **Small gaps between consecutive primes.** As this book was being completed, a breakthrough was announced in a long-standing question of the spacings of prime numbers [128]. The prime number theorem can be paraphrased as saying that the average size of $p_{n+1} - p_n$ is $\log p_n$, where p_n denotes the n-th prime, so that

$$\Delta \;=\; \liminf_{n \to \infty} \frac{p_{n+1} - p_n}{\log p_n} \leq 1.$$

In 1926 Hardy and Littlewood showed that assuming the generalized Riemann hypothesis, $\Delta \leq 2/3$. Other researchers lowered this figure, until 1986 when Maier established the result $\Delta \leq 0.2486$. In a dramatic new development, Daniel Goldston and Cem Yildirim announced in April 2003 that not only is $\Delta = 0$, but also

$$p_{n+1} - p_n \;<\; (\log p_n)^{8/9}$$

holds for infinitely many n. What's more, for any fixed integer r, the inequality

$$p_{n+r} - p_n \;<\; (\log p_n)^{(8r)/(8r+1)}$$

holds for infinitely many n. They anticipated that even stronger results of this type can be established (such as decreasing 8/9 to 4/5). This report was widely hailed in the mathematical community and the press, after being circulated in preprint and described at conferences.

Unfortunately, Andrew Granville and K. Soundararajan subsequently found a problem in one of the arguments—some "small" error terms are actually of the same order of magnitude as the main term. As of this date, the difficulty remains unresolved.

32. **More from Littlewood's miscellany.** [160] We conclude the chapter with a series of quotes and observations from Littlewood.

   (a) Page 56.

      "(A. S. Bescovitch) A mathematician's reputation rests on the number of bad proofs he has given. (Pioneer work is clumsy.)"

   (b) Page 60.

      "A precisian professor had the habit of saying '... quartic polynomial $ax^4 + bx^3 + cx^2 + dx + e$, where e need not be the base of the natural logarithms.' (It might be.)"

   (c) Page 61.

      "I read in the proof-sheets of Hardy on Ramanujan: 'As someone said, each of the positive integers was one of his personal friends.' My reaction was, 'I wonder who said that; I wish I had.' In the next proof-sheets I read what now stands, 'It was Littlewood who said ...'
      (What had happened was that Hardy had received the remark in silence and with poker face, and I wrote it off as a dud. I later taxed Hardy with this habit; on which he replied: 'Well, what is one to do, is one always to be saying "damned good"?' To which the answer is 'yes.')"

   (d) Pages 118–120. *Random Jottings on G. H. Hardy* (after a 35-year collaboration):

      "His spelling was not immaculate." ... "He preferred the Oxford atmosphere and said they took him seriously, unlike Cambridge." ... "He took a sensual pleasure in "calligraphy" and it would have been a deprivation if he didn't make the final copy of a joint paper.

(My standard role in a joint paper was to make the logical skeleton, in shorthand—no distinction between r and r^2, 2π and 1, etc., etc. But when I said 'Lemma 17' it stayed Lemma 17.)" ... "He was indifferent to noise; very rare in creative workers at least when no longer young."

(e) Page 149.

"Creative workers need drink at night, 'Roses and dung'. (Or: mathematicians read 'rubbish'.) An experimentalist, having spent the day looking for a leak, has had a perfect mental rest by dinner time, and overflows with minor mental activity."

(f) Page 164.

" 'Always verify references.' This is so absurd in mathematics that I used to say provocatively: 'never ...'."

6 | The Power of Constructive Proofs II

Mathematical proofs, like diamonds, are hard as well as clear, and
will be touched with nothing but strict reasoning.

John Locke, 1690,
from William Dunham, *The Mathematical Universe*

As we noted in [43], a computational or constructive approach can often
make a proof significantly more understandable. What's more, the key concepts involved are made more apparent, and are more likely to be remembered later. Here we present some additional examples of computer-aided
proofs, including some examples of "variational" methods of proof.

6.1 A More General AGM Iteration

We first study a more general form of the arithmetic-geometric mean iteration. This material has been adapted from [63]. Let a and b be real
numbers, $a > b > 0$, and let N be an integer greater than 1. By the *iteration* AG_N, we mean the following two-term recursion: $a_0 = a$, $b_0 = b$, and,
for any $k \geq 0$,

$$a_{k+1} = \frac{a_k + (N-1)b_k}{N} \tag{6.1}$$

$$b_{k+1} = \frac{\sqrt[N]{(a_k + (N-1)b_k)^N - (a_k - b_k)^N}}{N}. \tag{6.2}$$

In the case $N = 2$, we get the standard arithmetic-geometric mean (AGM)
iteration discussed in the previous section. The case $N = 3$ was studied in
detail in [46], [49], and [45]. We have

$$a_{k+1}^N - b_{k+1}^N = \left(\frac{a_k - b_k}{N}\right)^N \qquad \text{for any } k \geq 0. \tag{6.3}$$

This shows both the global convergence and local Nth-order convergence of the iteration. So there is a common limit of (a_k) and (b_k),

$$M_N(a_0, b_0) \quad = \quad \lim_{k \to \infty} a_k = \lim_{k \to \infty} b_k. \tag{6.4}$$

Let * be the involution on $[0, 1]$ defined by $x^* = \sqrt[N]{1 - x^N}$. Whenever we use the function symbol *, the respective value of N is clear from the context. Because of the homogeneity

$$M_N(\lambda a, \lambda b) \quad = \quad \lambda M_N(a, b) \qquad \text{for } \lambda > 0, \tag{6.5}$$

it is enough to investigate

$$A_N(x^N) \quad = \quad \frac{1}{M_N(1, x^*)}, \tag{6.6}$$

for $0 < x < 1$.

From (6.3), we can argue that $a_k^N(z)$ and $b_k^N(z)$ are analytic in the unit disk and converge uniformly to $M_N(1, z)$ therein. Thus $A_N(x^N)$ is analytic in a neighborhood of zero. Notice that $A_N(0) = 1$.

6.1.1   The Functional Equation for the AGM

For any $N > 1$ and any $0 < x < 1$ we have, applying one step of the AG_N iteration,

$$M_N(1 + (N - 1)x, 1 - x) \quad = \quad M_N(1, x^*). \tag{6.7}$$

Further, by (6.5) we have

$$M_N(1 + (N - 1)x, 1 - x) \quad = \quad (1 + (N - 1)x)M_N\left(1, \frac{1 - x}{1 + (N - 1)x}\right). \tag{6.8}$$

Combining the right-hand sides of (6.8) and (6.7), we arrive at the following functional equation for A_N:

$$(1 + (N - 1)x) \cdot A_N(x^N) \quad = \quad A_N\left(1 - \left(\frac{1 - x}{1 + (N - 1)x}\right)^N\right). \tag{6.9}$$

Notice that (6.9) uniquely determines the Taylor series for A_N at 0 and hence has a *unique* analytic solution

$$A_N(x^N) \quad = \quad \frac{1}{M_N(1, x^*)}. \tag{6.10}$$

6.1.2 The Quadratic Case Recovered

For $N = 2$, the equation (6.9) specializes to

$$(1 + x) \cdot A_2(x^2) = A_2\left(\frac{4x}{(1+x)^2}\right), \tag{6.11}$$

whose solution was found by Gauss (see [44]) in the form

$$A_2(x^2) = \sum_{i=0}^{\infty} \binom{2i}{i}^2 \left(\frac{x}{4}\right)^{2i} = {}_2F_1\left(\begin{array}{c} 1/2, 1/2 \\ 1 \end{array} \middle| x^2\right). \tag{6.12}$$

Gauss discovered this closed form after having been inspired by a great number of experimental numerical calculations (see [44, page 5, 7]).

Another way of expressing the AG_2 limit is by means of the following definite integral:

$$I_2(a_0, b_0) = \int_0^\infty \frac{dt}{\sqrt{(t^2 + a_0^2)(t^2 + b_0^2)}}. \tag{6.13}$$

Then it is straightforward to prove (see [44]) that

$$M_2(a_0, b_0) = \frac{I_2(1,1)}{I_2(a_0, b_0)} = \frac{\pi}{I_2(a_0, b_0)}. \tag{6.14}$$

The core step of this proof, namely showing the AG_2-invariance

$$I_2(a, b) = I_2\left(\frac{a+b}{2}, \sqrt{ab}\right) \tag{6.15}$$

follows fairly easily by the substitution $u = (t - ab/t)/2$. We leave this as a *Maple* or *Mathematica* exercise.

Yet another way of proving the AG_2 limit formula is by identifying $M_2(1, x)$ as a solution of a second-order linear differential equation (see [44]).

6.1.3 The Cubic Case Solved

The closed form for $M_3(1, x^*)$ was identified and proved in several ways in [46, 49, 45]. Nowadays, the discovery of the closed form for $A_3(x^3)$ as a formal power series again is a routine task using any computer mathematics software. We obtain

$$\frac{1}{M_3(1, x^*)} = A_3(x^3) = {}_2F_1\left(\begin{array}{c} 1/3, 2/3 \\ 1 \end{array} \middle| x^3\right). \tag{6.16}$$

The cubic counterpart of (6.13) is

$$I_3(a_0, b_0) \quad = \quad \int_0^\infty \frac{t\, dt}{\sqrt[3]{(t^3 + a_0^3)(t^3 + b_0^3)^2}}. \tag{6.17}$$

We have (see [40, 42])

$$M_3(a_0, b_0) \quad = \quad \frac{I_3(1, 1)}{I_3(a_0, b_0)} \quad = \quad \frac{2\pi}{\sqrt{27}} \cdot \frac{1}{I_3(a_0, b_0)}. \tag{6.18}$$

Again, the crucial part of the proof of (6.18) is to show that the integral is invariant with respect to the AG_3 iteration, that is,

$$I_3(a, b) \quad = \quad I_3\left(\frac{a + 2b}{3}, \sqrt[3]{b\frac{a^2 + ab + b^2}{3}}\right), \tag{6.19}$$

for all $a > b > 0$. The proof of (6.19) was proposed by one of the present authors [40] as Part (a) of the *American Mathematical Monthly* Problem 10281. The solutions published by the *Monthly* are concluded with the editorial comment, "... There is still no self-contained proof that avoids exploiting the identification with a hypergeometric function" ([42, pg. 183]).

Recently, John A. Macdonald found a proof of (6.19) that consists of a chain of variable substitutions, together with the split of the integration range at the point 1. Here we present a variation of this proof, in which all integral substitutions have been simplified to the extent that they can be checked by a computer. The proof was announced in the "Revivals" section of the Monthly [41].

A large portion of the proof of (6.19) is encapsulated in the following lemma.

Lemma 6.1. *For any* $\gamma \in (0, 1]$,

$$I_3(1, \sqrt[3]{1 - \gamma^3}) \quad = \quad \int_1^\infty \frac{dx}{\sqrt{(x - 1)((x + 3)x^2 - 4\gamma^3)}}. \tag{6.20}$$

Proof. We first note that

$$I_3(1, \sqrt[3]{1 - \gamma^3}) \quad = \quad \int_0^\infty \frac{u\, du}{\sqrt{(u^3 - 1)^2 + 4(1 - \gamma^3)u^3}}, \tag{6.21}$$

using the substitution

$$\boxed{u^3 = t^3(1 + t^3)/((1 - \gamma^3) + t^3)}.$$

Further, the equality

$$\int_0^\infty \frac{u\,du}{\sqrt{(u^3-1)^2+4(1-\gamma^3)u^3}} = \int_1^\infty \frac{(u+1)\,du}{\sqrt{(u^3-1)^2+4(1-\gamma^3)u^3}}$$

follows from

$$\int_0^\infty \frac{u\,du}{\sqrt{(u^3-1)^2+4(1-\gamma^3)u^3}} = \int_0^\infty \frac{du}{\sqrt{(u^3-1)^2+4(1-\gamma^3)u^3}}$$

and

$$\int_0^1 \frac{(u+1)\,du}{\sqrt{(u^3-1)^2+4(1-\gamma^3)u^3}} = \int_1^\infty \frac{(u+1)\,du}{\sqrt{(u^3-1)^2+4(1-\gamma^3)u^3}},$$

which both are easily verified by substituting $\boxed{u = 1/u.}$

Now

$$\int_1^\infty \frac{dx}{\sqrt{(x-1)((x+3)x^2-4\gamma^3)}} = \int_1^\infty \frac{(u+1)\,du}{\sqrt{(u^3-1)^2+4(1-\gamma^3)u^3}}$$

using the substitution $\boxed{x = u + 1/u - 1}$. This completes the proof of the lemma. $\square$

For any $y, z > 0$, we have

$$I_3(y,z) = \frac{1}{y}\cdot I_3\left(1,\frac{z}{y}\right). \tag{6.22}$$

Let $c = b/a$. It can be verified easily that, in view of (6.22), we can rewrite (6.19) as

$$\frac{2c+1}{3}\cdot I_3(1,c) = I_3(1,c^\wedge), \tag{6.23}$$

where $x \mapsto x^\wedge$ is defined by

$$x^\wedge = \sqrt[3]{\frac{9x(1+x+x^2)}{(1+2x)^3}}, \tag{6.24}$$

for any $x \in [0,1]$.

We will denote function composition in the obvious way, e.g., by $c^{*\wedge}$ we mean $(c^*)^\wedge$. One can check easily that, for any $c \in [0,1]$,

$$(2c+1)(2c^{\wedge*}+1) = 3, \tag{6.25}$$

or, in other words,

$$c^{\wedge *} = \frac{1-c}{1+2c}. \tag{6.26}$$

Recall that $*$ is an involution on $[0,1]$. From (6.26), it follows that also $^{\wedge *}$ is an involution on $[0,1]$. Thus,

$$^{\wedge * \wedge} = {}^{*} \tag{6.27}$$

as functions on $[0,1]$.

By an inspection of the function $^{\wedge *}$, we can see that the statement

$$\forall c \in (0,1] \qquad \frac{2c+1}{3} \cdot I_3(1,c) = I_3(1,c^{\wedge}) \tag{6.28}$$

is equivalent to

$$\forall C \in [0,1) \qquad \frac{2C^{\wedge *}+1}{3} \cdot I_3(1,C^{\wedge *}) = I_3(1,C^{\wedge * \wedge}). \tag{6.29}$$

Taking into account (6.27), (6.25) and swapping the sides, we see that the last statement is equivalent to

$$\forall C \in [0,1) \qquad (2C+1) \cdot I_3(1,C^{*}) = I_3(1,C^{\wedge *}). \tag{6.30}$$

Using (6.20), the equality (6.30) translates to

$$(2C+1) \cdot \int_1^\infty \frac{dx}{\sqrt{(x-1)((x+3)x^2 - 4C^3)}} \tag{6.31}$$

$$= \int_1^\infty \frac{dz}{\sqrt{(z-1)\left((z+3)z^2 - 36\frac{C(1+C+C^2)}{(2C+1)^3}\right)}}, \tag{6.32}$$

which can be proven by the substitution

$$z = \frac{(x-C)^3(x-1)}{(x^3-C^3)(2C+1)} + 1.$$

This completes the proof of (6.23) and thus also the proof of (6.19).

6.2   Variational Methods and Proofs

Whenever one can formulate a mathematical problem in variational form (i.e., the solution is the minimum of some function with or without constraints), one has access to a variety of constructive tools.

A key and representative tool is the next result. It states that if a closed function (a function that is defined on a closed set, lower semicontinuous, somewhere finite, and nowhere negative infinite) attains a value close to its infimum at some point, then a nearby point minimizes a slightly perturbed function.

Theorem 6.2. (Ekeland variational principle). *Suppose the function* $f : \mathrm{R}^n \to (-\infty, +\infty]$ *is closed and the point* $x \in \mathrm{R}^n$ *satisfies* $f(x) \le \inf f + \epsilon$ *for some real* $\epsilon > 0$. *Then for any real* $\lambda > 0$, *there is a point* $v \in \mathrm{R}^n$ *satisfying the conditions*

(a) $\|x - v\| \le \lambda$,

(b) $f(v) \le f(x)$, *and*

(c) v *is the unique minimizer of the function* $f(\cdot) + (\epsilon/\lambda)\| \cdot -v\|$.

Proof. We can assume f is proper, and by assumption, it is bounded below. Since the function

$$f(\cdot) + \frac{\epsilon}{\lambda}\| \cdot -x\|$$

therefore has compact level sets, its set of minimizers $M \subset \mathrm{R}^n$ is nonempty and compact. Choose a minimizer v for f on M. Then for points $z \ne v$ in M, we know

$$f(v) \le f(z) < f(z) + \frac{\epsilon}{\lambda}\|z - v\|,$$

while for z not in M, we have

$$f(v) + \frac{\epsilon}{\lambda}\|v - x\| < f(z) + \frac{\epsilon}{\lambda}\|z - x\|.$$

Part (c) follows by the triangle inequality. Since v lies in M, we have

$$f(z) + \frac{\epsilon}{\lambda}\|z - x\| \ge f(v) + \frac{\epsilon}{\lambda}\|v - x\| \quad \text{for all } z \text{ in } \mathrm{R}^n.$$

Setting $z = x$ shows the inequalities

$$f(v) + \epsilon \ge \inf f + \epsilon \ge f(x) \ge f(v) + \frac{\epsilon}{\lambda}\|v - x\|.$$

Properties (a) and (b) follow. $\qquad\square$

An immediate counterpart, but far from easy before the advent of Ekeland's principle, is the following.

Proposition 6.3. *Let $f : \mathrm{R}^n \to \mathrm{R}$ be differentiable and bounded below. Let $\varepsilon > 0$ be given. Then f has a ε-critical point: a point v with*

$$\|\nabla f(v)\| \leq \varepsilon. \tag{6.33}$$

Proof. Since f is bounded below $f_\varepsilon = f + \varepsilon \|\cdot\|$ has bounded lower level sets and so achieves its infimum at some point v. Then we may check that

$$\|\nabla f(v)\| \leq \|\nabla f_\varepsilon(v)\| + \varepsilon = \varepsilon. \tag{6.34}$$

$\square$

In many cases, this idea allows one to ultimately establish that the infimum does exist. Nonetheless, it is instructive to apply this result to a function, such as exp, that does not achieve its infimum.

Given a set $C \subset \mathrm{R}^n$ and a continuous *self map* $f : C \to C$, we ask whether f has a *fixed point*: $f(x) = x$. Ekeland's principle also has important, constructive applications to fixed point theory. We call f a *contraction map* if there is a real constant $\gamma_f < 1$ such that

$$\|f(x) - f(y)\| \leq \gamma_f \|x - y\| \quad \text{for all } x, y \in C. \tag{6.35}$$

We are now able to painlessly establish a version of one of the most important theorems in applied analysis.

Theorem 6.4. (Banach contraction). *Any contraction on a closed subset of R^n has a unique fixed point.*

Proof. Suppose the set $C \subset \mathrm{R}^n$ is closed and the function $f : C \to C$ satisfies the contraction condition (6.35). We apply the Ekeland variational principle (6.2) to the function

$$z \in \mathrm{R}^n \mapsto \begin{cases} \|z - f(z)\| & \text{if } z \in C \\ +\infty & \text{otherwise} \end{cases}$$

at an arbitrary point x in C, with the choice of constants

$$\epsilon = \|x - f(x)\| \quad \text{and} \quad \lambda = \frac{\epsilon}{1 - \gamma_f}.$$

This shows there is a point v in C satisfying

$$\|v - f(v)\| < \|z - f(z)\| + (1 - \gamma_f)\|z - v\|,$$

for all points $z \neq v$ in C. Hence, v is a fixed point, since otherwise choosing $z = f(v)$ gives a contradiction. The uniqueness is easy.   □

With a little more work, we may estimate how far the fixed point is from our initial guess. Now, what if the map f is not a contraction? A very useful weakening of the notion is the idea of a *nonexpansive map*, which is to say a self map f satisfying

$$\|f(x) - f(y)\| \leq \|x - y\| \quad \text{for all } x, y.$$

A nonexpansive map on a nonempty compact set or a nonempty closed convex set may not have a fixed point, as simple examples like translations on R or rotations of the unit circle show. On the other hand, a straightforward argument using the Banach contraction theorem shows this cannot happen if the set is nonempty, compact, *and* convex. Indeed, it suffices to pick some $x_0 \in C$ and to consider the perturbed contraction mapping

$$f_\tau(x) \quad = \quad (1 - \tau)f(x) + \tau x_0. \tag{6.36}$$

This map must have a fixed point x_τ with

$$\|f(x_\tau) - x_\tau\| \quad \leq \quad \tau \, \text{diam}(C). \tag{6.37}$$

Taking limits completes the argument. In practice, this is often not a good way to find the fixed point, but it is a very good way to start. Another example where the variational method shines is:

Example 6.5. The Rayleigh Quotient.

Let A be a $n \times n$ symmetric matrix. Consider the function

$$f_A(x) \quad = \quad \frac{\langle Ax, x \rangle}{\|x\|^2}, \tag{6.38}$$

defined on the open set $x \neq 0$ in R^n. Consider the minimum (or maximum) value of this function, which must exist since f_A is positively homogeneous and is continuous on the unit sphere.

Now the upshot is that, using the quotient rule, any such maximal (minimal) point, v, is an eigen-vector: $Av = \lambda v$ for some real λ which is a maximal (minimal) eigenvalue. Alternatively, the reader familiar with Lagrange multipliers can reach the same conclusion from studying $\min\{f_A(x) : \|x\| = 1\}$. From this, the whole spectral theory of symmetric matrices can be obtained.   □

6.3   Maximum Entropy Optimization

Maximum entropy methods are widely used in fields including crystallography, image reconstruction and the like. We consider the convex function $p : \mathrm{R} \mapsto (-\infty, +\infty]$ given by

$$p(u) = \begin{cases} u \log u - u & \text{if } u > 0 \\ 0 & \text{if } u = 0 \\ +\infty & \text{if } u < 0, \end{cases}$$

and the associated convex function (the negative of the *Boltzmann-Shannon entropy*) $f : \mathrm{R}^n \mapsto (-\infty, +\infty]$ by

$$f(x) = \sum_{i=1}^{n} p(x_i).$$

Then f is strictly convex on R_+^n, with compact lower level sets. Moreover, it is differentiable on the interior of the orthant, while the directional derivative $f'(x; \hat{x} - x) = -\infty$ for any point x on the boundary of R_+^n. This "barrier" property makes the entropy highly effective in many variational settings [61].

We consider a linear map $G : \mathrm{R}^n \mapsto \mathrm{R}^m$ such that $G\hat{x} = b$ for some $\hat{x}$. Then for any vector c in R^n, the minimization problem

$$\inf\{f(x) + \langle c, x \rangle \mid Gx = b, \ x \in \mathrm{R}^n\}$$

has a unique optimal solution, $\bar{x}$, and all its coordinates are strictly positive. Moreover, some vector λ in R^m satisfies

$$\nabla f(\bar{x}) = G^*\lambda - c \quad \text{that is} \quad \bar{x}_i = \exp(G^*\lambda - c)_i, \tag{6.39}$$

for all i. This can be achieved by solving $G(\exp(G^*\lambda - c)) = b$ for $\bar{\lambda}$, using any nonlinear solver one wishes, and setting $\bar{x} = \exp(G^*\bar{\lambda}) - c$. Here G^* is the transpose matrix.

As a striking example of the variational method, we consider the problem of determining when a square matrix A can be pre- and post-multiplied by diagonal matrices D_1 and D_2, so that $D_1 A D_2$ is *doubly stochastic* (each row and column sums to one and all entries are non-negative). Such problems (called "DAD problems") arise in actuarial science and afford a fine example of variational methods at work.

Suppose the $k \times k$ matrix A has each entry a_{ij} nonnegative. We say A *has doubly stochastic pattern* if there is a doubly stochastic matrix with exactly the same zero entries as A. Define a set $Z = \{(i,j) | a_{ij} > 0\}$, and let R^Z denote the set of vectors with components indexed by Z, and let R_+^Z denote those vectors in R^Z with all nonnegative components.

Consider the maximum entropy problem (P):

$$\inf \quad \sum_{(i,j)\in Z} (p(x_{ij}) - x_{ij} \log a_{ij})$$

$$\text{subject to} \quad \sum_{i:(i,j)\in Z} x_{ij} = 1 \ \text{ for } j = 1, 2, \cdots, k$$

$$\sum_{j:(i,j)\in Z} x_{ij} = 1 \ \text{ for } i = 1, 2, \cdots, k$$

$$x \in R^Z.$$

The next result answers the question of when diagonalization is possible.

Theorem 6.6. (DAD). *Suppose A has doubly stochastic pattern. Then there is a point $\hat{x}$ in the interior of R_+^Z which is feasible for the problem above. Hence the problem has a unique optimal solution $\bar{x}$, and, for some vectors λ and μ in R^k, $\bar{x}$ satisfies*

$$\bar{x}_{ij} = a_{ij} \exp(\lambda_i + \mu_j) \ \text{ for } (i,j) \in Z.$$

Moreover, A has doubly stochastic pattern if and only if there are diagonal matrices D_1 and D_2 with strictly positive diagonal entries and $D_1 A D_2$ doubly stochastic.

Note that the best case is when all entries of A are strictly positive. It is good fun to formally use the classical method of Lagrange multipliers (λ and μ above) to obtain this result; the prior discussion legitimates the process. A very satisfactory byproduct is that we have an algorithm for diagonalization, either by directly minimizing (P) or by using the dual system implicit in (6.39). (See Exercise 23.)

6.4 A Magnetic Resonance Entropy

The *Hoch and Stern information measure*, or *neg-entropy*, arises in nuclear magnetic resonance (NMR) analysis. It is defined in complex $n-$space by

$$H(z) = \sum_{j=1}^{n} h(z_j/b),$$

where h is convex and given (for scaling b) by

$$\boxed{h(z) = |z| \log\left(|z| + \sqrt{1 + |z|^2}\right) - \sqrt{1 + |z|^2}}$$

for quantum theoretic reasons.

It is easy to check by hand or computer (as was indeed the case) that

$$h^*(z) = \cosh(|z|).$$

By comparison the *Boltzmann-Shannon entropy* is

$$(z \log z - z)^* = \exp(z).$$

Efficient *dual algorithms* may now be constructed that are not at all apparent from the original formulation. Knowing "closed forms" helps:

$$(\exp \exp)^*(y) = y \log(y) - y\{W(y) + W(y)^{-1}\},$$

where *Maple* or *Mathematica* knows the complex *Lambert W* function, which is the solution of

$$W(x)e^{W(x)} = x.$$

Thus, the conjugate's series is as well known to the computer algebra system as that for exp:

$$-1 + (\log(y) - 1)\, y - \frac{1}{2}y^2 + \frac{1}{3}y^3 - \frac{3}{8}y^4 + \frac{8}{15}y^5 + O\left(y^6\right).$$

The W function arises usefully in many places. It was only recently that the function was named and then began to have a literature, largely because it had a name and existed in *Maple* and *Mathematica*. Unnamed objects are unlikely to be studied in an organized fashion.

6.5   Computational Complex Analysis

We would be remiss not to indicate how complex analysis comes to life in the presence of symbolic and numeric computation. We start by stating three fundamental theorems, each of which is most useful heuristically and formally.

Theorem 6.7. (Argument principle). *Suppose f is meromorphic inside a bounded region containing a simple closed curve C which contains no poles or zeroes of f. Then*

$$\frac{1}{2\pi i} \int_C \frac{f'(z)}{f(z)}\, dz = N(C) - M(C), \tag{6.40}$$

where $N(C)$ is the number of zeroes, and $M(C)$ is the number of poles inside C, and the integral is taken counterclockwise.

This is directly accessible to computation. We illustrate this at length in Exercise 2 at the end of this chapter. A shorter example is:

Example 6.8. Find the number of zeroes of $P(z) = z^3 + 3z^2 + 6z + 4$ with $\text{Re}(z) > 0$.

We note that P has no zeroes on the positive real axis and has one or three on the negative real axis (by Déscartes' rule of signs). If we integrate over $C(r) = \{z : |z| \leq r\}$, to six places we get $N(C(1.5)) = 1.00000$ and $N(C(2.5)) = 3.00001$, so we have located two complex zeroes in the annulus between these two circles. Correspondingly, $N(91/24 + C(109/24)) = 0$ (to eight digits) and thus with three circles, we have answered the question, as in Figure 6.1. Of course, the final two circles would have been enough to answer the question, but one cannot know that before beginning the computation. In fact, the zeroes are at -1 and $-1 \pm \sqrt{3}i$. □

A fairly direct consequence of the *maximum principle*, namely that the maximum of the modulus of a nonconstant analytic function occurs only on the boundary of whatever region we are considering, is:

Theorem 6.9. (Rouché's theorem). *Let $f(z)$ and $g(z)$ be analytic inside and on a simple closed curve C. Suppose that $|g(z)| < |f(z)|$ on C. Then $f(z)$ and $f(z) - g(z)$ have the same number of zeroes inside C.*

Computation can assure us that f and g do indeed satisfy both the hypotheses and the conclusions of the theorem. We leave some examples as exercises. The third more specialized result is a very useful consequence of the Cauchy residue theorem due to Lindelöf (1905).

Theorem 6.10. (Cauchy-Lindelöf). *Let $k(z)$ and $r(z)$ be meromorphic in the complex plane. Suppose that $r(z)$ is a rational function which is $O(z^{-2})$ at infinity while $k(z)$ is $o(z)$ over an infinite set of circles $|z| = R_n \to \infty$. Then*

$$-\sum_{p \in \mathcal{P}} \text{Res}(r\,k(s), s = p) \;=\; \sum_{q \in \mathcal{Q}} \text{Res}(r\,k(s), s = q), \qquad (6.41)$$

where $\mathcal{P}$ denotes the poles of r and $\mathcal{Q}$ denotes the poles of k that are not poles of r.

Recall that the *residue*, $\text{Res}(f(s), s = a)$, is defined as the coefficient of $(x - a)^{-1}$ in the Laurent series expansion of f. Again, they are most accessible to assisted computation.

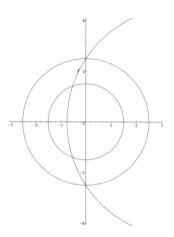

Figure 6.1. Trapping the zeroes.

Let $k(z) = \pi/\sin(\pi z)$ and $r(z) = 1/(z^2 + 1)$, which clearly satisfy the hypotheses of the Cauchy-Lindelöf theorem. Then $\mathcal{Q}$ consists of the zeroes of $\sin(\pi z)$, thus $\mathcal{Q} = \mathbb{Z}$, and $\mathcal{P}$ is $\{\pm i\}$. *Maple* computes the left side of (6.41) to be $\pi/\sinh(\pi)$. Correspondingly, we can easily check in *Mathematica* that the terms of the right side from -6 to 6 are

$$1/37, -1/26, 1/17, -1/10, 1/5, -1/2, 1, -1/2, 1/5, -1/10, 1/17, -1/26, 1/37,$$

so the general term is clearly $(-1)^n/(n^2 + 1)$, and we have evaluated

$$1 + 2 \sum_{n=1}^{\infty} \frac{(-1)^n}{n^2 + 1} = \frac{\pi}{\sinh(\pi)}.$$

Similarly, with $r(z) = 1/z^2$, we reobtain

$$-\frac{1}{6}\pi^2 = 2 \sum_{n=1}^{\infty} \frac{(-1)^n}{n^2}.$$

More examples are scattered through the book.

It is a nice exercise now to obtain the sin product from the Cauchy-Lindelöf theorem applied to $\cot(z) - 1/z$ and $1/((z - w)z)$ to deduce first that

$$w\cot(w) - 1 = 2w^2 \sum_{n=1}^{\infty} \frac{1}{n^2\pi^2 - w^2},$$

and to find corresponding formulas for other trigonometric functions.

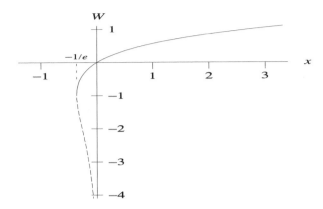

Figure 6.2. Real branches of Lambert's function that satisfy $W \exp W = x$

6.6 The Lambert Function

As noted in Section 6.4, the Lambert W function satisfies

$$W(x)e^{W(x)} = x \,. \tag{6.42}$$

We give a short primer on W in Section 6.6.3, and a survey of history, properties, and applications of W is to be found in [88, 90].

There is a branch point of W at $x = -1/e$, where $W(x) = -1$. See Figure 6.2, which can be produced in *Maple* by the command

```
> plot( [ t*exp(t), t, t=-5..1 ], -1..3, -4..1 );
```

The two real-valued branches of W are denoted $W_0(x)$ and $W_{-1}(x)$; we also refer to $W_0(x)$ as the *principal branch*. To understand the function near the branch point at $x = -1/e$, after various experiments, we decide to compute the series of $W_0\left(-e^{-1-z^2/2}\right)$. We obtain, very quickly, that

$$
\begin{aligned}
W_0 &\left(-\exp(-1-z^2/2)\right) \\
&= -1 + z - \frac{1}{3}z^2 + \frac{1}{36}z^3 + \frac{1}{270}z^4 + \frac{1}{4320}z^5 - \frac{1}{17010}z^6 \\
&\quad - \frac{139}{5443200}z^7 - \frac{1}{204120}z^8 - \frac{571}{2351462400}z^9 + O\left(z^{10}\right) \,. \tag{6.43}
\end{aligned}
$$

6.6.1 The Lambert Function and Stirling's Formula

We look up the sequence of denominators 1, 3, 36, 270, 4320, $\cdots$, in [200]. (Sometimes denominators have nontrivial common factors with numerators. Cancellation of these common factors makes any "guessing" procedure more difficult.) Thus, reference [163] would not easily be found by a normal citation search. We find out in [163] that Equation (6.43) gives coefficients needed in Stirling's formula for $n!$, which begins

$$n! \sim \sqrt{2\pi n}\, n^n e^{-n} \left(1 + \frac{1}{12n} + \frac{1}{288n^2} - \frac{139}{51840n^3} - \frac{571}{2488320n^4} + O\left(\frac{1}{n^5}\right)\right).$$

The connection we discover (without doing any work ourselves) is that if

$$W_0\left(-e^{-1-z^2/2}\right) = \sum_{k \geq 0} (-1)^{k-1} a_k z^k,$$

then

$$n! \sim \sqrt{2\pi n}\, n^n e^{-n} \sum_{k \geq 0} \frac{1 \cdot 3 \cdot 5 \cdots (2k+1)}{n^k} a_{2k+1}.$$

Moreover, there is a lovely (and useful!) recurrence relation for the a_k's, namely $a_0 = 1$, $a_1 = 1$, and

$$a_n = \frac{1}{n+1} \left(a_{n-1} - \sum_{k=2}^{n-1} k a_k a_{n+1-k}\right).$$

6.6.2   The Lambert Function and Riemann Surfaces

As we have seen, tools such as MATLAB and *Mathematica* permit easy accurate visualization of many objects, including Riemann surfaces [89, 209]. We now describe a simple technique for visualization of Riemann surfaces, namely to make three-dimensional plots of $\mathrm{Re}f(z)$ or $\mathrm{Im}f(z)$.

It is necessary to *prove* something—namely, that we do get a good picture of the Riemann surface and not just a three-dimensional plot of an imaginary or real part. The key point is that given $w = u + iv = f(z) = f(x + iy)$, then we get an accurate Riemann surface by plotting, say, (x, y, v) *if and only if* the missing piece of information (here, u) is completely determined once x, y, and v are given.

This is simple, if not obvious: Once we have a smooth three-dimensional surface, each point of which can be associated with a unique value (i.e., ordered pair) of the map $z \mapsto w = f(z)$, then we have a representation of the Riemann surface of f. The exact association is not automatic. For example, if $w = \log(z)$ and we plot (x, y, u), then we do *not* get a picture of the Riemann surface for logarithm, because the branch of $v = \Im(w) = \arg(z)$ is not determined from $u = \log(x^2 + y^2)/2$, x, and y. If we plot (x, y, v), we *do* recover the classical picture of the Riemann surface for

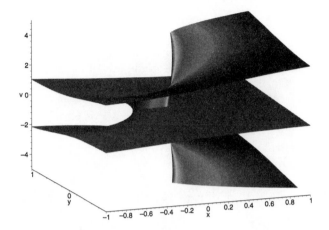

Figure 6.3. The Riemann surface for the Lambert function.

$\log(z)$, because given x, y, and v, we can easily find u. Figure 6.3 (see Color Plate V) gives a static representation of the Riemann surface for the Lambert W function. Many others are graphed in [89] and [209].

Here is a one-to-one correspondence proof. Given x, y, and v, we have to solve for u. Of course, one takes the existence of (u, v) for a given (x, y), [88]. We have

$$(u + iv)e^{u+iv} = x + iy,$$

which gives

$$ue^u + ive^u = (x + iy)e^{-iv} = (x + iy)(\cos v - i \sin v);$$

therefore,

$$ue^u + ive^u = (x \cos v + y \sin v) + i(y \cos v - x \sin v).$$

If $v \neq 0$, and moreover $y \cos v - x \sin v \neq 0$, then dividing the real part by the imaginary part gives u in terms of x, y, and v:

$$u = \frac{v(x \cos v + y \sin v)}{y \cos v - x \sin v}.$$

This solution is unique. Investigation of the exceptional conditions $v = 0$ or $y \cos v - x \sin v = 0$ leads to $u \exp u = x$, which has two solutions if and only if $-1/e \leq x < 0$, in the case $v = 0$, and to the singular condition $u = -\infty$ and $x = y = 0$.

This is precisely what we observe in the graph: Two sheets intersect only if $-1/e \leq x < 0$ (note in color, the colors are different and hence, the corresponding sheets on the Riemann surface do not "really" intersect), and all sheets have a singularity at the origin, except the central one, which contains $v = 0$. This is as good a representation of the Riemann surface for the Lambert W function as can be produced in three dimensions. However, Figure 6.3 is nowhere near as intelligible as a live plot. On a personal computer, the use of OpenGL by *Maple* allows the plot to be rotated by direct mouse control. This helps to give a good sense of what the surface is really like, in three dimensions.

6.6.3   The Lambert Function in Brief

If you have used *Maple* or *Mathematica* to solve transcendental equations, you may already have encountered the Lambert W function, defined by (6.42). The history and some of the properties of this remarkable function are described in [88]. The function provides a beautiful new look at much of undergraduate mathematics, and much of intrinsic interest. Here are some of the elementary properties of W.

1. On $0 \leq x < \infty$, there is one real-valued branch $W(x) \geq 0$ (see Figure 6.2). On $-1/e < x < 0$, there are two real-valued branches. We call the branch that has $W(0) = 0$ the principal branch. On this branch, it is easy to see that $W(e) = W(1 \cdot e^1) = 1$.

2. The derivative of W can be found by implicit differentiation to be

$$\frac{d}{dx}W(x) = \frac{1}{(1 + W(x))e^{W(x)}}$$
$$= \frac{W(x)}{(1 + W(x))x},$$

where the second formula follows on using $\exp W(x) = x/W(x)$, and holds if $x \neq 0$. We may use the first formula to find the value of the derivative at $x = 0$, and we see the singularity is just a removable one.

3. The function $y = W(\exp z)$ satisfies

$$y + \log y = z.$$

This function appears, for example, in convex optimization. Consider the convex conjugate, $f^*(s) = \sup_r rs - f(r)$, of the function $f(r) = r\log(r/(1-r)) - r$. Then we can show that $f^*(s) = W(\exp s)$.

4. $W(x)$ has a Taylor series about $x = 0$ with rational coefficients. Similarly, $W(\exp z)$ has a Taylor series with rational coefficients about $z = 1$. The first few terms are

$$W(e^z) = 1 + \frac{1}{2}(z-1) + \frac{1}{16}(z-1)^2 - \frac{1}{192}(z-1)^3 - \frac{1}{3072}(z-1)^4$$
$$+ \frac{13}{61440}(z-1)^5 - \frac{47}{1474560}(z-1)^6 - \frac{73}{41287680}(z-1)^7$$
$$+ \frac{2447}{1321205760}(z-1)^8 - \frac{16811}{47563407360}(z-1)^9$$
$$- \frac{15551}{1902536294400}(z-1)^{10} + O((z-1)^{11}).$$

5. There is an exact formula for the coefficients of the n-th derivative of $W(\exp z)$, in terms of second-order Eulerian numbers $\left\langle\!\!\left\langle \binom{n}{k} \right\rangle\!\!\right\rangle$ [131]. This formula comes from the following expression for the n-th derivative of $W(\exp z)$, which is stated in [88]. Once the answer is known, the proof is an easy induction, which we leave for the reader.

6. The derivatives of $W(\exp z)$ are

$$\frac{d^n}{dz^n}W(e^z) = \frac{q_n(W(e^z))}{(1+W(e^z))^{2n-1}}, \tag{6.44}$$

where $q_n(w)$ is a polynomial of degree n satisfying the recurrence relation

$$q_{n+1}(w) = -(2n-1)wq_n(w) + (w+w^2)q_n'(w), \qquad n > 1 \tag{6.45}$$

and having the explicit expression

$$q_n(w) = \sum_{k=0}^{n-1} \left\langle\!\!\left\langle \binom{n-1}{k} \right\rangle\!\!\right\rangle (-1)^k w^{k+1}. \tag{6.46}$$

If $n = 1$, we have $q_1(w) = w$, and it is convenient to put $q_0(w) = w/(1+w)$; this isn't a polynomial, but it makes things work out right. This means the series for $W(\exp z)$ about $z = 1$ is just

$$W(e^z) = \sum_{n \geq 0} \frac{q_n(1)}{n!2^{2n-1}}(z-1)^n. \tag{6.47}$$

6.7   Commentary and Additional Examples

1. **The Selberg integral.** The Selberg integral (see [11]) is an N-dimensional extension of Euler's beta integral

$$S_N(\lambda_1, \lambda_2, \lambda) = \left(\prod_{l=1}^{N} \int_0^1 dt_l \, t_l^{\lambda_1} (1 - t_l)^{\lambda_2}\right) \prod_{1 \le j < k \le N} |t_k - t_j|^{2\lambda}$$

(the beta integral is the case $N = 1$). Selberg evaluated this integral as a product of Gamma functions:

$$S_N(\lambda_1, \lambda_2, \lambda) = \prod_{j=0}^{N-1} \frac{\Gamma(\lambda_1 + 1 + j\lambda)\Gamma(\lambda_2 + 1 + j\lambda)\Gamma(1 + (j+1)\lambda)}{\Gamma(\lambda_1 + \lambda_2 + 2 + (N + j - 1)\lambda)\Gamma(1 + \lambda)}.$$

For example,

$$\int_0^1 \int_0^1 x^a (1 - x)^b \, y^a (1 - y)^b \, (|x - y|)^{2c} \, dx \, dy \tag{6.48}$$
$$= \frac{\Gamma(a + 1)\Gamma(b + 1)\Gamma(a + 1 + c)\Gamma(b + 1 + c)\Gamma(1 + 2c)}{\Gamma(a + b + 2 + c)\Gamma(a + b + 2 + 2c)\Gamma(c + 1)}.$$

The Selberg integral gives rise to many variations and to interesting tests of multidimensional integration routines. A beautiful and illustrative application is to evaluate Gaussian ensembles arising in statistical mechanics such as

$$\int_0^\infty \cdots \int_0^\infty \exp\left(-\frac{1}{2} \sum_{i=1}^n x_i^2\right) \prod_{1 \le i < j \le n} |x_i - x_j|^{2\gamma} \, dx = \prod_{j=1}^n \frac{\Gamma(\gamma j + 1)}{\Gamma(\gamma + 1)}.$$

2. **Establishing inequalities numerically via the argument principle.** This refers to Theorem 6.7. We consider the *logarithmic* and $2/3$-*power* means,

$$\mathcal{L}(x, y) = \frac{x - y}{\log(x) - \log(y)}, \quad \mathcal{M}(x, y) = \sqrt[\frac{3}{2}]{\frac{x^{\frac{2}{3}} + y^{\frac{2}{3}}}{2}}.$$

A delicate estimate of an *elliptic integral* was reduced to establishing the elementary inequalities:

$$\mathcal{L}^{-1}(\mathcal{M}(x, 1), \sqrt{x}) < \mathcal{L}^{-1}(x, 1) < \mathcal{L}^{-1}(\mathcal{M}(x, 1), 1) \text{ for } 0 < x < 1.$$

(a) To prove the right-hand inequality, we may draw on *graphic/symbolic* assistance to establish

$$\mathcal{F}(x) = \mathcal{L}^{-1}(\mathcal{M}(x,1),1) - \mathcal{L}^{-1}(x,1) > 0 \text{ for } 0 < x < 1.$$

Using any plotting routine, we will see that $\mathcal{M}$ is a mean (or in other words, $\min(a,b) \leq \mathcal{M}(a,b) \leq \max(a,b)$): as $x < \mathcal{M}(x,1) < 1$ illustrates. Also, we observe that $\mathcal{L}^{-1}$ is decreasing. These two graphical hints lead us directly to a proof of the right-hand inequality.

(b) The left-hand inequality is equivalent to

$$\mathcal{E}(x) = \mathcal{L}^{-1}(x,1) - \mathcal{L}^{-1}(\mathcal{M}(x,1),\sqrt{x}) > 0 \text{ for } 0 < x < 1.$$

This can be accomplished with a mix of *numeric/symbolic* methods:

i. establishing that $\lim_{x \to 0+} \mathcal{E}(x) = \infty$;

ii. using a Newton-like iteration to show that $\mathcal{E}(x) > 0$ on $[0.0, 0.9]$;

iii. using a Taylor series expansion to show $\mathcal{E}(x)$ has 4 zeroes at 1; and then

iv. using the Argument Principle to establish that there are no more zeroes inside $C = \{z : |z - 1| = \frac{1}{4}\}$:

$$\frac{1}{2\pi i} \int_C \frac{\mathcal{E}'}{\mathcal{E}} = \#(\mathcal{E}^{-1}(0); C)$$

(the number of zeroes inside C).

These steps can each be made effective, and so constitute a proof, the only one so far known. It is the last step (iv) that requires some care. In particular, a numerical quadrature scheme must be used that can ensure the integral is actually correct to at least one significant place as claimed; just because we can compute it correctly to twenty places does not ensure that.

3. **Two limit problems.** Evaluate:

(a)

$$\lim_{x \to \infty} \frac{x^4}{e^{3x}} \int_0^x \int_0^{x-u} e^{u^3 + v^3} \, dv \, du \qquad \left(= \frac{2}{9} \right).$$

(b)

$$\lim_{n\to\infty} \frac{1}{n^4} \prod_{i=1}^{2n} \left(n^2 + i^2\right)^{1/n} \qquad (= 25\exp(2\arctan(2) - 4)\,).$$

4. **Carleman's inequality.** Determine

$$\sup_{S} \frac{\sum_{n=1}^{\infty} (x_1 x_2 \cdots x_n)^{1/n}}{\sum_{n=1}^{\infty} x_n} \qquad (= e),$$

where S denotes all nonnegative nonzero sequences. *Hint*: Consider sequences of the form $n^n/(n+1)^{(n-1)}$. The recent survey [112] contains a fine selection of proofs and extensions of Carleman's inequality.

5. **The origin of the elliptic integrals.** The elliptic integrals take their name from the fact that E provides the arclength of an ellipse.

   (a) Show that the arclength, $\mathcal{L}$, of an ellipse with major semi-axis a and minor semi-axis b is

   $$\mathcal{L} = 4a\,\mathrm{E}'\left(\frac{b}{a}\right) = \int_{-\pi}^{\pi} \sqrt{a^2 \cos^2(\theta) + b^2 \sin^2(\theta)}\, d\theta.$$

   (b) The period, p, of a pendulum with amplitude α and length L is

   $$p = 4\sqrt{\frac{L}{g}}\,\mathrm{K}\left(\sin\left(\frac{\alpha}{2}\right)\right),$$

   where g is the gravitational constant. Deduce that for small amplitude $p \approx 2\pi\sqrt{L/g}$, as is the case of simple harmonic oscillation.

   (c) Use the AGM iteration

   $$\frac{\pi/2}{M(a,b)} = \int_{0}^{\pi/2} \frac{dt}{\sqrt{a^2 \cos^2 t + b^2 \sin^2 t}} \qquad (6.49)$$

   to write a fast algorithm to compute p to arbitrary precision. Combine this with Legendre's identity

   $$E(k)\,K(k') + K(k)\,E'(k) - K(k)\,K'(k) = \frac{\pi}{2}$$

   to compute the arclength of an ellipse.

Recall that a function is elementary if it can be realized in a finite number of steps from compositions of algebraic, exponential, and trigonometric functions and their inverses. In 1835, Liouville was able to show that E and K are nonelementary transcendental functions. By contrast, the formula for the area of an ellipse was known to Archimedes.

6. **Euler's integral for hypergeometric functions.** Establish Euler's integral for the hypergeometric function

$$F(a, b, c; x) = \frac{\Gamma(c)\Gamma(b)}{\Gamma(c-b)} \int_0^1 t^{b-1}(1-t)^{c-b-1}(1-tx)^{-a}\, dt \quad (6.50)$$

for $\mathrm{Re}(c) > \mathrm{Re}(b) > 0$. Appropriately interpreted, the right side of (6.50) is valid outside of $[1, \infty)$ and so provides an analytic continuation of F. *Hint*: Expand $(1 - xt)^a$ by the binomial theorem and observe that term-by-term one has a β integral which, as we saw, can be written in terms of Gamma functions.

7. **A hypergeometric function identity.** Show that

$$\frac{1}{x}F(1-y, 1, 1+x; -1) + \frac{1}{y}F(1-x, 1, 1+y; -1) = 2^{x+y-1}\beta(x, y).$$

Hint: Use (6.50).

8. **Multiple hypergeometric functions.** There are many extensions to hypergeometric functions in several complex variables. We list one as an example:

$$F(\alpha, \beta, \beta', \gamma; x, y) = \sum_{m,n \geq 0} \frac{(\alpha)_{m+n}(\beta)_m(\beta')_n}{(\gamma)_{m+n}\, m!\, n!} x^m y^n.$$

This function (and three similar ones) satisfies a pair of second order partial differential equations. For various choices of parameters, it reduces to the classical hypergeometric function

$$F(\alpha, \beta, \beta', \gamma; x, x) = F(\alpha, \beta + \beta', \gamma; x).$$

It also has a double integral of Euler type

$$F(\alpha, \beta, \beta', \gamma; x, y) = \frac{\Gamma(\gamma)}{\Gamma(\beta)\Gamma(\beta')\Gamma(\gamma - \beta - \beta')} \quad (6.51)$$

$$\times \int_\Omega u^{\beta-1} v^{\beta'-1}(1-u-v)^{\gamma-\beta-\beta'-1}(1-ux-vy)^{-\alpha}\, du\, dv,$$

where Ω is the positive quadrant of the unit circle, $\{(x, y) : x \geq 0, y \geq 0, x^2 + y^2 \leq 1\}$, and $\text{Re } \beta > 0$, $\text{Re } \beta' > 0$, $\text{Re } \gamma - \beta - \beta' > 0$. Attempt to numerically validate (6.51) and to prove the identity.

9. **Another hypergeometric function evaluation.** Following Gauss, evaluate

$$F(a, b, c; 1) = \frac{\Gamma(c)\Gamma(c - a - b)}{\Gamma(c - a)\Gamma(c - b)} \tag{6.52}$$

and determine the region of validity.

10. **Berkeley problem 2.2.13.** Find the singular matrix S nearest in Euclidean norm to

$$\begin{bmatrix} 1 & 0 \\ 0 & 2 \end{bmatrix}.$$

Hint: Directly solve the implied optimization problem using a computer algebra program.

Answer: $S = \begin{bmatrix} 0 & 0 \\ 0 & 2 \end{bmatrix}.$

11. **Kirchhoff's law.** Consider a finite, undirected, connected graph with vertex set V and edge set E. Suppose that α and β in V are distinct vertices and that each edge ij in E has an associated "resistance" $r_{ij} > 0$ in R. We consider the effect of applying a unit "potential difference" between the vertices α and β. Let $V_0 = V \setminus \{\alpha, \beta\}$, and for "potentials" x in R^{V_0}, we define the "power" $p : \text{R}^{V_0} \rightarrow$ R by

$$p(x) = \sum_{ij \in E} \frac{(x_i - x_j)^2}{2r_{ij}},$$

where we set $x_\alpha = 0$ and $x_\beta = 1$.

(a) Prove that the power function p has compact level sets.

(b) Deduce the existence of a solution to the following equations (describing "conservation of current"):

$$\sum_{j \,:\, ij \in E} \frac{x_i - x_j}{r_{ij}} = 0 \text{ for } i \text{ in } V_0$$

$$x_\alpha = 0$$

$$x_\beta = 1.$$

(c) Prove that the power function p is strictly convex.

(d) Deduce that the conservation of current equations in part (b) have a unique solution.

12. **Berkeley problem 7.1.14.** Show that the functions $t \mapsto \exp(\alpha_k t)$ are linearly independent over R when the corresponding real numbers $\alpha_1, \alpha_2, \cdots, \alpha_n$ are distinct. *Hint*: Use *Maple* or *Mathematica* to observe the relative growth of the exponential values, given some sample values of α_i.

13. **A central binomial series.** Show that

$$\arcsin^2 (x) = \frac{1}{2} \sum_{m=1}^{\infty} \frac{(2x)^{2m}}{m \binom{2m}{m}},$$

by showing that both sides satisfy the same differential equation.

14. **Branches of arcsin.** Consider

$$u(r) = \arcsin \left(4 \sqrt{\sin (r)} \, \frac{1 - \sin (r)}{1 + \sin^2 (r)} \right).$$

Determine the behavior of u and $u^{(2)}$ on $[0, \pi/2]$, and do the same for $v(r) = u(2r)$ on $[0, \pi/4]$. *Hint*: Start by plotting u and $u^{(2)}$ and the identity.

15. **Integrating the inverse of Gamma.**

(a) Evaluate

$$\int_0^{\infty} \frac{1}{x \left[\pi^2 + \log^2 (x)\right]} \, dx.$$

(b) Show

$$\int_0^{\infty} \frac{1}{\Gamma (x + 1)} \, dx = e - \frac{1}{\pi} \int_{-\frac{\pi}{2}}^{\frac{\pi}{2}} e^{-e^{\pi \tan(\theta)}} \, d\theta,$$

and

$$\int_0^{\infty} \frac{1}{\Gamma (x)} \, dx = e + \frac{1}{\pi} \int_{-\frac{\pi}{2}}^{\frac{\pi}{2}} e^{\pi \tan(\theta)} e^{-e^{\pi \tan(\theta)}} \, d\theta.$$

(c) For integer $n \geq 0$, show

$$\int_{-n}^{-n+1} \frac{1}{\Gamma(x)}\, dx = (-1)^n \int_0^\infty e^{-x} \frac{(x+1)\, x^{n-1}}{\pi^2 + \log^2(x)}\, dx$$

$$= \frac{(-1)^n}{\pi} \int_{-\frac{\pi}{2}}^{\frac{\pi}{2}} \left(e^{\pi\, \tan(\theta)} + 1 \right) \frac{e^{n\pi\, \tan(\theta)}}{e^{e^\pi\, \tan(\theta)}}\, d\theta$$

and

$$\int_{-n}^0 \frac{1}{\Gamma(x)}\, dx = \int_0^\infty e^{-x} \frac{(-x)^n - 1}{\pi^2 + \log^2(x)}\, dx.$$

Hence, estimate the size of $\int_{-n}^0 \Gamma(x)^{-1}\, dx$ asymptotically. In each case, the rightmost expression is much easier to compute. One way to do this is to use *Laplace's method*, which is suited to asymptotics of integrals of the form $\mathcal{I}(x) = \int_a^b e^{-xp(t)} q(t)\, dt$, for appropriate positive $p(t)$ and $q(t)$, for which the bulk of the value of $\mathcal{I}(x)$ is determined by the minimum value $p(c)$, and one estimates that $\mathcal{I}(x) \approx \int_a^b e^{-x(p(c)+(t-c)p'(c))} q(c)\, dt$, which can be integrated explicitly [175]. Note that it is easy to see that the value is "just less" than $n!$ (see [133, 158, 119]).

16. **Hankel loop integral for Gamma.** Show that for any complex z,

$$\Gamma(z) = \int_{-\infty}^{(0+)} e^t t^{-z}\, dt, \tag{6.53}$$

where the notation denotes the integral over a counterclockwise contour that starts at $t = -\infty$, circles around $t = 0$ once and returns to $t = -\infty$. (Here t^{-z} is taken to be the principal value where the contour crosses the positive real axis and to be continuous elsewhere.) Use (6.53) to show that

$$\int_0^\infty \frac{1}{\Gamma(x)}\, dx = e - \int_0^\infty \frac{e^{-t}}{\pi^2 + \log^2(t)}\, dt.$$

Compare part (b) of the previous exercise.

17. **A "gfun" proof for the AGM.** It is possible to entirely implement the evaluation of (6.12) and (6.16). We sketch Bruno Salvy's derivation within Murray, Salvy and Zimmermann's package "gfun" http://algo.inria.fr/libraries/software.html.

  (a) From (6.9), obtain the first dozen or so terms of the power series for $A_2(x^2)$.

(b) The "gfun" program will guess a differential equation for $A_2(x^2)$.

(c) It will then also provide a recursion for said differential equation.

(d) This recursion is solvable in *Maple* to produce

$$u_n = \frac{\Gamma^2\left(n + \frac{1}{2}\right)^2}{\Gamma^2(n+1)\,\pi},$$

and summing $\sum_{n \geq 0} u_n x^n$ produces the desired hypergeometric evaluation.

(e) Provide the details of a similar albeit a little more elaborate computer proof of hypergeometric evaluation for AG_3 of Section 6.1.3.

(f) Attempt to do likewise for $N = 4$ and $N = 7$.

18. **Some complex plots from Guenard and Lemberg.** Plot the following functions $\phi_k : (0,1] \mapsto \mathbb{C}$ for $k = 1, 2, \cdots, 6$ in random order on an interval of the origin and determine from the plots which function came from which plot [132].

$$
\begin{aligned}
\phi_1(t) &= t \sin\left(t^{-2}\right) e^{i\cos(t^4)/\sqrt{t}} \\
\phi_2(t) &= t \sin\left(t^{-1}\right) e^{i\cos(t^{-1})} \\
\phi_3(t) &= t \sin\left(t^{-2}\right) e^{i\cos(t^{-2})} \\
\phi_4(t) &= t \sin\left(t^{-1}\right) e^{i\cos^2(t^{-1})} \\
\phi_5(t) &= t \sin\left(t^{-1}\right) e^{i\sqrt{|\cos(t^{-1})|}} \\
\phi_6(t) &= \sqrt[3]{t} \sin^2\left(t^{-1/2}\right) e^{i\sin\left(\frac{1}{\sqrt{t}}\right)}.
\end{aligned}
$$

One of the functions is plotted in Figure 6.4. *Answer:* ϕ_3.

19. **Some ODE plots from Guenard and Lemberg.** Plot the families of solutions, in random order, to the following differential equations $\mathcal{E}_k$ for $k = 1, 2, \cdots, 5$ and determine from the plots which differential equation generated which plot. One of the solutions is plotted in Figure 6.5 [132].

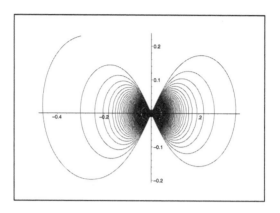

Figure 6.4. Which function ϕ_k is this?

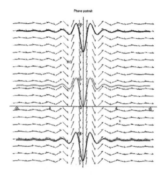

Figure 6.5. Which equation $\mathcal{E}_k$ generated this?

$$\mathcal{E}_1 : \quad y' \;=\; x \sin (xy)$$
$$\mathcal{E}_2 : \quad y' \;=\; \sin (x) \sin (y)$$
$$\mathcal{E}_3 : \quad y' \;=\; x \sin (y)$$
$$\mathcal{E}_4 : \quad y' \;=\; y \sin (x)$$
$$\mathcal{E}_5 : \quad y' \;=\; \frac{\sin (3x)}{1 + x^2}$$

Answer: $\mathcal{E}_3$.

20. **The Maclaurin series for the Lambert function.**

   (a) Show that the Lambert W function satisfies

   $$W(z) = \sum_{k=1}^{\infty} \frac{(-k)^{k-1}}{k!} z^k,$$

   and determine the radius of convergence.

   (b) Show that

   $$\sum_{k=1}^{\infty} \frac{k^k}{k!} x^k = \frac{1}{2}$$

   is solved by $x^* = 0.238843770191263\ldots$.

   (c) Determine the analytic form of this number. This evaluation was discovered by Harold Boas [28], using the Inverse Symbolic Calculator, while studying multidimensional versions of Harald Bohr's 1914 result that if $\left|\sum_{k=0}^{\infty} c_k z^k\right| < 1$ for $|z| < 1$, then $\sum_{k=0}^{\infty} \left|c_k z^k\right| < 1$ for $|z| < 1/3$ (the best possible radius).

21. **Berkeley problem 4.3.7.** Does there exist a continuous solution on $[0, 1]$ to $\mathcal{T}(f) = f$, where

   $$\mathcal{T}(g)(x) = \sin(x) + \int_0^1 \frac{g(y)}{e^{x+y+1}} dy?$$

   Answer: Yes. *Hint:* $\mathcal{T}$ is a Banach contraction with constant at most $\exp(-1) - \exp(-2)$. Try plotting $\mathcal{T}^n(0)$ for $n \le 3$.

22. **Torczon's multidirectional search.** As computing paradigms change, so do algorithms of choice—for good reasons, such as increased speed, storage and parallelism, and for less good reasons, such as vogue. Multidirectional search techniques were popular in the early days of modern computing because of their low overhead, but their relative lack of accuracy and speed drove them out of fashion. Nowadays, the need to handle extremely large problems, in which "noisy" gradients are either unavailable or very costly, coupled with the shift to parallel computation, has reawakened interest [108].

   (a) Standard methods in numerical optimization, such as the general classes of line search and trust region methods [173], use gradient calculations to determine *descent directions*, or directions in which the objective function decreases. A general class of methods that has (re)gained popularity is *direct search* or *pattern search* methods, which use only function evaluations to find stationary points.

The basic direct search method is the simplex method of Nelder and Mead [171], first published in 1965; this method starts with a *simplex* (a set of $n + 1$ nondegenerate points in R^n), and at each iteration replaces the simplex point with the largest objective function value with a new point with a smaller objective function value by reflecting across the centroid of the other simplex points, or contracting the entire simplex if such a point is not found.

Renewed interest in direct search methods began with the development in 1989 of the *multidirectional search* (MDS) algorithm of Virginia Torczon [207, 208], in which the update and contraction rules of Nelder/Mead are combined with additional rules to maintain the angles between the simplex points. The key achievement of the MDS algorithm over Nelder/Mead is a new set of convergence results; counterexamples exist where the Nelder/Mead algorithm converges to a nonstationary point [166]. Torczon generalized the MDS algorithm to a class of *pattern search* algorithms, where the set of gridpoints is expanded from a simplex to any set containing a *positive basis*, provided the new iterate, chosen to be the "center" of the grid, and the amount that the grid is expanded/contracted, both satisfy very general properties that are very easily verified.

Direct search algorithms benefit from being easy to implement and applicable to a wide range of problems for which gradient-based methods are inappropriate. In addition, direct search methods are also easy to demonstrate in practice, and provide an example of where visualization tools greatly aid the research and learning process.

(b) Figure 6.6 shows contour plots of three sets of three iterates (circles) of the Nelder/Mead, multidirectional search, and pattern search methods, each applied to the *Rosenbrock function*:

$$f_K(x, y) = K(y - x^2)^2 + (1 - x)^2,$$

where $K = 100$ is most frequently used ($K = 5$ is used in the illustrations). Using the general descriptions of each of these algorithms, match the algorithm to the set of iterations. When executed in parallel, it is instructive to color code the computations performed by different processors. Rosenbrock's function

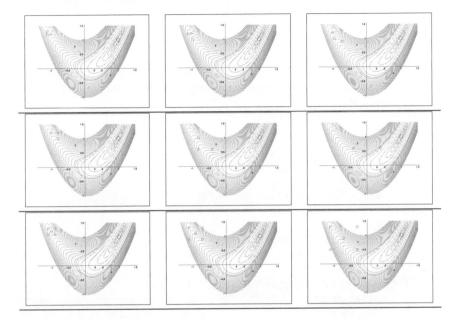

Figure 6.6. Comparison of three direct search methods for the Rosenbrock function.

causes pain for many minimization methods because of the shape of its contours.

23. **Computing with DAD problems.** Find diagonal matrices D_1 and D_2 with strictly positive diagonal entries such that $D_1 A D_2$ is doubly stochastic, where

$$A = \begin{pmatrix} 2 & 10 & 6 \\ 1 & 0 & 4 \\ 7 & 15 & 1 \end{pmatrix}.$$

Hint: Observe that such matrices D_1 and D_2 must exist since A has doubly stochastic pattern; for example, consider

$$\begin{pmatrix} 1/3 & 1/3 & 1/3 \\ 1/2 & 0 & 1/2 \\ 1/6 & 2/3 & 1/6 \end{pmatrix}.$$

(a) Construct the desired matrices by solving the minimization problem from Theorem 6.6.

(b) Alternatively, solve this minimization problem using one of the
    direct search methods outlined in Exercise 22. For example, the
    following *Maple* code applies Torczon's pattern search algorithm
    to this specific matrix, by considering convex combinations of
    the 3×3 permutation matrices with the same zero entry pattern
    as A.

```
p:=proc(u)
RETURN(piecewise(u>0,u*log(u)-u,u=0,0,u<0,infinity));end:

a:=proc(X,i,j)
RETURN(piecewise(A[i,j]>0,p(X[i,j])-X[i,j]*log(A[i,j]),0));
          end:

obj:=proc(X,n) local i,j;
RETURN(evalf(sum(sum(a(X,i,j),i=1..n),j=1..n))); end:

define_matrix:=proc(a) local X,an,i;
X[1]:=matrix(3,3,[1,0,0,0,0,1,0,1,0]);
X[2]:=matrix(3,3,[0,1,0,1,0,0,0,0,1]);
X[3]:=matrix(3,3,[0,1,0,0,0,1,1,0,0]);
X[4]:=matrix(3,3,[0,0,1,1,0,0,0,1,0]); an:=sum(a[i],i=1..4);
RETURN(evalm(sum(a[i]/an*X[i],i=1..4))); end:

inds:=[[1,0,0,0],[0,1,0,0],[0,0,1,0],[0,0,0,1],[0,0,0,0],
       [-1,0,0,0],[0,-1,0,0],[0,0,-1,0],[0,0,0,-1]];
numinds:=17; Digits:=50; numiters:=1000;
objvals:=[infinity\$numiters]; stepsize:=1.1; decfactor:=0.5;
incfactor:=4.0; minval:=infinity; oldminval:=infinity;
curriter:=[1,1,1,1]; for iter from 1 t to numiters do
  currind:=0;
  oldminval:=minval;
  for i from 1 to numinds do
    X1[i]:=define_matrix(evalm(stepsize*inds[i]+curriter));
    objvals[i]:=obj(X1[i],3);
    if objvals[i] < minval then
      minval:=objvals[i];
      currind:=i;
    fi;
  od;
  if currind > 0 then
    curriter:=stepsize*inds[currind]+curriter;
  fi;
  if minval = oldminval then
    stepsize:=decfactor*stepsize;
```

```
      else
         stepsize:=incfactor*stepsize;
      fi;
   od:
```

The solution is:

$$X = \begin{pmatrix} 1/5 & 1/2 & 3/10 \\ 1/3 & 0 & 2/3 \\ 7/15 & 1/2 & 1/30 \end{pmatrix}, \quad D_1 = \begin{pmatrix} 1/10 & 0 & 0 \\ 0 & 1/3 & 0 \\ 0 & 0 & 1/15 \end{pmatrix},$$

$$D2 = \begin{pmatrix} 1 & 0 & 0 \\ 0 & 1/2 & 0 \\ 0 & 0 & 1/2 \end{pmatrix}.$$

24. **Brouwer, eponymy and the fundamental theorem.** The Dutch
mathematician and philosopher Luitzen Egbertus Jan Brouwer (1881–
1966) is equally well known as one of the builders of modern topol-
ogy and as the father of *intuitionism* ("Mathematics is nothing more,
nothing less, than the exact part of our thinking." ... "The construction itself is an art, its application to the world an evil parasite.").
Intuitionism rejects the principle of the excluded middle ("A or not
A"), unless one can effectively determine which case happens, and
looks for fully analyzed proofs broken into "intuitively" compelling
steps. It shares certain concerns with Bishop's version of construc-
tivism.

Thus, Brouwer was influential in the development of both highly constructive mathematics and, apparently, far-from-constructive mathematics. We follow this up in the context of the fundamental theorem
of algebra—for which we gave a constructive proof in [43].

(i) Brouwer's fixed point theorem (1912), that *every continuous self-
map of a closed convex bounded set in* R^n *has a fixed point*, was certainly prefigured by others such as Hadamard and Poincaré, thereby
illustrating Stigler's principle of eponymy.

The $\sin(1/x)$ *circle.* One connects the segments in Figure 6.7 to the
$\sin(1/x)$ curve. This produces a connected, but not simply, highly
nonconvex set, γ, such that every continuous self-map g on γ has a
fixed point. Try to prove this by considering what happens if $g(0) \neq 0$.

(ii) The philosopher Brouwer soon disavowed much of the mathematical tool-set used in his and other early proofs of his eponymous
theorem.

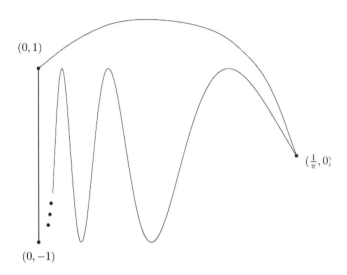

(0, 1)

($\frac{1}{\pi}$, 0)

(0, −1)

Figure 6.7. A nonconvex set with the fixed point property.

(iii) In modern effective incarnations due to Scarf (1973) and others, the theorem is used computationally in optimization and mathematical economics. Many wonderful results such as the von Neumann minimax theorem, the Nash equilibrium theorem and Arrow-Debreu's proof of the existence of market equilibria used Brouwer's theorem, or its extensions, in their first proofs.

(iv) There have been several false proofs—some by distinguished mathematicians—of the Fundamental Theorem of Algebra, based on Brouwer's theorem. In one case the result was applied to a discontinuous self-map! A correct proof, based on the existence of continuous n-th roots for continuous maps from the disc to the sphere, appears in [118], but citations of false ones continue.

(v) In this light, in her Simon Fraser Masters thesis, Tara Stuckless recently wrote:

> In a 1949 paper [12], B. H. Arnold wrote, "... it has been known for some time that the fundamental theorem of algebra could be derived from Brouwer's fixed point theorem." He continued to give a simple one page proof that a polynomial with degree $n \geq 1$ has at least one complex root. Unfortunately, he did this by applying Brouwer's theorem

to a discontinuous function. This error was spotted and a correction was published less than two years later [13]. In 1951, M. K. Fort followed up with a brief proof, using the existence of continuous nth roots of a continuous non zero function on a disk in the complex plane to show that Brouwer can be used to prove the fundamental theorem [118].

In this author's opinion, except for an ill-fated choice of titles, this would have probably been the end of the incorrect proof by Arnold. Arnold's paper was boldly called "A Topological Proof of the Fundamental Theorem of Algebra", whereas Fort's paper was modestly titled "Some Properties of Continuous Functions". Since 1949, citations of Arnold's paper have popped up from time to time as proof of the existence of a topological proof of the fundamental theorem of algebra, including in a four hundred page volume on fixed point theory published in 1981, a talk given in 2000 at a meeting of the Association for Symbolic Logic, as well as in newsgroup discussions that are at least as recent as 1998. On the other hand, this author had a difficult time tracking down Fort's paper, even knowing before hand that it did exist. No amount of searching electronic databases with relevant key words would produce it, though Arnold's paper was invariably returned. Perhaps it would have been useful for Fort to rename his work "A Correct Topological Proof of the Fundamental Theorem of Algebra". It seems certain that Arnold would have preferred this to having his blunder quoted so long after he made his apologies.

7 | Numerical Techniques II

Another thing I must point out is that you cannot prove a vague
theory wrong ... Also, if the process of computing the consequences
is indefinite, then with a little skill any experimental result can be
made to look like the expected consequences.

> Richard Feynman, 1964, from Gary Taubes, *The (Political)*
> *Science of Salt*, 1998

In this chapter, we will examine in more detail some additional underlying
computational techniques that are useful in experimental mathematics. In
particular, we shall briefly examine techniques for theorem proving, prime
number computations, polynomial root finding, numerical quadrature, and
infinte series summation. As in [43], we focus here on *practical* algorithms
and techniques. In some cases, there are known techniques that have su-
perior efficiencies or other characteristics, but for various reasons are not
considered suitable for practical implementation. We acknowledge the existence of such algorithms but do not, in most cases, devote space to them.

7.1 The Wilf-Zeilberger Algorithm

One fascinating non-numerical algorithm is the Wilf-Zeilberger (WZ) algorithm, which employs "creative telescoping" to show that a sum (with
either finitely or infinitely many terms) is zero. Below is an example of a
WZ proof of $(1 + 1)^n = 2^n$. This proof is from Doron Zeilberger's original
Maple program, which in turn is inspired by the proof in [215].

Let $F(n, k) = \binom{n}{k} 2^{-n}$. We wish to show that $L(n) = \sum_k F(n, k) =$
1 for every n. To this end, we construct, using the WZ algorithm, the
function

$$G(n, k) = \frac{-1}{2^{(n+1)}} \binom{n}{k-1} \left(= \frac{-k}{2(n - k + 1)} F(n, k) \right), \quad (7.1)$$

and observe that

$$F(n + 1, k) - F(n, k) \;=\; G(n, k + 1) - G(n, k). \qquad (7.2)$$

By applying the obvious telescoping property of these functions, we can write

$$\sum_k F(n + 1, k) - \sum_k F(n, k) \;=\; \sum_k (G(n, k + 1) - G(n, k))$$

$$=\; 0, \qquad\qquad (7.3)$$

which establishes that $L(n + 1) - L(n) = 0$. The fact that $L(0) = 1$ follows from the fact that $F(0, 0) = 1$ and 0 otherwise.

Obviously, this proof does not provide much insight, since the difficult part of the result is buried in the construction of 7.1. In other words, this is an instance where computers provide "proofs," but these "proofs" tend to be uninteresting. Nonetheless, the extremely general nature of this scheme is of interest. It possibly presages a future in which a wide class of such "proofs" can automatically be obtained in a computer algebra system. Details and additional applications of this algorithm are given in [215].

7.2   Prime Number Computations

In Chapter 2 of [43], we mentioned the connection between prime numbers and the Riemann zeta function. Prime numbers crop up in numerous other arenas of mathematical research, and often even in commercial applications, with the rise of RSA encryption methods on the Internet. Inasmuch as this research topic is certain to be of great interest for the foreseeable future, we mention here some of the techniques for counting, generating, and testing prime numbers.

The prime counting function $\pi(x)$ mentioned in Chapter 2 of [43] is of central interest in this research. It is clear from even a cursory glance at Table 2.2 of [43], that the researchers who have produced these counts are not literally testing every integer up to 10^{22} for primality—that would require much more computation than the combined power of all computers worldwide, even using the best known methods to test individual primes. Indeed, some very sophisticated techniques have been employed, which unfortunately are too technical to be presented in detail here. We refer interested readers to the discussion of this topic in the new book *Prime Numbers: A Computational Perspective* by Richard Crandall and Carl Pomerance [93, pg. 140–150]. Readers who wish to informally explore the behavior of $\pi(x)$ may use the following algorithm, which is a variant of a scheme originally presented by Eratosthenes of Cyrene about 200 BCE [93, pg. 114]:

Algorithm 7.1. *Blocked Sieve of Eratosthenes.*

We are given an interval (L, R), where L and R are even integers, where $L > P = \lceil \sqrt{R} \rceil$ and a blocksize B is assumed to divide $R - L$. We assume that a table of primes p_k up to P of size $Q = \pi(P)$ is available.

Initialize: For $k := 2$ to Q set $q_k := -(L + 1 + p_k)/2 \bmod p_k$.

Process blocks: For $T := L$ to $R - 1$ step $2 \cdot B$ do: for $j := 0$ to $B - 1$ set $b_j := 1$; for $k := 2$ to Q do: for $j := q_k$ to $B - 1$ step p_k do: set $b_j := 0$; enddo; set $q_k := (q_k - B) \bmod p_k$; enddo; for $j := 0$ to $B - 1$ do: if $b_j = 1$ then output $T + 2j + 1$; enddo; enddo.   □

The Sieve of Eratosthenes is only efficient if one wants to find all primes up to a given point, or all primes in a (fairly large) interval. For testing a single integer, or a few integers, faster means are available. We summarize some of these individual primality tests here.

The simplest scheme, from a programming point of view, of testing whether an integer n is prime is simply to generate a table of primes $(p_1, p_2, \cdots, p_n)$ up to $p_n > \sqrt{n}$, using the Sieve of Eratosthenes, and then to iteratively divide n by each p_i. This actually works quite well for modest-sized integers, but becomes infeasible for n beyond about 10^{16}.

If one does not require certainty, but only high probability that a number is prime, some very efficient probabilistic primality tests have been discovered in the past few decades. In fact, these schemes are now routinely used to generate primes for RSA encryption in Internet commerce. When you type in your Visa or Mastercard number in a secure web site to purchase a book or computer accessory, somewhere in the process it is quite likely that two large prime numbers have been generated, which were certified as prime using one of these schemes.

The most widely used probabilistic primality test is the following, which was originally suggested by Artjuhov in 1966, although it was not appreciated until it was rediscovered and popularized by Selfridge in the 1970s [93].

Algorithm 7.2. *Strong probable prime test.*

Given an integer $n = 1 + 2^s t$, for integers s and t (and t odd), select an integer a by means of a pseudorandom number generator in the range $1 < a < n - 1$.

1. Compute $b := a^t \bmod n$ using the binary algorithm for exponentiation (see Algorithm 3.2 in Chapter 3 of [43]). If $b = 1$ or $b = n - 1$ then exit (n is a strong probable prime base a).

2. For $j = 1$ to $s - 1$ do: Compute $b := b^2 \bmod n$; if $(b = n - 1)$ then exit (n is a strong probable prime base a).

3. Exit: n is composite.   □

This test can be repeated several times with different pseudo-randomly chosen a. In 1980 Monier and Rabin independently showed that an integer n that passes the test as a strong probable prime is prime with probability at least $3/4$, so that m tests increase this probability to $1 - 1/4^m$ [168, 183]. In fact, for large test integers n, the probability is even closer to unity. Damgard, Landrock and Pomerance showed in 1993 that if n has k bits, then this probability is greater than $1 - k^2 4^{2 - \sqrt{k}}$, and for certain k is even higher [104]. For instance, if n has 500 bits, then this probability is greater than $1 - 1/4^{28m}$. Thus a 500-bit integer that passes this test even once is prime with prohibitively safe odds—the chance of a false declaration of primality is less than one part in Avogadro's number (6×10^{23}). If it passes the test for four pseudo-randomly chosen integers a, then the chance of false declaration of primality is less than one part in a googol (10^{100}). Such probabilities are many orders of magnitude more remote than the chance that an undetected hardware or software error has occurred in the computation.

A number of more advanced probabilistic primality testing algorithms are now known. The current state-of-the-art is that such tests can determine the primality of integers with hundreds to thousands of digits. Additional details of these schemes are available in [93].

For these reasons, probabilistic primality tests are considered entirely satisfactory for practical use, even for applications such as large interbank financial transactions, which have extremely high security requirements. Nonetheless, mathematicians have long sought tests that remove this last iota of uncertainty, yielding a mathematically rigorous certificate of primality. Indeed, the question of whether there exists a "polynomial time" primality test has long stood as an important unsolved question in pure mathematics.

Thus it was with considerable elation that such an algorithm was recently discovered, by Manindra Agrawal, Neeraj Kayal, and Nitin Saxena (initials AKS) of the Indian Institute of Technology in Kanpur, India [2]. Their discovery sparked worldwide interest, including a prominent report in the New York Times [189]. Since the initial report in August 2002, several improvements have been made. We present here a variant of the original algorithm due to Lenstra [156], as implemented by Richard Crandall and Jason Papadopoulos, who note that the implementations of Daniel Bernstein already provide significant acceleration [102].

Algorithm 7.3. *Variant AKS provable primality test.*

Suppose we are given an integer p that we wish to establish as either prime or composite. This test assumes the existence of a table of primes covering integers up to roughly $(\log_2 p)^2$.

1. Establish that p is not a proper power, meaning that $p \neq a^b$ for $b > 1$. This can be done by computing $p^{1/c}$ for primes c up to $\lceil \log_2(p) \rceil$, using appropriately high-precision floating-point arithmetic. If p is a prime power, then of course it is composite.

2. Set $v = \lceil \log_2 p \rceil$. For integers r beginning with $v + 1$, test whether r is prime, and the multiplicative order of p modulo r is at least v. If not, increment r by one until these conditions are met.

3. For $a = 1$ to $r - 1$, test that the relation

$$(x - a)^p \quad \equiv \quad (x^p - a) \bmod p, (x^r - 1) \qquad (7.4)$$

holds in the polynomial ring $(Z/nZ)[x]$. The mod notation in (7.4) means to reduce every polynomial coefficient modulo p, and also to reduce modulo the polynomial $x^r - 1$. If Condition 3 holds for each a in the given range, then p is prime.   □

Arithmetic in the polynomial ring $(Z/nZ)[x]$ can be accelerated by noting that polynomial multiplication is simply another instance of acyclic convolution, which can be computed using fast Fourier transforms (FFTs), as in FFT-based multiprecision multiplication (which is discussed in the next section). Proof that this algorithm produces a certificate of primality are in Lenstra's manuscript [156], and a working implementation of this scheme (which uses Crandall's PrimeKit software) is available from Crandall's website http://www.perfsci.com. A brief review of AKS implementation issues is available at [102].

7.3   Roots of Polynomials

In Section 6.2.5 of [43], we showed how a relatively simple scheme involving Newton iterations can be used to compute high-precision square roots and even to perform high-precision division. This Newton iteration scheme is, in fact, quite general and can be used to solve many kinds of equations, both algebraic and transcendental. One particularly useful application, frequently encountered by experimental mathematicians, is to find roots of polynomials. This is done by using a careful implementation of the well-known version of Newton's iteration

$$x_{k+1} \;=\; x_k - \frac{p(x)}{p'(x)}, \tag{7.5}$$

where $p'(x)$ denotes the derivative of $p(x)$. As before, this scheme is most efficient if it employs a level of numeric precision that starts with ordinary double precision (16-digit) or double-double precision (32-digit) arithmetic until convergence is achieved at this level, then approximately doubles with each iteration until the final level of precision is attained. One additional iteration at the final or penultimate precision level may be needed to insure full accuracy.

Note that Newton's iteration can be performed, as written in (7.5), with either real or complex arithmetic, so that complex roots of polynomials (with real or complex coefficients) can be found almost as easily as real roots. Evaluation of the polynomials $p(x)$ and $p'(x)$ is most efficiently performed using Horner's rule: For example, the polynomial $p(x) = p_0 + p_1 x + p_2 x^2 + p_3 x^3 + p_4 x^4 + p_5 x^5$ is evaluated as $p(x) = p_0 + x(p_1 + x(p_2 + x(p_3 + x(p_4 + x p_5))))$.

There are two issues that arise here that do not arise with the Newton iteration schemes for division and square root. The first is the selection of the starting value—if it is not close to the desired root, then successive iterations may jump far away. If you have no idea where the roots are (or how accurate the starting value must be), then a typical strategy is to try numerous starting values, covering a wide range of likely values, and then make an inventory of the approximate roots that are found. If you are searching for complex roots, note that it is often necessary to use a two-dimensional array of starting values. These "exploratory" iterations can be done quite rapidly, since typically only a modest numeric precision is required—in almost all cases, just ordinary double precision (16 digits) or double-double precision (32 digits) arithmetic will do. Once the roots have been located in this fashion, then the full-fledged Newton scheme can be used to produce their precise high-precision values.

The second issue is how to handle repeated roots. The difficulty here is that, in such cases, convergence to the root is very slow, and instabilities may throw the search far from the root. In these instances, note that we can write $p(x) = q^2(x)r(x)$, where r has no repeated roots (if all roots are repeated, then $r(x) = 1$). Now note that $p'(x) = 2q(x)r(x) + q^2(x)r'(x) =$

$q(x)[2r(x) + q(x)r'(x)]$. This means that if $p(x)$ has repeated roots, then these roots are also roots of $p'(x)$, and, conversely, if $p(x)$ and $p'(x)$ have a common factor, then the roots of this common factor are repeated roots of $p(x)$. This greatest common divisor polynomial $q(x)$ can be found by performing the Euclidean algorithm (in the ring of polynomials) on $p(x)$ and $p'(x)$. The Newton iteration scheme can then be applied to find the roots of both $q(x)$ and $r(x)$. It is possible, of course, that $q(x)$ also has repeated roots, but recursive application of this scheme quickly yields all individual roots.

In the previous paragraph, we mentioned the possible need to perform the Euclidean algorithm on two polynomials, which involves polynomial multiplication and division. For modest-degree polynomials, a simple implementation of the schemes learned in high school algebra suffices—just represent the polynomials as strings of high-precision numbers. For high-degree polynomials, polynomial multiplication can be accelerated by utilizing fast Fourier transforms and a convolution scheme that is almost identical (except for release of carries) to the scheme, mentioned in Section 6.2 of [43], to perform high-precision multiplication. High-degree polynomial division can be accelerated by a Newton iteration scheme, similar to that mentioned above for high-precision division. See [93] for additional details on high-speed polynomial arithmetic.

We noted above that if the starting value is not quite close to the desired root, then successive Newton iterations may jump far from the root, and eventually converge to a different root than the one desired. In general, suppose we are given a degree-n polynomial $p(x)$ with m distinct complex roots r_k (some may be repeated roots). Define the function $Q_p(z)$ as the limit achieved by successive Newton iterations that start at the complex number z; if no limit is achieved, then set $Q_p(z) = \infty$. Then the m sets $\{z : Q_p(z) = r_k\}$ for $k = 1, 2 \cdots, m$ constitute a partition of the complex plane, except for a filamentary set of measure zero that separates the m sets. In fact, each of these m sets is itself an infinite collection of disconnected components.

The collection of these Newton-Julia sets and their boundaries form pictures of striking beauty, and are actually quite useful in gaining insight on both the root structure of the original polynomial and the behavior of Newton iteration solutions. Some of the most interesting graphics of this type are color-coded plots of the function $N_p(z)$, which is the number of iterations required for convergence (to some accuracy ϵ) of Newton's iteration for $p(x)$, beginning at z (if the Newton iteration does not converge at z, then set $N_p(z) = \infty$). A black-and-white plot for the cubic polynomial $p(x) = x^3 - 1$ is shown in Figure 7.1. Color Plate VI shows a color version.

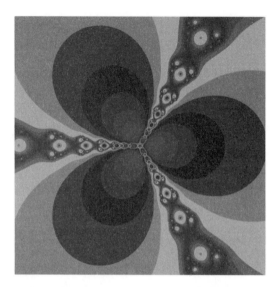

Figure 7.1. Newton-Julia set for $p(x) = x^3 - 1$.

7.4 Numerical Quadrature

Experimental mathematicians very frequently find it necessary to calculate definite integrals to high precision. Recall the examples given in Chapters 2 of [43] and Chapter 1 of this book, wherein we were able to experimentally identify certain definite integrals as analytic expressions, based only on their high-precision numerical value.

To briefly reprise one example, we were inspired by a recent problem in the *American Mathematical Monthly* [3]. By using one of the quadrature routines to be described below, together with a PSLQ integer relation detection program, we found that if $C(a)$ is defined by

$$C(a) = \int_0^1 \frac{\arctan(\sqrt{x^2 + a^2})\, dx}{\sqrt{x^2 + a^2}(x^2 + 1)}, \qquad (7.6)$$

then

$$
\begin{aligned}
C(0) &= \pi \log 2/8 + G/2 \\
C(1) &= \pi/4 - \pi\sqrt{2}/2 + 3\sqrt{2}\arctan(\sqrt{2})/2 \\
C(\sqrt{2}) &= 5\pi^2/96,
\end{aligned}
\qquad (7.7)
$$

where $G = \sum_{k \geq 0} (-1)^k/(2k+1)^2$ is Catalan's constant. The third of these results is the result from the *Monthly*. These particular results then led to the following general result, among others:

$$\int_0^\infty \frac{\arctan(\sqrt{x^2 + a^2})\, dx}{\sqrt{x^2 + a^2}(x^2 + 1)} = \tag{7.8}$$

$$\frac{\pi}{2\sqrt{a^2 - 1}} \left[2\arctan(\sqrt{a^2 - 1}) - \arctan(\sqrt{a^4 - 1}) \right].$$

The commercial packages *Maple* and *Mathematica* both include rather good high-precision numerical quadrature facilities. However, these packages do have some limitations, and in many cases much faster performance can be achieved with custom-written programs. And in general it is beneficial to have some understanding of quadrature techniques, even if you rely on software packages to perform the actual computation.

We describe here three state-of-the-art, highly efficient techniques for numerical quadrature. You can try programming these schemes yourself, or you can refer to the C++ and Fortran-90 programs available at http://www.expmath.info.

7.4.1 Gaussian Quadrature

Our first quadrature scheme is known as Gaussian quadrature, named after the famous mathematician who first discovered this method in the 19th century. Gaussian quadrature is particularly effective for functions that are bounded, continuous, and smooth on a finite interval. Unfortunately, many references only give tables of abscissas and weights, which are useless when we need hundreds or thousands of digits of precision. Thus we include here a fairly reasonably efficient scheme for computing these parameters to any desired accuracy (although, as we will see, the quadratic scaling of this scheme looms as an obstacle to very high-precision implementation).

Gaussian quadrature approximates the definite integral of the real function $f(x)$ on $(-1, 1)$ as

$$\int_{-1}^1 f(x)\, dx \approx \sum_{j=1}^n w_j f(x_j), \tag{7.9}$$

where the abscissas x_j are the zeroes of the degree-n Legendre polynomial $P_n(x)$ on $[-1, 1]$, and the weights w_j are given by

$$w_j = \frac{-2}{(n+1)P_n'(x_j)P_{n+1}(x_j)}, \tag{7.10}$$

for $1 \leq j \leq n$ (see [14, pg. 285]). The abscissas can be computed using a Newton iteration scheme, where the starting value for x_j is given by

$\cos[\pi(j-1/4)/(n+1/2)]$ [180, pg. 125]. The Legendre polynomial function values can be computed using an n-long iteration of the recurrence $P_0(x) = 0$, $P_1(x) = 1$ and

$$(k + 1)P_{k+1}(x) = (2k + 1)xP_k(x) - kP_{k-1}(x),$$

for $k \geq 2$. The derivative is computed as $P'_n(x) = n(xP_n(x) - P_{n-1}(x))/(x^2 - 1)$.

In a typical implementation of Gaussian quadrature, multiple "levels" or phases are employed, with a set of abscissas and weights precalculated for each level, starting with, say, $n = 3$, and then with n doubling with each level up to some maximum phase m. For many well-behaved functions, Gaussian quadrature typically exhibits quadratic convergence, in the sense that after a few initial levels, each subsequent level approximately doubles the number of correct digits. However, each level performed also approximately doubles the computation time, compared with the previous level.

We summarize this procedure as follows [180, pg. 125].

Algorithm 7.4. *Gaussian quadrature.*

Initialize (compute abscissas and weights):
For $k := 1$ to m do:
Set $n := 3 \cdot 2^k$;
For $j := 1$ to $n/2$ do:
Set $r := \cos[\pi(j - 1/4)/(n + 1/2)]$;
Iterate until $r = t_5$ to within the "epsilon" of the working precision level:
Set $t_1 := 1$ and $t_2 := 0$;
For $j_1 := 1$ to n do:
Set $t_3 := t_2$ and $t_2 = t_1$ and calculate $t_1 := [(2j_1 - 1)rt_2 - (j_1 - 1)t_3]/j_1$;
enddo
Calculate $t_4 = n(rt_1 - t2)/(r^2 - 1)$ and then set $t_5 := r$ and $r := r - t_1/t_4$;
end iterate
Set $x_{j,k} = r$ and $w_{j,k} = 2/[(1 - r^2)t_4^2]$;
enddo; enddo

Perform quadrature for a function $f(x)$ on $(-1, 1)$:
For $k := 1$ to m (or until successive instances yield results identical to within the working precision) do:
Set $n := 3 \cdot 2^k$ and set $s := 0$;
For $j = 1$ to $n/2$ do:
Set $s := s + w_{j,k}[f(-x_{j,k}) + f(x_{j,k})]$;

enddo
Set $S_k := s$;
enddo ☐

The Newton iteration scheme given in lines 4–11 above can be accelerated by utilizing a dynamic level of precision, as in other Newton-based algorithms. This means that the first few iterations are performed with modest precision, say double precision or double-double accuracy. Once convergence has been achieved at this level, additional iterations are performed with a level of precision that nearly doubles with each iteration, until the final level of precision has been achieved.

Error bounds are known for Gaussian quadrature, but since they rely on bounds for the n-th derivatives of $f(t)$, they are not very useful in this context, since we seek schemes that work for arbitrary functions, where such information is generally not available. See Section 7.4.4 for some alternative means to estimate errors and determine when results are sufficiently accurate.

7.4.2 Error Function Quadrature

The second scheme we will discuss here is known as "error function" or "erf" quadrature. While error function quadrature is not as efficient as Gaussian quadrature for continuous, bounded, well-behaved functions on finite intervals, it often produces highly accurate results even for functions with (integrable) singularities or vertical derivatives at one or both endpoints of the interval. In contrast, Gaussian quadrature typically performs very poorly in such instances.

The error function quadrature scheme and the tanh-sinh scheme to be described in the next section are based on the Euler-Maclaurin summation formula, which can be stated as follows [14, pg. 280]. Let $m \geq 0$ and $n \geq 1$ be integers, and define $h = (b - a)/n$ and $x_j = a + jh$ for $0 \leq j \leq n$. Further, assume that the function $f(x)$ is at least $(2m + 2)$-times continuously differentiable on $[a, b]$. Then

$$\int_a^b f(x)\, dx = h \sum_{j=0}^n f(x_j) - \frac{h}{2}\left(f(a) + f(b)\right)$$

$$- \sum_{i=1}^m \frac{h^{2i} B_{2i}}{(2i)!}\left(f^{(2i-1)}(b) - f^{(2i-1)}(a)\right) - E, \quad (7.11)$$

where B_{2i} denote the Bernoulli numbers, and

$$E = \frac{h^{2m+2}(b-a)B_{2m+2}f^{(2m+2)}(\xi)}{(2m+2)!}, \qquad (7.12)$$

for some $\xi \in (a, b)$.

In the circumstance where the function $f(x)$ and all of its derivatives are zero at the endpoints a and b, the second and third terms of the Euler-Maclaurin formula are zero. Thus the error in a simple step-function approximation to the integral, with interval h, is simply E. But since E is then less than a constant times $h^{2m+2}/(2m+2)!$, for any m, we conclude that the error goes to zero more rapidly than any power of h. In the case of a function defined on $(-\infty, \infty)$, the Euler-Maclaurin summation formula still applies to the resulting doubly infinite sum approximation, provided as before that the function and all of its derivatives tend to zero for large positive and negative arguments.

This principle is utilized in the error function and tanh-sinh quadrature scheme (see the next subsection) by transforming the integral of $f(x)$ on a finite interval, which we will take to be $(-1, 1)$ for convenience, to an integral on $(-\infty, \infty)$ using the change of variable $x = g(t)$. Here $g(x)$ is some monotonic function with the property that $g(x) \to 1$ as $x \to \infty$, and $g(x) \to -1$ as $x \to -\infty$, and also with the property that $g'(x)$ and all higher derivatives rapidly approach zero for large arguments. In this case we can write, for $h > 0$,

$$\int_{-1}^{1} f(x)\, dx = \int_{-\infty}^{\infty} f(g(t))g'(t)\, dt = h \sum_{-\infty}^{\infty} w_j f(x_j), \qquad (7.13)$$

where $x_j = g(hj)$ and $w_j = g'(hj)$. If the convergence of $g'(t)$ and its derivatives to zero is sufficiently rapid for large $|t|$, then even in cases where $f(x)$ has a vertical derivative or an integrable singularity at one or both endpoints, the resulting integrand $f(g(t))g'(t)$ will be a smooth bell-shaped function for which the Euler-Maclaurin summation formula applies, as described above. In such cases we have that the error in the above approximation decreases faster than any power of h. The summation above is typically carried out to limits $(-N, N)$, beyond which the terms of the summand are less than the "epsilon" of the multiprecision arithmetic being used.

The error function integration scheme uses the function $g(t) = \mathrm{erf}(t)$ and $g'(t) = (2/\sqrt{\pi})e^{-t^2}$. Note that $g'(t)$ is merely the bell-shaped probability density function, which is well known to converge rapidly to zero, together with all of its derivatives, for large arguments. The error function

$\text{erf}(x)$ can be computed to high precision as $1 - \text{erfc}(x)$, using the following formula given by Crandall [96, pg. 85] (who in turn attributes it to a 1968 paper by Chiarella and Reichel):

$$\text{erfc}(t) = \frac{e^{-t^2}\alpha t}{\pi} \left(\frac{1}{t^2} + 2 \sum_{k \geq 1} \frac{e^{-k^2\alpha^2}}{k^2\alpha^2 + t^2} \right) + \frac{2}{1 - e^{2\pi t/\alpha}} + E, \qquad (7.14)$$

where $|E| < e^{-\pi^2/\alpha^2}$. The parameter $\alpha > 0$ here is chosen small enough to ensure that the error E is sufficiently small. We summarize this scheme with the following algorithm statement. Here n_p is the precision level in digits, and ϵ is the "epsilon" level, which is typically 10^{-n_p}.

Algorithm 7.5. *Error function complement [erfc] evaluation.*

Initialize:
Set $\alpha := \pi/\sqrt{n_p \log(10)}$, and set $n_t := n_p \log(10)/\pi$.
Set $t_2 := e^{-\alpha^2}$, $t_3 := t_2^2$, and $t_4 := 1$.
For $k := 1$ to n_t do: set $t_4 := t_2 \cdot t_4$, $E_k := t_4$, $t_2 := t_2 \cdot t_3$; enddo.

Evaluation of function, with argument x:
Set $t_1 := 0$, $t_2 := x^2$, $t_3 := e^{-t_2}$ and $t_4 := \epsilon/(1000 \cdot t_3)$.
For $k := 1$ to n_t do: set $t_5 := E_k/(k^2\alpha^2 + t_2)$ and $t_1 := t_1 + t_5$.
If $|t_5| < t_4$ then exit do; enddo.
Set $\text{erfc}(x) := t_3\alpha x/\pi \cdot (1/t_2 + 2t_1) + 2/(1 - e^{2\pi x/\alpha})$.   □

We now state the algorithm for error function quadrature. As with the Gaussian scheme, m levels or phases of abscissas and weights are precomputed in the error function scheme. Then we perform the computation, increasing the level by one (each of which approximately doubles the computation, compared to the previous level), until an acceptable level of estimated accuracy is obtained (see Section 7.4.4). In the following, ϵ is the "epsilon" level of the multiprecision arithmetic being used.

Algorithm 7.6. *Error function quadrature.*

Initialize:
Set $h := 2^{2-m}$.
For $k := 0$ to $20 \cdot 2^m$ do:
Set $t := kh$, $x_k := 1 - \text{erf}(t)$ and $w_k := 2/\sqrt{\pi} \cdot e^{-t^2}$.
If $|x_k - 1| < \epsilon$ then exit do; enddo.
Set $n_t = k$ (the value of k at exit).

Perform quadrature for a function $f(x)$ on $(-1, 1)$:
Set $S := 0$ and $h := 4$.
For $k := 1$ to m (or until successive values of S are identical to within ϵ)
do: $h := h/2$.
For $i := 0$ to n_t step 2^{m-k} do:
If $(\text{mod}(i, 2^{m-k+1}) \neq 0$ or $k = 1)$ then
If $i = 0$ then $S := S + w_0 f(0)$ else $S := S + w_i(f(-x_i) + f(x_i))$ endif.
endif; enddo; endo.
Result $= hS$.   □

7.4.3   Tanh-Sinh Quadrature

The third scheme we will discuss here is known informally as "tanh-sinh"
quadrature. It is not well known, but based on the authors' experience, it
deserves to be taken seriously because of its speed, robustness, and ease
of implementation. Like the error function quadrature scheme, it is often
successful in producing high-precision quadrature values even for functions
with (integrable) singularities or vertical derivatives at endpoints. It was
first introduced by Hidetosi Takahasi and Masatake Mori [204, 169].
    The tanh-sinh scheme is very similar to the error function scheme, in
that it is based on the Euler-Maclaurin summation formula. The difference
here is that it employs the transformation $x = \tanh(\pi/2 \cdot \sinh t)$, where
$\sinh t = (e^t - e^{-t})/2$, $\cosh t = (e^t + e^{-t})/2$, and $\tanh t = \sinh t / \cosh t$.
As before, this transformation converts an integral on $(-1, 1)$ to an integral on the entire real line, which can then be approximated by means
of a simple step-function summation. In this case, by differentiating the
transformation, we obtain the abscissas x_k and the weights w_k as

$$
\begin{aligned}
x_j &= \tanh[\pi/2 \cdot \sinh(jh)] \\
w_j &= \frac{\pi/2 \cdot \cosh(jh)}{\cosh^2[\pi/2 \cdot \sinh(jh)]}.
\end{aligned}
\tag{7.15}
$$

Note that these functions involved here are compound exponential, so, for
example, the weights w_j converge very rapidly to zero. As a result, the
tanh-sinh quadrature scheme is often even more effective than the error
function scheme in dealing with singularities at endpoints.

Algorithm 7.7. *tanh-sinh quadrature.*

Initialize: Set $h := 2^{-m}$.
For $k := 0$ to $20 \cdot 2^m$ do:

Set $t := kh$, $x_k := \tanh(\pi/2 \cdot \sinh t)$ and $w_k := \pi/2 \cdot \cosh t / \cosh^2(\pi/2 \cdot \sinh t)$;
If $|x_k - 1| < \epsilon$ then exit do; enddo.
Set $n_t = k$ (the value of k at exit).

Perform quadrature for a function $f(x)$ on $(-1,1)$:
Set $S := 0$ and $h := 1$.
For $k := 1$ to m (or until successive values of S are identical to within ϵ)
do:
$h := h/2$.
For $i := 0$ to n_t step 2^{m-k} do:
If $(\mod(i, 2^{m-k+1}) \neq 0$ or $k = 1)$ then
If $i = 0$ then $S := S + w_0 f(0)$ else $S := S + w_i(f(-x_i) + f(x_i))$ endif.
endif; enddo; endo.
Result $= hS$.   □

7.4.4   Practical Considerations for Quadrature

Each of the schemes described above have assumed a function of one variable defined and continuous on the interval $(-1,1)$. Integrals on other finite intervals (a,b) can be found by applying a linear change of variable:

$$\int_a^b f(t)\, dt \;=\; \frac{b-a}{2} \int_{-1}^1 f\left(\frac{b+a}{2} + \frac{b-a}{2}x\right) dx. \qquad (7.16)$$

Note also that integrable functions on an infinite interval can, in a similar manner, be reduced to an integral on a finite interval, for example:

$$\int_0^\infty f(t)\, dt \;=\; \int_0^1 [f(x) + f(1/x)/x^2]\, dx. \qquad (7.17)$$

Integrals of functions with singularities (such as "corners" or step discontinuities) within the integration interval (i.e., not at the endpoints) should be broken into separate integrals.

The above algorithm statements each suggest increasing the level of the quadrature (the value of k) until two successive levels give the same value of S, to within some tolerance ϵ. While this is certainly a reliable termination test, it is often possible to stop the calculation earlier, with significant savings in runtime, by means of making reasonable projections of the current error level. In this regard, the authors have found the following scheme to be fairly reliable: Let S_1, S_2, and S_3 be the value of S at the current level, the previous level, and two levels back, respectively. Then set $D_1 := \log_{10} |S_1 - S_2|$, $D_2 := \log_{10} |S_1 - S_3|$, and $D_3 := \log_{10} \epsilon - 1$. Now we can estimate the error E at level $k > 2$ as 10^{D_4}, where $D_4 = \min(0, \max(D_1^2/D_2, 2D_1, D_3))$. These estimation calculations may be performed using ordinary double precision arithmetic.

All three quadrature schemes have been implemented in C++ and Fortran-90 programs, available at http://www.expmath.info.

7.4.5   Higher-Dimensional Quadrature

The error function and tanh-sinh quadrature schemes can be easily generalized to perform two-dimensional (2-D) and three-dimensional (3-D) quadrature. Runtimes are typically many times higher than with one-dimensional (1-D) integrals. However, if one is content with, say, 32-digit or 64-digit results (by using double-double or quad-double arithmetic, respectively), then many two-variable functions can be integrated in reasonable run time (say, a few minutes). One advantage that these schemes have is that they are very well suited to parallel processing. Thus even several-hundred digit values can be obtained for 2-D and 3-D integrals if one can utilize a highly parallel computer, such as a "Beowulf" cluster. One can even envision harnessing many computers on a geographically distributed grid for such a task, although the authors are not aware of any such attempts yet.

One sample computation of this sort, performed by one of the present authors, produced the following evaluation:

$$\int_{-1}^{1} \int_{-1}^{1} \frac{dx\, dy}{\sqrt{1 + x^2 + y^2}} \;=\; 4\log(2 + \sqrt{3}) - \frac{2\pi}{3}. \qquad (7.18)$$

7.5   Infinite Series Summation

We have already seen numerous examples in previous chapters of mathematical constants defined by infinite series. In experimental mathematics work, it is usually necessary to evaluate such constants to say several hundred digit accuracy. The commercial software packages *Maple* and *Mathematica* include quite good facilities for the numerical evaluation of series. However, as with numerical quadrature, these packages do have limitations, and in some cases better results can be obtained using custom-written computer code. In addition, even if one relies exclusively on these commercial packages, it is useful to have some idea of the sorts of operations that are being performed by such software.

Happily, in many cases of interest to the experimental mathematician, infinite series converge sufficiently rapidly that they can be numerically evaluated to high precision by simply evaluating the series directly as written, stopping the summation when the individual terms are smaller than the "epsilon" of the multiprecision arithmetic system being used. All of the BBP-type formulas, for instance, are of this category. But other types of infinite series formulas present considerable difficulties for high-precision evaluation. Two simple examples are Gregory's series for $\pi/4$ and a similar series for Catalan's constant:

$$\begin{aligned} \pi/4 &= 1 - 1/3 + 1/5 - 1/7 + \cdots \\ G &= 1 - 1/3^2 + 1/5^2 - 1/7^2 + \cdots . \end{aligned} \quad (7.19)$$

We describe here one technique that is useful in many such circumstances. In fact, we have already been introduced to it in an earlier section of this chapter: It is the Euler-Maclaurin summation formula. The Euler-Maclaurin formula can be written in somewhat different form than before, as follows [14, page 282]. Let $m \geq 0$ be an integer, and assume that the function $f(x)$ is at least $(2m + 2)$-times continuously differentiable on $[a, \infty)$, and that $f(x)$ and all of its derivatives approach zero for large x. Then

$$\sum_{j=a}^{\infty} f(j) = \int_a^{\infty} f(x)\,dx + \frac{1}{2}f(a) - \sum_{i=1}^{m} \frac{B_{2i}}{(2i)!} f^{(2i-1)}(a) + E, \quad (7.20)$$

where B_{2i} denote the Bernoulli numbers, and

$$E = \frac{B_{2m+2} f^{(2m+2)}(\xi)}{(2m + 2)!}, \quad (7.21)$$

for some $\xi \in (a, \infty)$.

This formula is not effective as written. The strategy is instead to evaluate a series manually for several hundred or several thousand terms, then to use the Euler-Maclaurin formula to evaluate the tail. Before giving an example, we need to describe how to calculate the Bernoulli numbers B_{2k}, which are required here. The simplest way to compute them is to recall that [1, page 807]

$$\zeta(2k) = \frac{(2\pi)^{2k} |B_{2k}|}{2(2k)!}, \quad (7.22)$$

(see also Equation (3.17)), which can be rewritten as

$$\frac{B_{2k}}{(2k)!} = \frac{2(-1)^{k+1} \zeta(2k)}{(2\pi)^{2k}}. \quad (7.23)$$

The Riemann zeta function at real arguments s can, in turn, be computed using the formula [64]

$$\zeta(s) \;=\; \frac{-1}{2^n(1-2^{1-s})} \sum_{j=0}^{2n-1} \frac{e_j}{(j+1)^s} + E_n(s), \tag{7.24}$$

where

$$e_j \;=\; (-1)^j \left(\sum_{k=0}^{j-n} \frac{n!}{k!(n-k)!} - 2^n \right) \tag{7.25}$$

(the summation is zero when its index range is null), and $|E_n(s)| < 1/(8^n|1-2^{1-s}|)$. This scheme is encapsulated in the following algorithm.

Algorithm 7.8. *Zeta function evaluation.*

Initialize: Set $n = P/3$, where P is the precision level in bits, and set $t_1 := -2^n$, $t_2 := 0$, $S := 0$, and $I = 1$.

For $j := 0$ to $2n - 1$ do: If $j < n$ then $t_2 := 0$ elseif $j = n$ then $t_2 := 1$ else $t_2 := t_2 \cdot (2n - j + 1)/(j - n)$ endif.
Set $t_1 := t_1 + t_2$, $S := S + I \cdot t_1/(j+1)^s$ and $I := -I$; enddo.
Return $\zeta(s) := -S/[2^n \cdot (1 - 2^{1-s})]$. $\square$

A more advanced method to compute the zeta function in the particular case of interest here, where we need the zeta function evaluated at all even integer arguments up to some level m, is described in [16].

We illustrate the above by calculating Catalan's constant using the Euler-Maclaurin formula. We can write

$$
\begin{aligned}
G \;&=\; (1 - 1/3^2) + (1/5^2 - 1/7^2) + (1/9^2 - 1/11^2) + \cdots \\
&=\; 8 \sum_{k=0}^{\infty} \frac{2k+1}{(4k+1)^2(4k+3)^2} \\
&=\; 8 \sum_{k=0}^{n} \frac{2k+1}{(4k+1)^2(4k+3)^2} + 8 \sum_{k=n+1}^{\infty} \frac{2k+1}{(4k+1)^2(4k+3)^2} \\
&=\; 8 \sum_{k=0}^{n} \frac{2k+1}{(4k+1)^2(4k+3)^2} + 8 \int_{n+1}^{\infty} f(x)\,dx + 4f(n+1) \\
&\quad\; -8 \sum_{i=1}^{m} \frac{B_{2i}}{(2i)!} f^{(2i-1)}(n+1) + 8E,
\end{aligned}
\tag{7.26}
$$

where $f(x) = (2x + 1)/[(4x + 1)^2(4x + 3)^2]$ and $|E| < 3/(2\pi)^{2m+2}$. Using $m = 20$ and $n = 1000$ in this formula, we obtain a value of G correct to 114 decimal digits. We presented the above scheme for Catalan's constant because it is illustrative of the Euler-Maclaurin method. However serious computation of Catalan's constant can be done more efficiently using the Boole summation formula (see Item 16 at the end of this chapter), the recently discovered BBP-type formula (given in Table 3.5 of [43]), Ramanujan's formula (given in Item 7 of Chapter 6 in [43]), or Bradley's formula (also given in Item 7 of Chapter 6 in [43]).

One less-than-ideal feature of the Euler-Maclaurin approach is that high-order derivatives are required. In many cases of interest, successive derivatives satisfy a fairly simple recursion and can thus be easily computed with an ordinary handwritten computer program. In other cases, these derivatives are sufficiently complicated that such calculations are more conveniently performed in a symbolic computing environment such as *Mathematica* or *Maple*. In a few applications of this approach, a combination of symbolic computation and custom-written numerical computation is required to produce results in reasonable runtime [18].

7.5.1   Computation of Multiple Zeta Constants

As we saw in Chapter 3, one class of mathematical constants that has been of particular interest to experimental mathematicians in the past few years are multiple zeta constants. Research in this arena has been facilitated by the discovery of methods that permit the computation of these constants to high precision. While Euler-Maclaurin-based schemes can be used (and in fact were used) in these studies, they are limited to 2-order sums. We present here an algorithm that permits even high-order sums to be evaluated to hundreds or thousands of digit accuracy. We will limit our discussion here to multiple zeta constants of the form

$$\zeta(s_1, s_2, \cdots, s_n) = \sum_{n_1 > n_2 > \cdots > n_k} \frac{1}{n_1^{s_1} n_2^{s_2} \cdots n_k^{s_k}}, \qquad (7.27)$$

for positive integers s_k and n_k, although in general the technique we describe here has somewhat broader applicability.

This scheme is as follows (see the discussion in Chapter 3, as well as [36] and [94] for further details). For $1 \leq j \leq m$, define the numeric strings

$$a_j = \{s_j + 2, \{1\}_{r_j}, s_{j+1}, \{1\}_{r_{j+1}}, \cdots, s_m + 2, \{1\}_{r_m}\} \qquad (7.28)$$

$$b_j = \{r_j + 2, \{1\}_{s_j}, r_{j-1}, \{1\}_{s_{j-1}}, \cdots, r_1 + 2, \{1\}_{s_1}\}, \qquad (7.29)$$

where by the notation $\{1\}_n$, we mean n repetitions of 1. For convenience, we will define a_{m+1} and b_0 to be the empty string. Define

$$\kappa(s_1, s_2, \cdots, s_k) \;=\; \sum_{n_j > n_{j+1} > 0} 2^{-n_1} \prod_{j=1}^{k} n_j^{-s_j}. \qquad (7.30)$$

Then we have the *Hölder convolution*

$$\zeta(a_1) \;=\; \sum_{j=1}^{m} \left[\sum_{t=0}^{s_j+1} \kappa(s_j + 2 - t, \{1\}_{r_j}, a_{j+1}) \kappa(\{1\}_t, b_{j-1}) \right.$$
$$\left. + \sum_{u=1}^{r_j} \kappa(\{1\}_u, a_{j+1}) \kappa(r_j + 2 - u, \{1\}_{s_j}, b_{j-1}) \right] + \kappa(b_m). \quad (7.31)$$

An online tool that implements this procedure is available at http://www.cecm.sfu.ca/projects/ezface+. This procedure has also been implemented as part of the Experimental Mathematician's Toolkit, available at http://www.expmath.info.

7.6 Commentary and Additional Examples

1. **Primes in pi.** $\pi_1 = 3, \pi_2 = 31, \pi_6 = 314159$, and π_{38} are the only prime initial strings in the first 500 digits of π. The next probable prime string is π_{16208}. This suspected prime would be a superb candidate for the new AKS-class of rigorous, deterministic algorithms discussed in Section 7.2. For more information on primes in π, see http://mathpages.com/home/kmath184/kmath184.htm.

2. **Adaptation of Putnam problem 1993–A5.** Consider the integral

$$J(a, b) = \int_a^b \frac{\left(x^2 - x\right)^2}{\left(x^3 - 3\,x + 1\right)^2} \, dx.$$

Show that

$$J\left(-100, -10\right) + J\left(\frac{1}{101}, \frac{1}{11}\right) + J\left(\frac{101}{100}, \frac{11}{10}\right)$$

is rational.

Hint: Numerically observe that the given expression is rational. Then consider the changes of variables $x \to 1 - 1/x$ and $x \to 1/(1 - x)$, respectively, to move the second and third integrals to the interval $[-100, -10]$.

Answer: $11131110/107634259$.

3. **Berkeley problem 5.10.26.** Evaluate

$$\mathcal{I} = \int_0^{2\pi} \frac{\cos^2(3t)}{5 - 4\cos(2t)}\, dt.$$

Answer: $3\pi/8$.

Hint: $\mathcal{I} = \frac{1}{2}\mathrm{Re}\left(i \int_{|z|=1}(z^3 + 1)/(2z^2 - 5z + 2)\, dz\right) = -\pi\,\mathrm{Res}((z^3 + 1)/(2z^2 - 5z + 2), z = \frac{1}{2})$.

4. **Berkeley problem 5.10.27.** Evaluate

$$\mathcal{I} = \int_{|z|=1} \frac{\cos^3(z)}{z^3}\, dz$$

taken counterclockwise.

Answer: $3\pi i$.

Hint: $2\pi i\,\mathrm{Res}\left(\cos^3(z)/z^3, z = 0\right)$.

5. **Berkeley problem 5.11.4.** Evaluate

$$\int_0^\infty \frac{\sin^2(t)}{t^2}\, dt.$$

Answer: $\pi/2$.

Hint: These integrals can be evaluated using residues, or using numerical methods. Although this exercise has appeared before, we present it here as an introduction to the problems that follow.

6. **Berkeley problem 5.11.5.** Evaluate

$$\int_{-\infty}^\infty \frac{\sin^3(t)}{t^3}\, dt \qquad \left(= \frac{3\pi}{4}\right).$$

Hint: $4\sin^3(z) = \mathrm{Re}\,3\exp(iz) - \exp(3iz)$. Use Cauchy's theorem on a contour C which consists of semicircles of radius $\varepsilon \to 0$ and $R \to \infty$ along with the intervals $[-R, -\varepsilon]$ and $[\varepsilon, R]$.

7. **Berkeley problem 5.11.8.** Show

$$\int_0^\infty \frac{\sin(x)}{x(x^2 + a^2)}\, dx = \frac{\pi(1 - e^{-a})}{2a^2}.$$

8. **Berkeley problem 5.11.14.** Evaluate

$$\int_0^\infty \frac{1+x^2}{1+x^4}\,dx \qquad \left(=\frac{\pi}{\sqrt{2}}\right).$$

Hint: Again use residues or consider $\int_0^t (1+x^2)/(1+x^4)\,dx$.

9. **Berkeley problem 5.11.22.** Show

$$\int_0^\infty \frac{x}{\sinh(x)}\,dx = \frac{\pi^2}{4}.$$

Then determine

$$\int_0^t \frac{x}{\sinh(x)}\,dx.$$

10. **Berkeley problem 5.11.25.** Show

$$\int_0^\infty \frac{\log^2(x)}{1+x^2}\,dx = \frac{\pi^3}{8}.$$

11. **Evaluation of integrals.** Evaluate the following integrals, by numerically computing them and then trying to recognize the answers, either by using the Inverse Symbolic Calculator at http://www.cecm. sfu.ca/projects/ISC, or by using a PSLQ facility, such as that built into the Experimental Mathematician's Toolkit, available at http://www.expmath.info.

These examples are taken from Gradsteyn and Ryzhik [130]. All of the answers are simple one- or few-term expressions involving familiar mathematical constants such as π, e, $\sqrt{2}$, $\sqrt{3}$, $\log 2$, $\zeta(3)$, G (Catalan's constant), and γ (Euler's constant). We recognize that many of these can be evaluated analytically using symbolic computing software (depending on the available versions). The intent here is to provide exercises for numerical quadrature and constant recognition facilities.

(a) $$\int_0^1 \frac{x^2\,dx}{(1+x^4)\sqrt{1-x^4}} \qquad (7.32)$$

(b) $$\int_0^\infty xe^{-x}\sqrt{1-e^{-2x}}\,dx \qquad (7.33)$$

(c) $$\int_0^\infty \frac{x^2\,dx}{\sqrt{e^x-1}} \qquad (7.34)$$

(d) $$\int_0^{\pi/4} x\tan x\,dx \qquad (7.35)$$

$$(e) \qquad \int_0^{\pi/2} \frac{x^2\,dx}{1-\cos x} \qquad\qquad (7.36)$$

$$(f) \qquad \int_0^{\pi/4} (\pi/4 - x\tan x)\tan x\,dx \qquad\qquad (7.37)$$

$$(g) \qquad \int_0^{\pi/2} \frac{x^2\,dx}{\sin^2 x} \qquad\qquad (7.38)$$

$$(h) \qquad \int_0^{\pi/2} \log^2(\cos x)\,dx \qquad\qquad (7.39)$$

$$(i) \qquad \int_0^1 \frac{\log^2 x\,dx}{x^2+x+1} \qquad\qquad (7.40)$$

$$(j) \qquad \int_0^1 \frac{\log(1+x^2)\,dx}{x^2} \qquad\qquad (7.41)$$

$$(k) \qquad \int_0^\infty \frac{\log(1+x^3)\,dx}{1-x+x^2} \qquad\qquad (7.42)$$

$$(l) \qquad \int_0^\infty \frac{\log x\,dx}{\cosh^2 x} \qquad\qquad (7.43)$$

$$(m) \qquad \int_0^1 \frac{\arctan x}{x\sqrt{1-x^2}} \qquad\qquad (7.44)$$

$$(n) \qquad \int_0^{\pi/2} \sqrt{\tan t}\,dt \qquad\qquad (7.45)$$

Answers: (a) $\pi/8$, (b) $\pi(1+2\log 2)/8$, (c) $4\pi(\log^2 2 + \pi^2/12)$, (d) $(\pi\log 2)/8 + G/2$, (e) $-\pi^2/4 + \pi\log 2 + 4G$, (f) $(\log 2)/2 + \pi^2/32 - \pi/4 + (\pi\log 2)/8$, (g) $\pi\log 2$, (h) $\pi/2(\log^2 2 + \pi^2/12)$, (i) $8\pi^3/(81\sqrt{3})$, (j) $\pi/2 - \log 2$, (k) $2(\pi\log 3)/\sqrt{3}$, (l) $\log\pi - 2\log 2 - \gamma$, (m) $[\pi\log(1+\sqrt{2})]/2$, (n) $\pi\sqrt{2}/2$.

12. **Evaluation of infinite series.** Evaluate the following infinite series, by numerically computing them and then trying to recognize the answers, either by using the Inverse Symbolic Calculator at http://www.cecm.sfu.ca/projects/ISC, or else by using a PSLQ facility, such as that built into the Experimental Mathematician's Toolkit, available at http://www.expmath.info.

These examples have been provided to the authors by Gregory and David Chudnovsky of the Institute for Mathematics and Supercomputing at Brooklyn Polytechnic College. All of the answers are simple one- or few-term expressions involving familiar mathematical constants such as π, e, $\sqrt{2}$, $\sqrt{3}$, $\log 2$, $\zeta(3)$, G (Catalan's constant), and γ (Euler's constant).

$$\text{(a)} \qquad \sum_0^\infty \frac{50n - 6}{2^n \binom{3n}{n}} \qquad\qquad (7.46)$$

$$\text{(b)} \qquad \sum_0^\infty \frac{2^{n+1}}{\binom{2n}{n}} \qquad\qquad (7.47)$$

$$\text{(c)} \qquad \sum_0^\infty \frac{12n2^{2n}}{\binom{4n}{2n}} \qquad\qquad (7.48)$$

$$\text{(d)} \qquad \sum_0^\infty \frac{(4n)!(1 + 8n)}{4^{4n}n!^4} \qquad\qquad (7.49)$$

$$\text{(e)} \qquad \sum_0^\infty \frac{(4n)!(19 + 280n)}{4^{4n}n!^4 99^{2n+1}} \qquad\qquad (7.50)$$

$$\text{(f)} \qquad \sum_0^\infty \frac{(2n)!(3n)!4^n(4 + 33n)}{n!^5 108^n 125^n} \qquad\qquad (7.51)$$

$$\text{(g)} \qquad \sum_0^\infty \frac{(-27)^n(90n + 177)}{16^n \binom{3n}{n}} \qquad\qquad (7.52)$$

$$\text{(h)} \qquad \sum_0^\infty \frac{275n - 158}{2^n \binom{3n}{n}} \qquad\qquad (7.53)$$

$$\text{(i)} \qquad \sum_0^\infty \frac{8^n(520 + 6240n - 430n^2)}{\binom{4n}{n}} \qquad\qquad (7.54)$$

$$\text{(j)} \qquad \sum_0^\infty \frac{\binom{2n}{n}}{n^2 4^n} \qquad\qquad (7.55)$$

$$\text{(k)} \qquad \sum_0^\infty \frac{(-1)^n}{n^3 2^n \binom{2n}{n}} \qquad\qquad (7.56)$$

$$\text{(l)} \qquad \sum_0^\infty \frac{8^n(338 - 245n)}{3^n \binom{3n}{n}} \qquad\qquad (7.57)$$

$$\text{(m)} \qquad \sum_1^\infty \frac{(-9)^n \binom{2n}{n}}{6n^2 64^n} - \sum_1^\infty \frac{3^n \binom{2n}{n}}{n^2 16^n} \qquad\qquad (7.58)$$

Answers: (a) π, (b) $\pi + 4$, (c) $3\pi + 8$, (d) $2/(\pi\sqrt{3})$, (e) $2/(\pi\sqrt{11})$, (f) $15\sqrt{3}/(2\pi)$, (g) $120 - 64\log 2$, (h) $6\log 2 - 135$, (i) $45\pi - 1164$, (j) $\pi^2/6 - 2\log^2 2$, (k) $\log^3 2/6 - \zeta(3)/4$, (l) $162 - 6\pi\sqrt{3} - 18\log 3$, (m) $\pi^2/18 + \log^2 2 - \log^3 3/6$.

13. **Pisot and Salem numbers.** A real algebraic integer $\alpha > 1$ is a
Pisot number (respectively, *Salem number*) if all other roots of its
monic polynomial lie inside (respectively, inside or on) the unit disc.
An algebraic number α is a Pisot number if and only if, for $n \in \mathbb{N}$,
the fractional parts $\{n\alpha\} \to 0$ as $n \to \infty$. The smallest Pisot number
is the largest root of $z^3 - z - 1$ and is approximately $1.3247179\ldots$
The smallest Salem number is *conjectured* to be the largest root of
Lehmer's polynomial

$$p(z) = z^{10} + z^9 - z^7 - z^6 - z^5 - z^4 - z^3 + z + 1,$$

which has two real roots $0.850137130927042\ldots$ and $1.176280818259917\ldots$
The conjecture is called *Lehmer's problem (1933)* [65].

(a) Show that for a Pisot number, the fractional parts $\{n\alpha\} \to 0$ as
$n \to \infty$. *Hint*: use Newton's formula for

$$\sigma_n = \sum_{p(\beta)=0} \beta^n,$$

where p is the polynomial associated with α.

(b) Computationally establish Lehmer's conjecture up to as high a
degree as possible (this has been done to about degree 40).

14. **Halley's method.** Suppose f is a suitably smooth real function and
$N \geq 0$ is integer. Consider the iteration starting at an initial value
$y_0 = y$ and iterating

$$y_{n+1} = y_n + (N+1)\frac{(1/f)^{(N)}(y)}{(1/f)^{(N+1)}(y)}\bigg|_{y=y_n}.$$

For $N = 0$, this is Newton's method to find a zero of f. For $N = 1$,
this is a locally cubically convergent method due to the astronomer
and mathematician Halley. Show in general that the method converges of order $N + 2$ for the initial guess close enough to a (nondegenerate) zero of f.

15. **Bender's continued fraction for the Bernoulli numbers.** The
Bernoulli numbers as exact rational numbers can be generated from
a continued fraction, due to Bender, which makes a nice equally

divergent counterpart to Exercise 56 of Chapter 7 of [43]. Formally,

$$S(a) = 1 + \cfrac{b_0\,a^2}{1 + \cfrac{b_1\,a^2}{1 + \cfrac{b_2\,a^2}{1 + \cfrac{b_3\,a^2}{1 + \cfrac{}{\ddots}}}}} = \sum_{n=0}^{\infty} B_{2n} a^{2n},$$

with $b_0 = 1/6$ and

$$b_n = \frac{n(n+1)^2(n+2)}{4(2n+1)(2n+3)},$$

for $n > 0$.

(a) While the series does not converge in any obvious sense, it gives
a very satisfactory symbolic expansion. Decide whether $S(a)$
coincides with the subsequent three equivalent expressions:

$$S(a) \overset{?}{=} a \int_0^{\infty} t \coth\left(\frac{ta}{2}\right) e^{-t}\, dt$$

$$= a \int_0^1 \frac{\log(s)}{s^a - 1}\, ds - \frac{a}{2} = 1 + \frac{a}{2} + a \sum_{n=1}^{\infty} \frac{1}{(an+1)^2}.$$

(b) Evaluate $S(4)$, $S(1/4)$, $S(3)$, and $S(1/3)$.

(c) Show that $2n \int_0^{\infty} x^{2n-1} e^{-\pi x} \operatorname{cosech}(\pi x)\, dx = |B_{2n}|$ for $n = 1, 2, \cdots$. (What are the even moments?) Deduce, for all real a,
that

$$\Psi'\left(1 + a^{-1}\right)/a + a/2 = 1 + \frac{a}{2} + a \sum_{n=1}^{\infty} \frac{1}{(an+1)^2}$$

$$= \int_0^{\infty} \coth\left(\frac{\pi x}{a}\right) \frac{2x}{(1+x^2)^2}\, dx,$$

and that

$$\sum_{n=1}^{N} B_{2n} a^{2n} =$$

$$4a \int_0^{\infty} \frac{ya\left(1 - (-1)^N (ya)^{2N}\left\{1 + N(1 + (ya)^2)\right\}\right)}{(e^{2\pi y} - 1)(1 + (ya)^2)^2}\, dy,$$

for $N > 0$.

(d) In consequence,

$$\left| \int_0^\infty \coth\left(\frac{\pi x}{a}\right) \frac{2x}{(1+x^2)^2}\, dx - \sum_{n=0}^{N-1} B_{2n} a^{2n} \right| < |B_{2N}|\, a^{2N},$$

and we obtain a genuine asymptotic expansion.

(e) Show, for $a > 1$, that

$$S(a) = \frac{a\left(3 + 2a + a^2\right)}{2(1+a)^2} - \sum_{n=1}^\infty \frac{n}{(-a)^n}\left(\zeta(n+1) - 1\right)$$

$$= \frac{1}{2a}\pi^2 \csc^2\left(\frac{\pi}{a}\right) - 2\frac{a^2}{(1-a^2)^2} - 2\sum_{n=1}^\infty \frac{n}{a^{2n}}\left(\zeta(2n+1) - 1\right).$$

For what a are these valid equalities?

(f) Show that

$$T(a) = \int_0^\infty \tanh(ax)\, e^{-x}\, dx = \int_0^1 \frac{1 - y^{2a}}{1 + y^{2a}}\, dy$$

generates the continued fraction

$$\cfrac{a}{1 + \cfrac{2a^2}{1 + \cfrac{6a^2}{1 + \cfrac{12a^2}{1 + \cfrac{20a^2}{1 + \ddots}}}}} = -\sum_{n\geq 1} T_{2n+1} a^{2n-1}.$$

Here the T_n are, as before, tangent numbers.

(g) Provide a justification for the last identity, along the lines given above for the Bernoulli numbers.

(h) Show

$$\frac{T(a) + 1}{2} = F\left(\frac{1}{2a}, 1; \frac{1}{2a} + 1; -1\right)$$

is of the form for Gauss's continued fraction.

16. **The Boole summation formula.** Boole summation is a counterpart of Euler-Maclaurin summation especially fitted to alternating series. Based upon Euler numbers, rather than Bernoulli numbers,

it is described in detail in [37], and in the revised version of Abromowitz and Stegun, available at http://dlmf.nist.gov. We recall the *Euler polynomials* are defined by

$$\frac{2e^{tx}}{e^t + 1} = \sum_{n=0}^{\infty} E_n(x) \frac{t^n}{n!}$$

for $|t| < \pi$, and let the *periodic Euler function* be defined by $\overline{E}_n(x) = E_n(x)$ for $0 < x \le 1$ and $\overline{E}_n(x+1) = -\overline{E}_n(x)$ for all x. Then the Euler numbers are recovered as $E_n = 2^n E_n \left(\frac{1}{2}\right)$. The version of Boole summation we need tells us that if f has m derivatives on $t \ge x$, and $f^{(k)}(t) \to 0$ as $t \to \infty$ for $0 \le k \le m$, then

$$\sum_{n=0}^{\infty} (-1)^n f(x+h+n) = \sum_{k=0}^{m} \frac{E_k(h)}{2k!} f^{(k)}(x) + R_m \qquad (7.59)$$

where

$$R_m = \frac{1}{2(m-1)!} \int_0^{\infty} \overline{E}_{m-1}(h-t) f^{(m)}(x+t) \, dt.$$

(a) Use (7.59) to prove the formulas for the "errors" in the Gregory series for π and the classical series for $\log 2$ given in Section 2.2 of [43].

(b) Apply the same technique to Catalan's constant and to $\zeta(3)$.

17. **Value recycling, zeta, and gamma.** This material is culled from [53]. The notion of *recycling* is that previously calculated ζ-values—or initialization tables of those calculations—are re-used to aid in the extraction of other ζ-values, or that many ζ-values are somehow simultaneously determined, and so on. So by value recycling, we intend that the computation of a collection of ζ-values is more efficient than establishment of independent values. We record

$$\pi t \cot (\pi t) = -2 \sum_{m=0}^{\infty} \zeta(2m) t^{2m}, \qquad (7.60)$$

and the *incomplete Gamma function*, given (at least for $\operatorname{Re}(z) > 0$) by:

$$\Gamma(a, z) = \int_z^{\infty} t^{a-1} e^{-t} dt = \frac{2z^a e^{-z}}{\Gamma(1-a)} \int_0^{\infty} \frac{t^{1-2a} e^{-t^2}}{t^2 + z} dt,$$

where the integral representation is valid for $\operatorname{Re}(a) < 1$. Evaluation of $\Gamma(a, z)$ is not as problematic as it may seem; many computer systems have suitable machinery. We note the special cases $\Gamma(s,0) = \Gamma(s)$ and $\Gamma(1, z) = e^{-z}$, and the recursion

$$a\Gamma(a, z) = \Gamma(a+1, z) - z^a e^{-z}. \tag{7.61}$$

Also,

$$\zeta(s)\Gamma\left(\tfrac{1}{2}s\right) = \frac{\pi^{s/2}}{s(s-1)} + \sum_{n=1}^{\infty} n^{-s}\Gamma\left(\tfrac{1}{2}s, \pi n^2\right)$$

$$+\pi^{s-1/2} \sum_{n=1}^{\infty} n^{s-1}\Gamma\left(\tfrac{1}{2}(1-s), \pi n^2\right), \tag{7.62}$$

$$\zeta(s)\Gamma(s) = -\frac{\lambda^s}{2s} + \frac{\lambda^{s-1}}{s-1} + \sum_{n=0}^{\infty} n^{-s}\Gamma(s, \lambda n)$$

$$-2\lambda^{s-1} \sum_{n=1}^{\infty} \left(\frac{\lambda}{2\pi i}\right)^{2n} \frac{\zeta(2n)}{2n+s-1}. \tag{7.63}$$

(a) One can use either (7.62) or (7.63) to efficiently evaluate zeta at each of N arguments $\{s, s+2, s+4, \cdots, s+2(N-1)\}$ for any complex s. This approach is fruitful for obtaining a set of zeta values at odd positive integers, for example. The idea is to exploit the recursion relation (7.61) for the incomplete Gamma function and, when N is sufficiently large, effectively remove incomplete Gamma evaluations via $\Gamma(\{s/2\}, x), \Gamma(\{(1-s)/2\}, x)$, where $\{z\}$ here denotes the fractional part of z, over a collection of x-values, and then use the above recursion either backward or forward to rapidly evaluate series terms for the whole set of desired zeta values.

Given the initial $\Gamma(\{s/2\}, x)$ evaluations, further effort is sharply reduced. When $\{s + 2k\}$ are odd integers, precomputation involve only $\Gamma(0, x)$ and $\Gamma(1/2, x)$ values—known classically as exponential-integral and error-function values. Reference [96] contains explicit pseudocode for a recycling evaluation of $\zeta(3), \zeta(5), \cdots, \zeta(L)$ via (7.62), and one initializes error function and exponential-integral values by

$$\{\Gamma(1/2, \pi n^2) : n \in [1, \lfloor D \rfloor]\}, \tag{7.64}$$

$$\{\Gamma(0, \pi n^2) : n \in [1, \lfloor D \rfloor]\},$$

when D decimal digits precision is ultimately desired for each ζ value. The notion of "recycling" takes its purest form in this method, for the incomplete-Gamma evaluations above are reused for every $\zeta(\text{odd})$.

(b) A second approach, relevant for *even* integer arguments, involves
a method of series inversion used by J. P. Buhler for numerical
analysis on Fermat's Last Theorem and on the Vandiver conjecture [73, 74]. This uses a generating function for Bernoulli
numbers, and invokes Newton's method for series inversion of
the key elementary functions. To obtain values at even positive
integers, one may use an expansion related to (7.60). One has

$$\frac{\sinh(2\pi\sqrt{t})}{4\pi\sqrt{t}}\frac{2\pi^2 t}{\cosh(2\pi\sqrt{t})-1} = -\sum_{n=0}^{\infty}(-1)^n\zeta(2n)t^n,$$

which we have written this way for Algorithm 7.9. We have
split the left-hand side into two series, each in t: one of the form
$(\sinh\sqrt{z})/\sqrt{z}$, and the other like $(\cosh\sqrt{z}-1)/z$. The idea is to
invert the latter series via a fast polynomial inversion algorithm
(Newton method). Using t as a placeholder, one reads off the ζ-
values as coefficients in a final polynomial. In Algorithm 7.9, we
assume that $\zeta(2), \zeta(4), \cdots \zeta(2N-2)$ are desired. The polynomial
arithmetic is most efficient when truncation of large polynomials
occurs as needed. We denote by $q(t)$ mod t^k the truncation
of polynomial q through t^{k-1}. Also, polynomial multiplication
operation is signified by "$*$."

Algorithm 7.9.   *Recycling scheme for* $\zeta(0), \zeta(2), \zeta(4), ...,$
$\zeta(2(N-1))$.

(1) [Denominator set-up] Create the polynomial $f(t) =$
$(\cosh(2\pi\sqrt{t})-1)/(2\pi^2 t)$, through degree N (i.e., through power
t^N inclusive);
(2) [Newton polynomial inversion, to obtain $g := f^{-1}$] Set $p =$
$g = 1$;
    while$(p < \deg(f))$ do begin
        $p = \min(2p, \deg(f))$; $h = f$ mod t^p; $g = (g + g * (1 - h *$
$g))$ mod t^p; end;
(3) [Numerator set-up] Create the polynomial $k(t) =$
$\sinh(2\pi\sqrt{t})/(4\pi\sqrt{t})$, through degree N;
    $g = g * k$ mod t^{2N-1}; for $n \in [0, 2N-2]$, read off $\zeta(2n)$ as
$-(-1)^n$ times the coefficient of t^n in polynomial $g(t)$.   □

In step (1) the polynomial can have floating point or symbolic
coefficients with their respective powers of π and so on. If used

symbolically, the ζ values of the indicated finite set are exact, through $\zeta(2N-2)$ inclusive. The method has been used, numerically so that fast Fourier transform methods may also be applied, to calculate the relevant ζ-values for high-precision values of the Khintchine constant [16].

If memory storage is an issue, there is a powerful technique called *multisectioning*, whereby one calculates all the $\zeta(2k)$ for k lying in some congruence class (mod 4, 8, or 16, say), using limited memory for that calculation, then moving on to the next congruence class, and so on. Observe that, by looking only at even-indexed Bernoulli numbers in the previous algorithm, we have effectively multisectioned by 2 already. To multisection by 4, observe:

$$\frac{x\cosh x\sin x \pm x\cos x\sinh x}{\sinh x\sin x} = 2\sum_{n\in S^\pm}\frac{B_n}{n!}(2x)^n,$$

where the sectioned sets are $S^+ = \{0,4,8,12,...\}$ and $S^- = \{2,6,10,14,...\}$. The key is that the denominator $(\sinh x\sin x)$ is, perhaps surprisingly, x^2 times a series in x^4; namely, we have the attractive series

$$\sinh x\sin x = \sum_{n\in S^-}(-1)^{(n-2)/4}\,2^{n/2}\,\frac{x^n}{n!},\qquad(7.65)$$

so that the key Newton inversion of a polynomial approximant to the denominator only has *one-fourth* the terms of the standard Bernoulli denominator (e^x-1). Thus, reduced memory is used to establish a congruence class of Bernoulli indices, then that memory is reused for the next congruence class, and so on. Thus, these methods function well in parallel or serial environments.

Multisectioning was used by Buhler and colleagues—as high as level-16 sections—to verify Fermat's Last Theorem to exponent 8,000,000. They desired Bernoulli numbers modulo primes, and so employed integer arithmetic. A detailed analysis of multisectioning is to be found in Kevin Hare's 1999 Master's thesis [139].

(c) A third approach is to contemplate continued fraction representations that yield ζ-values. For example, the well known fraction for $\sqrt{z}\coth\sqrt{z}$ gives:

$$\cfrac{\pi^2 z}{3+\cfrac{\pi^2 z}{5+\cfrac{\pi^2 z}{7+\ddots}}} = 2\sum_{n=1}^{\infty}(-1)^{n-1}\zeta(2n)z^n.$$

This is advantageous if one already has an efficient continued fraction engine. As a final alternative for fast evaluation at even positive integer arguments, there is an interesting approach, due to Plouffe and Fee, in which the Von-Staudt-Clausen formula for the fractional part of B_n is invoked, then asymptotic techniques are used to determine the integer part. In this way, the number $B_{200,000}$ has been calculated in exact, rational form.

18. **Pseudospectra.** A fine example of the changing nature of numerical mathematics is afforded by the use of *pseudospectra*. One considers the ε-pseudospectrum of a complex $n \times n$ matrix A for $\varepsilon > 0$

$$\Lambda_\varepsilon(A) = \{z \in \mathbb{C} : z \in \Lambda(B), \text{ for some } \|B - A\| \le \varepsilon\},$$

which consists of the eigenvalues of all nearby matrices (in operator norm), and coincides with the spectrum of A, $\Lambda(a)$, for $\varepsilon = 0$. This is much more stable than the spectrum and can be graphed very informatively for various values of ϵ as illustrated in Figure 7.2 for the matrix $A(0.5)$.

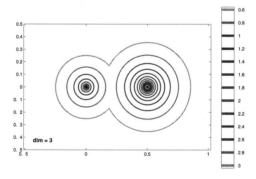

Figure 7.2. The pseudospectra of matrix $A(0.5)$.

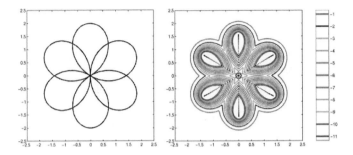

Figure 7.3. Daisy matrix pseudospectra.

Here,

$$A(t) = \begin{bmatrix} t & -1/4 & 0 \\ 0 & t & 0 \\ 0 & 0 & t - 1/2 \end{bmatrix}$$

$$A^{-1}(t) = \begin{bmatrix} t^{-1} & 1/4\,t^{-2} & 0 \\ 0 & t^{-1} & 0 \\ 0 & 0 & 2\,(2\,t - 1)^{-1} \end{bmatrix}.$$

Thus, $A(t)$ has eigenvalues $t, t, t - 1/2$, and the inverse exists for $t \neq 0, 1/2$. Try to correlate the graphic information with the behavior of $A^{-1}(t)$.

See http://web.comlab.ox.ac.uk/projects/pseudospectra for a host of tools and examples, from which the picture (Figure 7.3) of a 200-dimension Toeplitz matrix with "symbol" $t^{-5} - t$ is taken (thanks to Nick Trefethen). Similarly, the ε-spectral radius of a matrix, $\alpha_\varepsilon(A) = \{\operatorname{Re} z : z \in \Lambda_\varepsilon(A)\}$, is more robust than the spectral radius ($\varepsilon = 0$), and is important in engineering control problems. The stability of the pseudospectrum is, in large part, explained by its variational formulation.

19. **Square-full numbers and one bits.** A *square-full number*, also known as a squareful or powerful number, is a natural number all of whose prime factors occur to multiplicity greater than one. In [17] it is shown that the number of 1's in the first N bits of the theta

function value $\sigma = \left(\sum_{n \geq 0} 2^{-n^2} \right)^2 = (\theta_3(1/2)/2 + 1/2)^2$ behaves like

$$C_0 \frac{N}{\sqrt{\log N}}, \text{ where } C_0 = \frac{16 L^3}{\pi^2} \sum_g \frac{b(\tau(g))}{\psi(g)}, \tag{7.66}$$

and g runs over the square-full integers divisible solely by primes that are congruent to 1 (mod 4). Here b denotes the number of bits, τ the number of divisors, $\psi(g) = g \prod_{p|g}(1 + 1/p)$, and L is the *Landau constant*

$$L = \left(\frac{1}{2} \prod_{p \equiv 3 \mod 4} \left(1 - \frac{1}{p^2} \right)^{-1} \right)^{1/2} = 0.764223653\ldots$$

(a) Compute the predicted density (7.66) of ones in the binary expansion of σ, for $N = 10^n$, $n = 1, \cdots, 8$.

(b) Compare the empirical density for the same values of N.

20. **Converting series into products.** Use the ideas in Exercise 8 of Chapter 4 to produce two algorithms for converting series into products of the form in that exercise; compare the efficiency of the two methods.

21. **Self-reference and Grelling's paradox of heterologicality.** One of the classic linguistic (or predicate) paradoxes, which like Russell's naive set theory paradox of all non-self containing sets prefigures Gödel's theorem and Chaitin's work, is as follows. Say that a word or property is *homological* if it describes itself and *heterological* otherwise. Thus, "short" is short and, arguably, "ugly" is ugly and so both are homological. By contrast, "long" is not long and so is heterological. A problem occurs when we try to decide if "heterological" is a heterological property. This is usually dealt with by restricting the range of predicates or the class of sets. It would be amusing if we could show that Khintchine's constant did not respect Khintchine's constant.

22. **Progress and silence.** "It's generally the way with progress that it looks much greater than it really is" is the epigraph that Ludwig Wittgenstein (1889–1951) ("whereof one cannot speak, thereof one must be silent") had wished for an unrealized joint publication of *Tractatus Logico-Philosophicus* (1922) and *Philosophical Investigations* (1953). This suggests the two volumes are not irreconcilable, as often described. Wittgenstein was one of the influential members of the Vienna Circle from which Gödel took many ideas.

23. **Quantum error-correcting codes.** An intriguing paper entitled "Quantum Error Correction Via Codes over $GF(4)$," authored by A. R. Calderbank, E. M. Raines, P. W. Shor, and N. J. A. Sloane, translates the problem of finding quantum-error-correcting codes into the problem of finding additive codes over the field $GF(4)$ that are self-orthogonal with respect to a certain trace inner product. The authors describe the origin of this paper as follows [79].

> (xv) There is a remarkable story behind this paper. About two years ago one of us (P.W.S.) was studying fault tolerant quantum computation, and was led to investigate a certain group of 8×8 orthogonal matrices. P.W.S. asked another of us (N.J.A.S.) for the best method of computing the order of this group. N.J.A.S. replied by citing the computer algebra system MAGMA [67, 69, 68], and gave as an illustration the MAGMA commands needed to specify a certain matrix group that had recently arisen in connection with packings in Grassmannian spaces. This group was the symmetry group of a packing of 70 4-dimensional subspaces of R^8 that had been discovered by computer search [85]. It too was an 8-dimensional group, of order 5160960. To our astonishment the two groups turned out to be identical (not just isomorphic)! We then discovered that this group was a member of an infinite family of groups that played a central role in a joint paper [76] of another of the authors (A.R.C.). This is the family of real Clifford groups L_R, described in Section 2 (for $n = 3$, L_R has order 5160960).

> This coincidence led us to make connections which further advanced both areas of research (fault tolerant quantum computing and Grassmannian packings [195, 194]).

> While these three authors were pursuing these investigations, the fourth author (E.M.R.) happened to be present for a job interview and was able to make further contributions to the Grassmannian packing problem [77]. As the latter involved packings of 2^k-dimensional subspaces in 2^n dimensional space, it was natural to ask if the same techniques could be used for constructing quantum error-correcting codes, which are also subspaces of 2^n dimensional space. This question led directly to [78] and the present paper. (Incidentally, he got the job.)

24. **Chaitin on randomness.** It seems apropos to end with Greg
    Chaitin's views in "The Creative Life: Science vs Art," an article
    available at the URL http://www.cs.umaine.edu/~chaitin/cdg.html.

> The message is that mathematics is quasi-empirical, that
> mathematics is not the same as physics, not an empirical
> science, but I think it's more akin to an empirical science
> than mathematicians would like to admit.

> Mathematicians normally think that they possess absolute
> truth. They read God's thoughts. They have absolute cer-
> tainty and all the rest of us have doubts. Even the best
> physics is uncertain, it is tentative. Newtonian science was
> replaced by relativity theory, and then—wrong!—quantum
> mechanics showed that relativity theory is incorrect. But
> mathematicians like to think that mathematics is forever,
> that it is eternal. Well, there is an element of that. Cer-
> tainly a mathematical proof gives more certainty than an
> argument in physics or than experimental evidence, but
> mathematics is not certain. This is the real message of
> Gödel's famous incompleteness theorem and of Turing's
> work on uncomputability.

> You see, with Gödel and Turing the notion that mathe-
> matics has limitations seems very shocking and surprising.
> But my theory just measures mathematical information.
> Once you measure mathematical information you see that
> any mathematical theory can only have a finite amount of
> information. But the world of mathematics has an infi-
> nite amount of information. Therefore it is natural that
> any given mathematical theory is limited, the same way
> that as physics progresses you need new laws of physics.
> Mathematicians like to think that they know all the laws.
> My work suggests that mathematicians also have to add
> new axioms, simply because there is an infinite amount of
> mathematical information. This is very controversial. I
> think mathematicians, in general, hate my ideas. Physi-
> cists love my ideas because I am saying that mathematics
> has some of the uncertainties and some of the characteris-
> tics of physics. Another aspect of my work is that I found
> randomness in the foundations of mathematics. Mathe-
> maticians either don't understand that assertion or else it
> is a nightmare for them ...

Bibliography

[1] Milton Abramowitz and Irene A. Stegun. *Handbook of Mathematical Functions*. Dover Publications, New York, 1970.

[2] Manindra Agrawal, Neeraj Kayal, and Nitin Saxena. PRIMES is in P. http://www.cse.iitk.ac.in/news/primality.pdf, 2002.

[3] Zafar Ahmed. Definitely An Integral. *American Mathematical Monthly*, 109:670–671, 2002.

[4] D. Albers, G. Alexanderson, and C. Reid. *More Mathematical People*. Harcourt Brace Jovanovich, Orlando, 1990.

[5] W. R. Alford, A. Granville, and C. Pomerance. There Are Infinitely Many Carmichael Numbers. *Annals of Mathematics*, 140:1–20, 1994.

[6] G. Almkvist and A. Granville. Borwein and Bradley's Apery-like Formulae for Zeta(4n+3). *Experimental Mathematics*, 8:197–204, 1999.

[7] G. Almkvist, C. Krattenthaler, and J. Petersson. Some New Formulas for Pi. *Manuscript*, 2001.

[8] Gerg Almkvist and Herbert S. Wilf. On the Coefficients in the Hardy-Ramanujan-Rademacher Formula for p(n). *Journal of Number Theory*, 50:329–334, 1995.

[9] George E. Andrews. *The Theory of Partitions*. Addison-Wesley, Boston, 1976.

[10] George E. Andrews. *q-Series: Their Development and Application in Analysis, Number Theory, Combinatorics, Physics and Computer Algebra*. C.B.M.S. Regional Conference Series in Math, No. 66, American Mathematical Society, Providence, 1986.

[11] George E. Andrews, Richard Askey, and Ranjan Roy. *Special Functions*. Cambridge University Press, Cambridge, 1999.

[12] B. H. Arnold. A Topological Proof of the Fundamental Theorem of Algebra. *American Mathematical Monthly*, 56:465–466, 1949.

[13] B. H. Arnold and Ivan Niven. A Correction. *American Mathematical Monthly*, 58:104, 1951.

[14] Kendall E. Atkinson. *An Introduction to Numerical Analysis*. John Wiley and Sons, Hoboken, NJ, 1989.

[15] C. Bachoc, R. Coulangeon, and G. Nebe. Designs in Grassmannian Spaces and Lattices. *Journal of Algebraic Combinatorics*, 16:5–19, 2002.

[16] David H. Bailey, Jonathan M. Borwein, and Richard E. Crandall. On the Khintchine Constant. *Mathematics of Computation*, 66:417–431, 1997.

[17] David H. Bailey, Jonathan M. Borwein, Richard E. Crandall, and Carl Pomerance. On the Binary Expansions of Algebraic Numbers. *Journal of Number Theory Bordeaux*, to appear.

[18] David H. Bailey, Jonathan M. Borwein, and Roland Girgensohn. Experimental Evaluation of Euler Sums. *Experimental Mathematics*, 3:17–30, 1994.

[19] David H. Bailey and David J. Broadhurst. A Seventeenth-Order Polylogarithm Ladder. http://crd.lbl.gov/~dhbailey/dhbpapers/ladder.pdf, 1999.

[20] Alexander Barvinok. *A Course in Convexity*. American Mathematical Society, 2002.

[21] P. Bateman. On the Representation of a Number as the Sum of Three Squares. *Transactions of the AMS*, 71:70–101, 1951.

[22] E. Bedocchi. Nota ad una congettura sui numeri primi. *Rivista di Matematica della Università di Parma*, 11:229–236, 1985.

[23] Michael Berry. Why Are Special Functions Special? *Physics Today*, 54(4):11–12, 2001.

[24] Frits Beukers. A Rational Approach to Pi. *Nieuw Archief voor Wiskunde*, 5:372–379, 2000.

[25] Rajendra Bhatia. *Fourier Series*. Hindustan Book Agency, Delhi, India, 1993.

[26] Patrick Billingsley. *Probability and Measure*. John Wiley, Hoboken, NJ, 1986.

[27] L.M. Blumenthal. *Theory and Applications of Distance Geometry*. Chelsea, New York, 1970.

[28] Harold P. Boas. Majorant Series. *Journal of the Korean Mathematical Society*, 37:321–337, 2000.

[29] D. Borwein, J. M. Borwein, P. B. Borwein, and R. Girgensohn. Giuga's Conjecture on Primality. *American Mathematical Monthly*, 103:40–50, 1996.

[30] D. Borwein, J. M. Borwein, and R. Girgensohn. Explicit Evaluation of Euler Sums. *Proceedings of the Edinburgh Mathematical Society*, 38:273–294, 1995.

[31] D. Borwein, J. M. Borwein, and C. Pinner. Convergence of Madelung-Like Lattice Sums. *Transactions of the American Mathematical Society*, 350:3131–3167, 1998.

[32] D. Borwein, J. M. Borwein, and K. Taylor. Convergence of Lattice Sums and Madelung's Constant. *Journal of Mathematical Physics*, 26:2999–3009, 1985.

[33] David Borwein and Jonathan M. Borwein. Some Remarkable Properties of Sinc and Related Integrals. *The Ramanujan Journal*, 5:73–90, 2001.

[34] David Borwein, Jonathan M. Borwein, and Bernard A. Mares Jr. Multi-Variable Sinc Integrals and Volumes of Polyhedra. http://www.cecm.sfu.ca/preprints/2001pp.html, 2001.

[35] J. M. Borwein and G. De Barra. Nested Radicals. *Americn Mathematical Monthly*, 98:735–739, 1991.

[36] J. M. Borwein, D. M. Bradley, D. J. Broadhurst, and P. Lisoněk. Special Values of Multidimensional Polylogarithms. *Transactions of the American Mathematical Society*, 353:907–941, 2001.

[37] Jonathan Borwein, Peter Borwein, and Karl Dilcher. Pi, Euler Numbers and Asymptotic Expansions. *American Mathematical Monthly*, 96:681–687, 1989.

[38] Jonathan Borwein and Kwok-Kwong (Stephen) Choi. On Dirichlet Series for Sums of Squares. *Ramanujan Journal*, to appear.

[39] Jonathan Borwein, Richard Crandall, and Greg Fee. On Ramanujan's AGM fraction. *Experimental Mathematics*, to appear.

[40] Jonathan M. Borwein. Problem 10281. *American Mathematical Monthly*, 100:76–77, 1993.

[41] Jonathan M. Borwein. Revivals. Problem 10281. *American Mathematical Monthly*, 103:911–912, 1996.

[42] Jonathan M. Borwein. Solutions to Problem 10281. *American Mathematical Monthly*, 103:181–183, 1996.

[43] Jonathan M. Borwein and David H. Bailey. *Mathematics by Experiment: Plausible Reasoning in the 21st century*. A K Peters Ltd, Natick, MA, 2003.

[44] Jonathan M. Borwein and Peter B. Borwein. *Pi and the AGM: A Study in Analytic Number Theory and Computational Complexity*. CMS Series of Mongraphs and Advanced books in Mathematics, John Wiley, Hoboken, NJ, 1987.

[45] Jonathan M. Borwein and Peter B. Borwein. A Remarkable Cubic Mean Iteration. In *Lecture Notes in Mathematics, vol. 1435: Proceedings of Computational Methods and Function Theory*, pages 27–31. Springer-Verlag, Heidelberg, 1990.

[46] Jonathan M. Borwein and Peter B. Borwein. A Cubic Counterpart of Jacobi's Identity and the AGM. *Transactions of the American Mathematical Society*, 323:691–701, 1991.

[47] Jonathan M. Borwein and Peter B. Borwein. Strange Series Evaluations and High Precision Fraud. *American Mathematical Monthly*, 99:622–640, 1992.

[48] Jonathan M. Borwein and Peter B. Borwein. On the Generating Function of the Integer Part: [na+b]. *Journal of Number Theory*, 43:293–318, 1993.

[49] Jonathan M. Borwein, Peter B. Borwein, and Frank G. Garvan. Some Cubic Identities of Ramanujan. *Transactions of the American Mathematical Society*, 343:35–47, 1994.

[50] Jonathan M. Borwein, Peter B. Borwein, Roland Girgensohn, and Sheldon Parnes. Experimental Mathematical Investigation of Decimal and Continued Fraction Expansions of Selected Constants. *Manuscript*, 1995.

[51] Jonathan M. Borwein, David M. Bradley, and David J. Broadhurst. Evaluations of k-fold Euler/Zagier Sums: A Compendium of Results for Arbitrary k. *The Wilf Festschrift, Electronic Journal of Combinatorics*, 4(2), 1997.

[52] Jonathan M. Borwein, David M. Bradley, David J. Broadhurst, and Petr Lisoněk. Combinatorial Aspects of Multiple Zeta Values. *Electronic Journal of Combinatorics*, 5, R38, 1998.

[53] Jonathan M. Borwein, David M. Bradley, and Richard E. Crandall. Computational Strategies for the Riemann Zeta Function. *Journal of Computational and Applied Mathematics*, 121:247–296, 2000.

[54] Jonathan M. Borwein, David J. Broadhurst, and Joel Kamnitzer. Central Binomial Sums, Multiple Clausen Values and Zeta Values. *Experimental Mathematics*, 10:25–41, 2001.

[55] Jonathan M. Borwein and Kwok-Kwong Stephen Choi. On the Representations of $xy + xz + yx$. *Experimental Mathematics*, 9:153–158, 2000.

[56] Jonathan M. Borwein, Kwok-Kwong Stephen Choi, and Wilfrid Pigulla. Continued Fractions as Accelerations of Series. http://www.cecm.sfu.ca/preprints/2003pp.html, 2003.

[57] Jonathan M. Borwein and Roland Girgensohn. Addition Theorems and Binary Expansions. *Canadian Journal of Mathematics*, 47:262–273, 1995.

[58] Jonathan M. Borwein and Roland Girgensohn. Solution to AMM Problem 10901. *American Mathematical Monthly*, 108, 2001.

[59] Jonathan M. Borwein and Roland Girgensohn. Evaluations of Binomial Series. http://www.cecm.sfu.ca/preprints/2002pp.html, 2002.

[60] Jonathan M. Borwein and Adrian S. Lewis. Partially-Finite Convex Programming in L^1: Entropy Maximization. *SIAM Journal on Optimization*, 3:248–267, 1993.

[61] Jonathan M. Borwein and Adrian S. Lewis. *Convex Analysis and Nonlinear Optimization. Theory and Examples.* CMS books in Mathematics, Springer-Verlag, vol. 3, Heidelberg, 2000.

[62] Jonathan M. Borwein and Mark A. Limber. Maple as a High Precision Calculator. *Maple News Letter*, 8:39–44, 1992.

[63] Jonathan M. Borwein, Petr Lisoněk, and John A. Macdonald. Arithmetic-Geometric Means Revisited. *MapleTech*, pages 20–27, 1997.

[64] Peter Borwein. An Efficient Algorithm for the Riemann Zeta Function. http://www.cecm.sfu.ca/preprints/1995pp.html, 1995.

[65] Peter B. Borwein. *Computational Excursions in Analysis and Number Theory*. Springer-Verlag, CMS Books in Mathematics, Heidelberg, 2002.

[66] Peter B. Borwein and Christopher Pinner. Polynomials with (0, +1, -1) Coefficients and Roots Close to a Given Point. *Canadian Journal of Mathematics*, 49:887–915, 1997.

[67] W. Bosma and J. Cannon. *Handbook of Magma Functions*. School of Mathematics and Statistics, Sydney, 1995.

[68] W. Bosma, J. Cannon, and C. Playout. The Magma Algebra System I: The User Language. *Journal of Symbolic Computing*, 24:235–265, 1998.

[69] W. Bosma, J. J. Cannon, and G. Matthews. Programming with Algebraic Structures: Design of the Magma Language. In M. Giesbrecht, editor, *Proceedings of the 1994 International Symposium on Symbolic and Algebraic Computation*, pages 52–57. Association of Computing Machinery, 1994.

[70] Douglas Bowman and David M. Bradley. Multiple Polylogarithms: A Brief Survey. In Bruce C. Berndt and Ken Ono, editors, *Proceedings of a Conference on q-Series with Applications to Combinatorics, Number Theory and Physics*, volume 291 of *Contemporary Mathematics*, pages 71–92. American Mathematical Society, 2001.

[71] Douglas Bowman and David M. Bradley. On Multiple Polylogarithms and Certain Generalizations Possessing Periodic Argument Lists. http://germain.umemat.maine.edu/faculty/bradley/papers/pub.html, 2003.

[72] Douglas Bowman and David M. Bradley. Resolution of Some Open Problems Concerning Multiple Zeta Evaluations of Arbitrary Depth. *Compositio Mathematica*, to appear.

[73] J. Buhler, R. Crandall, R. Ernvall, and T. Metsänkylä. Irregular Primes to Four Million. *Mathematics of Computation*, 61:151–153, 1993.

[74] J. Buhler, R. Crandall, and R. W. Sompolski. Irregular Primes to One Million. *Mathematics of Computation*, 59:717–722, 1992.

[75] P.L. Butzer and R.J. Nessel. *Fourier Analysis and Approximation*. Birkhäuser, Basel, 1971.

[76] A. R. Calderbank, P. J. Cameron, W. M. Kantor, and J. J. Seidel. $Z4$ Kerdock Codes, Orthogonal Spreads, and Extremal Euclidean Line-Sets. *Proceedings of the London Mathematical Society*, 75:436–480, 1997.

[77] A. R. Calderbank, R. H. Hardin, E. M. Rains, P. W. Shor, and N. J. A. Sloane. A Group-Theoretic Framework for the Construction of Packings in Grassmannian Spaces. *Journal of Algebraic Combinatorics*, 9:129–140, 1999.

[78] A. R. Calderbank, E. M. Rains, P. W. Shor, and N. J. A. Sloane. Quantum Error Correction and Orthogonal Geometry. *Physics Review Letters*, 78:405–409, 1997.

[79] A. R. Calderbank, E. M. Rains, P. W. Shor, and N. J. A. Sloane. Quantum Error Correction Via Codes over $GF(4)$. *IEEE Transactions on Information Theory*, 44:1369–1387, 1998.

[80] E. Rodney Canfield. Engel's Inequality for Bell Numbers. *Journal of Combinatorial Theory, Series A*, 72:184–187, 1995.

[81] L. Carleson. On Convergence and Growth of Partial Sums of Fourier Series. *Acta Mathematica*, 116:135–157, 1966.

[82] Julien Cassaigne and Steven R. Finch. A Class of 1-Additive Sequences and Quadratic Recurrences. *Experimental Mathematics*, 4:49–60, 1995.

[83] K. Chandrasekharen. *Arithmetical Functions*. Springer-Verlag, Heidelberg, 1970.

[84] Man-Duen Choi. Tricks or Treats with the Hilbert Matrix. *American Mathematical Monthly*, 90:301–312, 1983.

[85] J. H. Conway, R. H. Hardin, and N. J. A. Sloane. Packing Lines, Planes, etc.: Packings in Grassmannian Space. *Experimental Mathematics*, 5:139–159, 1996.

[86] J. H. Conway and N. J. A. Sloane. *Sphere Packings, Lattices and Groups*. Springer-Verlag, Heidelberg, 1999.

[87] Shaun Cooper. Sums of Five, Seven and Nine Squares. *Ramanujan Journal*, 6:469–490, 2002.

[88] Robert M. Corless, Gaston H. Gonnet, David E. G. Hare, David J. Jeffrey, and Donald E. Knuth. On the Lambert W function. *Advances in Computational Mathematics*, 5:329–359, 1996.

[89] Robert M. Corless and David J. Jeffrey. Graphing Elementary Riemann Surfaces. *ACM SIGSAM Bulletin: Communications in Computer Algebra*, 32:11–17, 1998.

[90] Robert M. Corless, David J. Jeffrey, and Donald E. Knuth. A Sequence of Series for the Lambert W function. In W. Kuchlin, editor, *Proceedings of ISSAC'97*, pages 197–204, 1997.

[91] David A. Cox, Donal O'Shea, and John B. Little. *Using Algebraic Geometry*. Springer-Verlag, Heidelberg, 1998.

[92] David A. Cox, John B. Little, and Donal O'Shea. *Ideals, Varieties and Algorithms*. Springer-Verlag, Heidelberg, 1996.

[93] Richard Crandall and Carl Pomerance. *Prime Numbers: A Computational Perspective.* Springer-Verlag, Heidelberg, 2001.

[94] Richard E. Crandall. Fast Evaluation of Multiple Zeta Sums. *Mathematics of Computation,* 67:1163–1172, 1996.

[95] Richard E. Crandall. On the Quantum Zeta Function. *Journal of Physics A: Mathematical and General,* 29:6795–6816, 1996.

[96] Richard E. Crandall. *Topics in Advanced Scientific Computation.* Springer-Verlag, Heidelberg, 1996.

[97] Richard E. Crandall. New Representations for the Madelung Constant. *Experimental Mathematics,* 8:367–379, 1999.

[98] Richard E. Crandall and Joseph P. Buhler. Elementary Function Expansions for Madelung Constants. *Journal of Physics A: Mathematics and General,* 20:5497–5510, 1987.

[99] Richard E. Crandall and Joseph P. Buhler. On the Convergence Problem for Lattice Sums. *Journal of Physics A: Mathematics and General,* 23:2523–2528, 1990.

[100] Richard E. Crandall and J. F. Delord. The Potential within a Crystal Lattice. *Journal of Physics A: Mathematics and General,* 20:2279–2292, 1987.

[101] Richard E. Crandall, Karl Dilcher, and Carl Pomerance. A Search for Wieferich and Wilson Primes. *Mathematics of Computation,* 66:433–449, 1997.

[102] Richard E. Crandall and Jason S. Papadopoulos. On the Implementation of AKS-Class Primality Tests. http://developer.apple.com/hardware/ve/acgresearch.html, 2003.

[103] D. P. Dalzell. On 22/7 and 355/113. *Eureka,* 34:10–13, 1971.

[104] I. Damgard, P. Landrock, and C. Pomerance. Average Case Error Estimates for the Strong Probable Prime Test. *Mathematics of Computation,* 61:177–194, 1993.

[105] Ingrid Daubechies and Jeffrey C. Lagarias. Two-Scale Difference Equations. I. Existence and Global Regularity of Solutions. *SIAM Journal of Mathematical Analysis,* 22:1338–1410, 1991.

[106] Ingrid Daubechies and Jeffrey C. Lagarias. Two-Scale Difference Equations. II. Local Regularity, Infinite Products of Matrices, and Fractals. *SIAM Journal of Mathematical Analysis,* 23:1031–1079, 1992.

[107] P. Delsarte, J.-M. Goethals, and J. J. Seidel. Spherical Codes and Designs. *Geometriae Dedicata,* 6:363–388, 1977.

[108] John E. Dennis Jr. and Victoria Torczon. Direct Search Methods on Parallel Machines. *SIAM Journal of Optimization,* 1(1):448–474, 1991.

[109] Keith Devlin. *Mathematics: The Science of Patterns: The Search for Order in Life, Mind and the Universe.* W. H. Freeman, New York, 1997.

[110] Leonard Eugene Dickson. *History of the Theory of Numbers*, volume 1–3. Chelsea Publishing (1952), Carnegie Institute (1991), American Mathematical Society, Providence, 1999.

[111] Tomslav Doslic and Darko Velian. Calculus Proofs of Some Combinatorial Identities. *Mathematical Inequalities and Applications*, 6:197–209, 2003.

[112] John Duncan and Colin M. McGregor. Carleman's Inequality. *American Mathematical Monthly*, 110:424–430, 2003.

[113] William Dunham. *Euler: The Master of Us All*. Mathematical Association of America (Dolciani Mathematical Expositions), Washington, 1999.

[114] Harold M. Edwards. *Riemann's Zeta Function*. Academic Press, New York, 1974.

[115] Noam D. Elkies. On the Sums $\sum_{k=-\infty}^{\infty} (4k + 1)^{-n}$. *American Mathematical Monthly*, 110:561–573, 2003.

[116] Paul Erdős. On a Family of Symmetric Bernoulli Convolutions. *Transactions of the American Mathematical Society*, 61:974–976, 1939.

[117] Philippe Flajolet and Bruno Salvy. Euler Sums and Contour Integral Representations. *Experimental Mathematics*, 7:15–34, 1997.

[118] M. K. Fort Jr. Some Properties of Continuous Functions. *American Mathematical Monthly*, 59:372–375, 1952.

[119] L. Gårding. An Inequality for Hyperbolic Polynomials. *Journal of Mathematics and Mechanics*, 8:957–965, 1959.

[120] Adriano Garsia. Arithmetic Properties of Bernoulli Convolutions. *Transactions of the American Mathematical Society*, 102:409–432, 1962.

[121] Roland Girgensohn. Functional Equations and Nowhere Differentiable Functions. *Aequationes Mathematicae*, 46:243–256, 1993.

[122] Roland Girgensohn. Nowhere Differentiable Solutions of a System of Functional Equations. *Aequationes Mathematicae*, 47:89–99, 1994.

[123] Roland Girgensohn. A Unified View of Classical Summation Kernels. *Resultate der Mathematik (Results in Mathematics)*, 38:48–57, 2000.

[124] G. Giuga. Su una presumibile proprietà caratteristica dei numeri primi. *Rendiconti Scienze Matem tiche e Applicazioni A*, 83:511–528, 1950.

[125] J.-M. Goethals and J. J. Seidel. Spherical designs. In D. K. Ray-Chaudhuri, editor, *Relations Between Combinatorics and Other Parts of Mathematics, Proceedings of the Symposium on Pure Mathematics*, volume 34, pages 255–272. American Mathematical Society, 1979.

[126] J.-M. Goethals and J. J. Seidel. Cubature Formulae, Polytopes and Spherical Designs. In C. Davis et al., editor, *The Geometric Vein: The Coxeter Festschrift*, pages 203–218. Springer-Verlag, Heidelberg, 1981.

[127] J.-M. Goethals and J. J. Seidel. The Football. *Nieuw Archief Wisk,* 29:50–58, 1981.

[128] Daniel Goldston and Cem Yildirim. Small Gaps Between Primes. http://www.aimath.org/goldston_tech, 2003.

[129] R. William Gosper. Unpublished Research Announcement. *Manuscript,* 1974.

[130] I. S. Gradshteyn and I. M. Ryzhik. *Table of Integrals, Series, and Products, Fifth Edition.* Academic Press, New York, 1994.

[131] Ronald L. Graham, Donald E. Knuth, and Oren Patashnik. *Concrete Mathematics.* Addison-Wesley, Boston, 1994.

[132] Francois Guénard and Henri Lemberg. *La Méthode Expérimentale en Mathématiques.* Scopos, Springer-Verlag, Heidelberg, 2001.

[133] O. Guler. Hyperbolic Polynomials and Interior Point Methods for Convex Programming. *Mathematics of Operations Research,* 22:350–377, 1997.

[134] Richard K. Guy. Aliquot Sequences. In *Unsolved Problems in Number Theory,* pages 60–62. Springer-Verlag, Heidelberg, 1994.

[135] R. H. Hardin and N. J. A. Sloane. New Spherical 4-Designs. *Discrete Mathematics,* 106/107:255–264, 1992.

[136] R. H. Hardin and N. J. A. Sloane. McLaren's Improved Snub Cube and Other New Spherical Designs in Three Dimensions. *Discrete Computational Geometry,* 15:429–441, 1996.

[137] Godfrey H. Hardy. *Ramanujan.* Chelsea, New York, 1978.

[138] Godfrey H. Hardy and Edward M. Wright. *An Introduction to the Theory of Numbers.* Oxford University Press, Oxford, 1985.

[139] Kevin Hare. Multisectioning, Rational Poly-exponential Functions and Parallel Computation. http://www.cecm.sfu.ca/preprints/1999pp.html, 1999.

[140] J. Herman, R. Kučera, and J. Šimša. *Equations and Inequalities: Elementary Problems and Theorems in Number Theory,* volume 1. CMS Books, Springer-Verlag, Heidelberg, 2000.

[141] J. Holt. Mistaken Identity Theory: Why Scientists Always Pick the Wrong Man. *Lingua Franca,* page 60, March 2000.

[142] R. Hooke and T. A. Jeeves. Direct Search Solution of Numerical and Statistical Problems. *Journal of the Association for Computing Machinery,* 8:212–229, 1961.

[143] John H. Hubbard, Jean Marie McDill, Anne Noonburg, and Beverly H. West. A New Look at the Airy Equation with Fences and Funnels. In *The Organic Mathematics Project Proceedings,* volume 20, Ottawa, 1997. Canadian Mathematical Society.

[144] R. A. Hunt. On the Convergence of Fourier Series. In *Orthogonal Expansions and Their Continuous Analogues*, pages 235–255, Carbondale, IL, 1968. Southern Illinois University Press.

[145] Borge Jessen and Aurel Wintner. Distribution Functions and the Riemann Zeta Function. *Transactions of the American Mathematical Society*, 38:48–88, 1935.

[146] G. S. Joyce and I. J. Zucker. Evaluation of the Watson integral and Associated Logarithmic Integral for the d-Dimensional Hypercubic Lattice. *Journal of Physics A*, 34:7349–7354, 2001.

[147] G. S. Joyce and I. J. Zucker. On the Evaluation of Generalized Watson Integrals. *Proceedings of the AMS*, to appear.

[148] Anatolij A. Karatsuba. *Basic Analytic Number Theory*. Springer-Verlag, Heidelberg, 1993.

[149] Yitzhak Katznelson. *An Introduction to Harmonic Analysis*. Wiley, Hoboken, NJ, 1968.

[150] Akio Kawauchi. *A Survey of Knot Theory*. Birkhäuser Verlag, Basel, 1996.

[151] R. Kershner and A. Wintner. On Symmetric Bernoulli Convolutions. *American Journal of Mathematics*, 57:541–548, 1935.

[152] Donald E. Knuth. *The Art of Computer Programming*, volume 2. Addison-Wesley, Boston, 1998.

[153] Pavel P. Korovkin. On Convergence of Linear Positive Operators in the Space of Continuous Functions. *Doklady Akademii Nauk. Rossijskaya Akademiya Nauk*, 90:961–964, 1953.

[154] George Lamb. Problem 11000. *American Mathematical Monthly*, 110:240, 2003.

[155] Andrzej Lasota and Michael C. Mackey. *Chaos, Fractals and Noise: Stochastic Aspects of Dynamics*. Springer-Verlag, Heidelberg, 1994.

[156] H. W. Lenstra Jr. Primality Testing with Cyclotomic Rings. *preprint*, 2002.

[157] A. S. Lewis, P. A. Parrilo, and M. V. Ramana. The Lax Conjecture Is True. *Proceedings of the American Mathematical Society*, to appear.

[158] A. S. Lewis and H. S. Sendov. Self-Concordant Barriers for Hyperbolic Means. *Mathematical Programming*, 91, Ser. A:1–10, 2001.

[159] Peter Lisoněk and Robert Israel. Metric Invariants of Tetrahedra via Polynomial Elimination. *Proceedings of the 2000 International Symposium on Symbolic and Algebraic Manipulation*, pages 217–219, 2000.

[160] John E. Littlewood. *Littlewood's Miscellany*. Cambridge University Press, Cambridge, second edition, edited by Béla Bollobás edition, 1988.

[161] I. G. Macdonald. *Symmetric Functions and Hall Polynomials*. Oxford University Press, Oxford, 1995.

[162] K. Mahler. Arithmetische Eigenschaften der Lösungen einer Klasse von Funktionalgleichungen. *Mathematik Annalen*, 101:342–366, 1929.

[163] George Marsaglia and John C. Marsaglia. A New Derivation of Stirling's Approximation to $n!$. *American Mathematical Monthly*, 97:826–829, 1990.

[164] R. Daniel Mauldin and Karoly Simon. The Equivalence of Some Bernoulli Convolutions to Lebesgue Measure. *Proceedings of the American Mathematical Society*, 126:2733–2736, 1998.

[165] M. Mazur. Problem 10717. *American Mathematical Monthly*, 106:167, 1999.

[166] K. I. M. McKinnon. Convergence of the Nelder-Mead Simplex Method to a Nonstationary Point. *SIAM Journal on Optimization*, 9:148–158, 1998.

[167] J. Milnor. On the Betti Numbers of Real Algebraic Varieties. *Proceedings of the American Mathematical Society*, 15:275–280, 1964.

[168] L. Monier. Evaluation and Comparison of Two Efficient Probabilistic Primality Testing Algorithms. *Theoretical Computer Science*, 12:97–108, 1980.

[169] M. Mori. Developments in the Double Exponential Formula for Numerical Integration. In *Proceedings of the International Congress of Mathematicians*, pages 1585–1594, Heidelberg, 1991. Springer-Verlag.

[170] David Mumford, Caroline Series, and David Wright. *Indra's Pearls: The Vision of Felix Klein*. Cambridge University Press, Cambridge, 2002.

[171] J. A. Nelder and R. Mead. A Simplex Method for Function Minimization. *The Computer Journal*, 7:308–313, 1965.

[172] Ivan Niven, Herbert S. Zuckerman, and Hugh L. Montgomery. *An Introduction to the Theory of Numbers*. John Wiley, Hoboken, NJ, 1991.

[173] Jorge Nocedal and Stephen J. Wright. *Numerical Optimization*. Springer-Verlag, Heidelberg, 1999.

[174] Yasuo Ohno. A Generalization of the Duality and Sum Formulas on the Multiple Zeta Values. *Journal of Number Theory*, 74:39–43, 1999.

[175] Frank W. J. Olver. *Asymptotics and Special Functions*. Academic Press, New York, 1974.

[176] Yuval Peres, Wilhelm Schlag, and Boris Solomyak. Sixty Years of Bernoulli Convolutions. In C. Bandt, S. Graf, and M. Zaehle, editors, *Progress in Probability vol. 46: Fractals and Stochastics II*, pp. 39–65, Birkhäuser, Boston, 2000.

[177] Yuval Peres and Boris Solomyak. Absolute Continuity of Bernoulli Convolutions, a Simple Proof. *Mathematical Research Letters*, 3:231–239, 1996.

[178] Kenneth A. Perko. On the Classifications of Knots. *Proceedings of the American Mathematical Society*, 45:262–266, 1974.

[179] Oskar Perron. *Die Lehre von den Kettenbrüchen*. Chelsea, New York, 1950.

[180] William H. Press, Brian P. Flannery, Saul A. Teukolsky, and William T. Vetterling. *Numerical Recipes*. Cambridge University Press, Cambridge, 1986.

[181] P. Pritchard, A. Moran, and A. Thyssen. Twenty-Two Primes in Arithmetic Progression. *Mathematics of Computation*, 64:1337–1339, 1995.

[182] A.P. Prudnikov, Yu.A. Brychkov, and O.I. Marichev. *Integrals and Series* (Volumes 1 and 2). Taylor and Francis, London, 1986.

[183] M. Rabin. Probabilistic Algorithms for Testing Primality. *Journal of Number Theory*, 12:128–138, 1980.

[184] Bruce Reznick. *Sums of Even Powers of Real Linear Forms*. American Mathematical Society, 1992.

[185] Bruce Reznick. Some Constructions of Spherical 5-Designs. *Linear Algebra and Its Applications*, 226/228:163–196, 1995.

[186] Paulo Ribenboim. *The Little Book of Big Primes*. Springer-Verlag, Heidelberg, 1991.

[187] Bernhard Riemann. Über die Anzahl der Primzahlen unter einer Gegebenen Grösse. In Hermann Weyl, editor, *Das Kontinuum und Andere Monographien*. Chelsea Publishing Co., New York, 1972.

[188] John Riordan. *Combinatorial Identities*. John Wiley, Hoboken, NJ, 1968.

[189] Sara Robinson. New Method Said to Solve Key Problem In Math. *New York Times*, Aug. 8 2002.

[190] Dale Rolfsen. *Knots and Links*. Publish or Perish, Inc., Houston, 1976.

[191] Raphael Salem. A Remarkable Class of Algebraic Integers, Proof of a Conjecture by Vijayaraghavan. *Duke Mathematical Journal*, 11:103–108, 1944.

[192] Byron Schmuland. Random Harmonic Series. *American Mathematical Monthly*, 110:407–416, 2003.

[193] P. D. Seymour and T. Zaslavsky. Averaging Sets: A Generalization of Mean Values and Spherical Designs. *Advances in Mathematics*, 52:213–240, 1984.

[194] P. W. Shor and N. J. A. Sloane. A Family of Optimal Packings in Grassmannian Manifolds. *Journal of Algebraic Combinatorics*, 7:157–163, 1998.

[195] Peter W. Shor. Fault-Tolerant Quantum Computation. In *Proceedings of the 37th Symposium on the Foundations of Computer Science*, pages 56–65. IEEE Computer Society, 1996.

[196] Carl L. Siegel. Über Riemanns Nachlaß zur Analytischen Zahlentheorie. *Quellenstudien zur Geschichte der Math. Astron. und Phys.*, 2:45–80, 1932.

[197] L. E. Sigler. *Fibonacci's Liber Abaci: A Translation into Modern English of Leonardo Pisano's Book of Calculation.* Springer-Verlag, Heidelberg, 2002.

[198] N. J. A. Sloane, R. H. Hardin, and P. Cara. Spherical Designs in Four Dimensions. In *Proceedings of the Workshop on Information and Coding.* IEEE Press, New York, 2003.

[199] N. J. A. Sloane, R. H. Hardin, and P. Cara. Spherical Designs in Three and Four Dimensions. *preprint*, 2003.

[200] N. J. A. Sloane and Simon Plouffe. *The Encyclopedia of Integer Sequences.* Academic Press, New York, 1995.

[201] Boris Solomyak. On the Random Series $\sum \pm \lambda^i$ (an Erdős Problem). *Annals of Mathematics*, 142:611–625, 1995.

[202] Richard Stanley. *Enumerative Combinatorics*, volume 1 and 2. Cambridge University Press, Cambridge, 1999.

[203] Karl R. Stromberg. *An Introduction to Classical Real Analysis.* Wadsworth, Thompson Learning, Florence, KY, 1981.

[204] Hidetosi Takahasi and Masatake Mori. Double Exponential Formulas for Numerical Integration. *Publications of RIMS, Kyoto University,* 9:721–741, 1974.

[205] Radu Theodorescu. Problem 10738. *American Mathematical Monthly,* 106:471, 1999.

[206] Edward C. Titchmarsh. *The Theory of the Riemann Zeta-Function.* Oxford University Press, 1951.

[207] Virginia Torczon. *Multi-Directional Search: A Direct Search Algorithm for Parallel Machines.* PhD thesis, Rice University, May 1989.

[208] Virginia Torczon. On the Convergence of Pattern Search Algorithms. *SIAM Journal on Optimization*, 7:1–25, Feb. 1997.

[209] Michal Trott. Visualization of Riemann Surfaces of Algebraic functions. *Mathematica in Education and Research*, 6(4):15–36, 1997.

[210] Mitsuru Utchiyama. Proofs of the Korovkin Theorems via Inequalities. *American Mathematical Monthly*, 110:334–336, 2003.

[211] Alfred J. van der Poorten and Xuan Chuong Tran. Quasi-Elliptic Integrals and Periodic Continued Fractions. *Monatshefte für Mathematik,* 131:155–169, 2000.

[212] H. S. Wall. *Analytic Theory of Continued Fractions.* Chelsea, New York, 1948.

[213] George N. Watson. The Final Problem: An Account of the Mock Theta Functions. *Journal of the London Mathematical Society*, 11:55–80, 1936.

[214] Aurel Wintner. On Convergent Poisson Convolutions. *American Journal of Mathematics*, 57:827–838, 1935.

[215] D. Zeilberger. Identities in Search of Identity. *Theoretical Computer Science*, 117:23–38, 1993.

[216] Antoni Zygmund. *Trigonometric Series*. Cambridge University Press, Cambridge, 1959.

Index

---

[1]In this Index, ‡ denotes a figure.

[2]In this Index, † denotes a quote.